Topics in
Calculus

Topics in Calculus

second edition

Morton Lowengrub Indiana University

Joseph G. Stampfli Indiana University

JOHN WILEY & SONS   New York   •   London   •   Sydney   •   Toronto

ISBN 0 471 01088 X
Library of Congress Catalog Card Number: 74–78064
Printed in the United States of America.

5 6 7 8 9 10

To Carol and Linda

Preface

This second edition, like the first, is a basic introduction to the tools and techniques of calculus that are useful to students of business and economics, the social sciences, and the biological sciences. Only a minimal knowledge of high school algebra is assumed.

There are several significant differences between this book and its predecessor. Many examples from the nonphysical sciences have been added: elasticity of demand, pricing policies for monopolies, how bugs climb incline planes, nerve physiology, the mechanics of coughing, and inventory analysis, to name a few. Nearly every section in the text now contains exercises emphasizing such applications. In addition, the section on inverse functions has been moved from Chapter 6 to Chapter 1 in order to present a more coherent discussion of root functions, logarithms, and exponentials. Chapter 5 has been extensively rewritten. The reorganization enables one to get to area and other applications of integration more quickly. A section on the Riemann integral has been added and integration by parts has been moved to the end of the chapter. Finally, many exercises have been changed and over 100 new problems have been added.

The spirit of the first edition has been retained. To quote from the original preface, "If this book can be said to have a philosophy or strategy, it is the following: First, we quickly try to give the student the tools and techniques he needs to solve interesting problems. Second, we do not worry about rigor, but rather attempt to develop the student's understanding of the concepts and ideas of calculus from a geometric or intuitive point of view."

We urge instructors to emphasize examples and applications rather than the theoretical aspects of the material. Although the text does contain some statements of theorems and proofs, these are, in effect, tools or labor-saving devices, and this fact should be stressed. Simply because a proof is included does not mean it should be presented. In fact, if a blanket rule were to be adopted, it would be to omit all proofs. There are, however, situations where a proof reveals how simple and natural a theorem is, or acts as a mnemonic device. In situations where the proof "explains" the theorem, one might comment on it or assign the proof for reading. In general, it is best to introduce theorems as problem-solving devices and then quickly move on to the examples.

The first five chapters form the basis for a one-semester course and deserve some specific comments.

Chapter 1 is mainly background material and can be covered quickly if the student has had an adequate course in high school algebra. The main emphasis should be on the graphs and elementary properties of functions.

In any beginning calculus course there is a real question about how much emphasis should be placed on limits. Three reasonable alternatives to Chapter 2 are

1. Skip Chapter 2 and go right to Chapter 3.
2. Cover Sections 2.1 and 2.2 carefully and then skim over the remainder.
3. Cover the entire chapter.

If one adopts the first approach, some extra time and effort will probably be needed when the derivative is introduced. The third alternative seems overly intensive in a one-semester course. Our own choice is to favor 2 or some variation thereof. The primary purpose of the text is, after all, to introduce calculus as a problem-solving device, and in this context the need for limits is minimal. On the other hand, limits are the crucial underlying concept in calculus and necessary to its understanding and appreciation.

It is essential to cover all of Chapter 3 so that the students have a firm grounding in the techniques of differentiation. It is impossible to teach applications to students who cannot differentiate (or integrate).

Chapter 4 contains the greatest number of applications. Hopefully the instructor will stress the idea of setting up a mathematical model for a "simple" real-life situation. The examples in Sections 4.2 and 4.4 offer an ample opportunity to do this.

In Chapter 5 we have added a section on the Riemann integral. This discussion provides a natural introduction to the Newton integral and is included for that reason. It may also be considered for its own sake. Our own preference would be to omit Section 5.4 and proceed immediately to Section 5.5. On historical grounds the Newton integral predates the Riemann integral by over 100 years.

It has been our experience that the material in these five chapters is sufficient for a one-semester course. If time permits, we suggest that the instructor discuss material from Chapter 8, where again there is a wealth of examples on nontechnical applications of differential equations.

Chapters 6–9 are independent and can be covered in any order. In a two-semester or three-quarter course it is possible to cover all of the text.

As in the first edition, the trigonometric functions have been relegated to starred (optional) sections. In general these starred sections contain interesting but peripheral material that is not intended for a one-semester course but could be included in a two-semester course.

We would like to renew our thanks to those who helped in the first edition: We are grateful to George Springer for suggesting we write such a book, and in aiding us in many ways after we had started. Our colleagues have been a source of valuable suggestions and criticism, and we single out John Chadam, Peter Fillmore, and Glenn Schober in particular. Several classes at Indiana had to live with the preliminary (or phone book) edition, and did so cheerfully, for which we salute them.

We would like to thank Christopher Nevison, Morton Harris, and Stanley Luka-wecki for their extremely valuable and detailed critiques of the new edition and the many others who made suggestions for improving the book based on their experience in teaching from it. Arlen Brown and Brian Schmidt of Indiana University contributed many useful suggestions at various stages. We are very grateful to Jean Clydesdale of Glasgow, Scotland, for her excellent job of typing the manuscript. Finally we would like to express our appreciation to Marret McCorkle and Arthur Evans of Xerox College Publishing for their dedicated aid and assistance in the preparation of the second edition.

M.L.
J.G.S.

Skeleton Syllabus

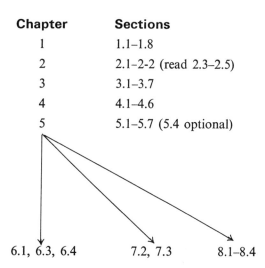

Chapter	Sections
1	1.1–1.8
2	2.1-2-2 (read 2.3–2.5)
3	3.1–3.7
4	4.1–4.6
5	5.1–5.7 (5.4 optional)

6.1, 6.3, 6.4   7.2, 7.3   8.1–8.4

Chapters 1–5 are the minimum backbone for a one-semester course. There is time to add additional topics.

Contents

9

**Functions of
Several Variables**

Topics in
Calculus

Introduction

1.1 Real Numbers

We begin our study of calculus with a discussion of the real numbers. The real numbers are basic to many branches of mathematics. In particular, they are of primary importance to our further study of calculus. The answer to the question "What is a real number?" is not easy. It would be nice if we could simply state a one-sentence definition and let it go at that. However, this is not possible. We assume that you already know something about numbers—namely addition, subtraction, multiplication, and division (by numbers other than 0).

The numbers most familiar to you, 1, 2, 3, . . . , are called *positive integers*. If we add or multiply two positive integers, we again obtain a positive integer. What happens if we subtract two positive integers? First, observe that if we subtract 3 from 5, we again obtain a positive integer, 2. However, if we subtract 5 from 3 (this is certainly possible), then we do not obtain a positive integer. Clearly if you owe the bookstore $5 and only pay $3, you still owe them $2. Thus, we see the need for considering numbers of the form $-1, -2, -3, ...$ These numbers are called *negative integers*. The collection of positive integers, negative integers, and the number 0 is called the *set of integers*. It should also be clear from real life situations that we need to consider fractions as part of our number system. We define a *rational number* to be any number that can be expressed as the quotient of two integers, where it is understood that the integer in the denominator cannot be zero. Thus, $\frac{1}{3}$, $\frac{17}{2}$, $-\frac{2}{3}$, $\frac{4}{2}$, $\frac{6}{8}$ are all rational numbers. Note that the integers themselves form part of the rational numbers. (An integer can be considered as a rational number with the number 1 in the denominator.)

Now, one might think that we have listed all the real numbers above, namely the collection of all integers and rational numbers. Unfortunately, this is not the case.

There are other numbers which should also belong to the collection of real numbers. Suppose we are given a right isosceles triangle whose legs are of unit length (see Figure 1.1.1). The problem is to determine the length of the hypotenuse. Recalling

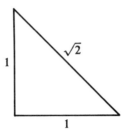

Figure 1.1.1

the Pythagorean theorem,* we see that the length is given by the number $\sqrt{2}$. Such numbers as $\sqrt{2}$, $\sqrt{3}$, and π must also have a place in our number system. These numbers cannot, however, be represented as the quotient of two integers. (See Exercise 13 at the end of this section.) We call these numbers *irrational* numbers. The collection of all integers, rational numbers, and irrational numbers composes the *real number* system.

A geometrical interpretation of the number system may be provided by the following construction. On a straight line, called the "coordinate line," we mark off a segment 0 to 1 as in Figure 1.1.2. This procedure establishes our unit of length,

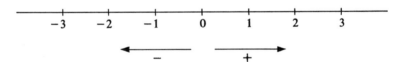

Figure 1.1.2

which may be chosen at will. The positive and negative integers are then represented as a set of equidistant points on the coordinate line, with positive integers to the right of 0 and negative integers to the left of 0. To represent a fraction with denominator n, we simply divide each of the unit segments into n equal parts and represent the fractions with denominator n as the points of the subdivision. The placement of the irrational numbers is accomplished by an approximation procedure. An irrational number can be represented by an infinite decimal expansion; for example, $\sqrt{2} = 1.41214\ldots$. Thus we can say that $\sqrt{2}$ must be somewhere between the numbers $1.4 = 1\frac{2}{5}$ and $1.5 = 1\frac{1}{2}$. This enables us to determine approximately how each of the irrational numbers fits into our coordinate line. In fact, if we carry this procedure further, we can determine precisely the location of an irrational number on the line.

---

* Given a right triangle with sides of length a, b, c, then $a^2 + b^2 = c^2$ where c is the length of the hypotenuse.

Sometimes the coordinate line is referred to as the real line or the real axis.

Recall that division by 0 is not permitted. Here is one reason why. Suppose $a \neq 0$ and we assume that a can be divided by 0. This would mean that $a/0 = b$, where b is some real number. However, since $a/0 = b$ means that $a = b \cdot 0$, it follows that $a = 0$, contrary to our assumption that $a \neq 0$. For different reasons we do not attempt to define $0/0$.

Let us make one more remark concerning the coordinate axis. Observe that if we are given any real number, we can determine a point on the line corresponding to this number. Conversely, given any point on the line, we can determine a number to assign to this point. (This fact should be evident from our geometric construction.) Thus we see that corresponding to every point on the line there exists one and only one real number we can assign to it; conversely, given any real number, there exists one and only one point on the line representing this number. This kind of correspondence (one number for one point and conversely) is called a *one-to-one* correspondence.

It is clear that we need some order properties for our number system. That is, we must know what we mean by saying that a number a is greater than or less than a number b. Geometrically, this will mean that any number to the right of a given number on the line is greater than the given number and any number to the left of the given number on the line is less than that number. It will be advantageous for us to make these ideas more precise and to introduce some symbols to describe them.

We already have the positive numbers placed to the right of 0 on our real axis. We may therefore say a number a is greater than 0 if and only if a is positive. We represent this relation symbolically by writing

$$a > 0,$$

where the symbol $>$ is read "is greater than." We may now define precisely the term "a is greater than b," written $a > b$.

Notation. If a is a real number then by $-a$ we mean the number $(-1)a$.

DEFINITION 1. Let a and b be any two numbers. We say that $a > b$ if $a - b$ is positive; that is, $a - b > 0$.

Thus, $5 > 2$ means that $5 - 2 > 0$, which is certainly obvious.

We now define the concept "less than" in terms of our new symbol $>$. A number a is said to be less than 0, or negative, if and only if $-a > 0$. Introducing a new symbol $<$, read "less than," we say that $a < 0$ if $-a > 0$. We say that $a < b$, where a and b are two numbers, if $b - a = -(a - b) > 0$, or, using our symbol $<$, if $a - b < 0$. Thus $2 < 5$ means that $2 - 5 < 0$ or $-(2 - 5) > 0$. When we write the number a, it is not necessarily positive. For example, if $a = -3$, then a is negative and $-a = -(-3) = 3$ is positive. Note that $a < b$ if and only if $b > a$.

Let us now state some rules concerning the symbols $<$ and $>$.

RULE 1. If $a > 0$ and $b > 0$, then $ab > 0$ and $a + b > 0$.

3

RULE 2. If *a* is any real number different from zero, then either $a > 0$ or $-a > 0$ but not both.

RULE 3. If $a < 0$ and $b < 0$, then $ab > 0$. (Thus the product of two negative numbers is positive.)

RULE 4. If $a < 0$ and $b > 0$, then $ab < 0$. (Thus the product of a positive number and a negative number is negative.)

Rules 1–4 are useful and should be remembered. It is possible to deduce Rules 3 and 4 from Rules 1 and 2, but for our purposes we accept these rules as given. As an example, Rule 3 tells us that $(-2)(-5) > 0$; certainly $+10 > 0$. In like manner, Rule 4 tells us that $(-2)(5) < 0$; certainly $-10 < 0$ since $-(-10) > 0$. Try some examples to convince yourself that Rules 1–4 are plausible.

From Rules 1–4 we can establish many other results concerning $<$ and $>$. We state the most useful ones here. If *a*, *b*, and *c* are any three real numbers, then we have the following:

RULE 5. If $a > b$, then $a + c > b + c$; and if $a > b$ and $c > 0$, then $ac > bc$.

RULE 6. If $a > b$, then $(-a) < (-b)$. In particular, if $c < 0$, then $ac < bc$.

RULE 7. If $a > b > 0$, or $0 > a > b$, then $1/a < 1/b$.

▶ Example 1. If $5 > 2$, then $-5 < -2$. Just look at the coordinate line to establish this. In like manner, if $5 > 2$, then $(-2)5 < (-2)2$, which implies that $-10 < -4$. Rule 7 tells us that if $5 > 2$, then $\frac{1}{5} < \frac{1}{2}$.

Rules 1–7 are simply computational rules, which you will use over and over again.

A relation involving the symbols $<$ and $>$ is called an *inequality*. All of our discussions so far have involved inequalities. We may often write the symbol $a \geq b$ when we mean the number *a* is greater than or equal to the number *b*. In like manner we may use the symbol $\leq$ to designate less than or equal.

In order to define the concept of distance between points on the real axis we need the idea of the absolute value of a number.

DEFINITION 2. Let a be any real number. We define the absolute value of a, written $|a|$*, as follows:*

$$|a| = \begin{cases} a & \text{if } a > 0, \\ 0 & \text{if } a = 0, \\ -a & \text{if } a < 0. \end{cases}$$

For example, we observe that $|2| = 2$ while $|-2| = -(-2) = 2$, since $-2 < 0$. It is a simple matter to show that if *b* is any real number, then $|b| = |-b|$. If *b* is positive, then $|b| = b$ and $|-b| = -(-b) = b$. Hence, for $b \geq 0$, $|b| = |-b|$. If $b < 0$, $|b| = -b$, while $|-b| = -b$ since $-b > 0$. Thus, for $b < 0$, $|b| = |-b|$. We see that in all cases $|b| = |-b|$. It is clear from the definition that $|b| \geq 0$, and

$|b| = 0$ if and only if $b = 0$. This proof illustrates something very useful about dealing with absolute values. One must always consider the separate cases $b \geq 0$, $b < 0$ when verifying statements concerning $|b|$. Another example illustrates this idea.

▶ **Example 2.** For what real numbers x is $|x + 1| = 3$?

Solution. We must solve the above equation for x. First consider the case $x + 1 > 0$; then $|x + 1| = x + 1$ and $x + 1 = 3$, which implies $x = 2$. Next, consider the case $x + 1 < 0$; here $|x + 1| = -(x + 1)$ and thus $-(x + 1) = 3$. Hence $x + 1 = -3$, which yields $x = -4$. It is clear that the above equation is satisfied for $x = -4$ and $x = 2$.

In the next section we shall see that $|x|$ simply represents the distance from the number x to the number 0 on the coordinate axis, and $|x - a|$, where a is any real number, denotes the distance from x to the real number a. Thus, $|x - 3|$ is the distance from x to 3.

We shall now state some properties of absolute values which will be useful for our further work. All of the results can be verified from the definition, and we leave it to the interested reader to do this.

PROPERTY 1. If x is any real number, then $|x|^2 = x^2$ and $\sqrt{x^2} = |x|$.

PROPERTY 2. If x, y are any real numbers, then $|xy| = |x|\,|y|$.

PROPERTY 3. If x, y are any real numbers, then $|x + y| \leq |x| + |y|$.

Property 3 is called the *triangle inequality*. Let us illustrate the Properties 1–3 with some examples.

▶ **Example 3.** $|5|^2 = 5^2 = 25$, $\sqrt{5^2} = |5| = 5$; $\sqrt{(-4)^2} = |-4| = -(-4) = 4$.

▶ **Example 4.** $|5(-2)| = |5|\,|-2| = (5)(-(-2)) = 5 \cdot 2 = 10$.

▶ **Example 5.** Let $x = 3$, $y = -4$; then $|3 + (-4)| = |3 - 4| \leq |3| + |-4| = 3 + 4 = 7$. Since $|3 - 4| = |-1| = 1$, it is clear that in this case $|3 - 4| < 7$. If we let $x = 3$ and $y = 4$, then $|3 + 4| = |3| + |4| = 7$.

You should construct other numerical examples to help convince yourself that Properties 1–3 are valid.

Let us pose the following problem: Find all those real numbers x such that $|x| < 4$. In order to answer the question we must return to the definition of absolute value. First, if $x > 0$ we recall that $|x| = x$, and hence those numbers for which $x < 4$ and $x \geq 0$ certainly satisfy the inequality. Now, we must see what happens if $x < 0$. In this case, $|x| = -x$ and $-x < 4$. It follows from Rule 6 that $-x < 4$ is equivalent to the statement $x > -4$. Combining these two results, we see that all those numbers x such that $x < 4$ *and* $x > -4$ satisfy $|x| < 4$. We write this in a combined form as follows: those numbers x such that $-4 < x < 4$ satisfy the inequality $|x| < 4$. In Figure 1.1.3 we see how this can be interpreted geometrically on the real axis.

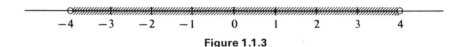

Figure 1.1.3

You can easily convince yourself that all those numbers x represented on the line between -4 and 4 (the crosshatched area) are those numbers for which $|x| < 4$. The open circles at -4 and 4 indicate that these two numbers are *not* included. If the numbers -4 and 4 were meant to be included, then the inequality would be of the form $|x| \leq 4$.

We can consider inequalities such as the one just given in a more general way. The verification of the results proceeds in much the same way as the preceding example. We then add the following properties of inequalities, which will be very useful for future work:

PROPERTY 4. Let a be any positive number. The inequality $|x| \leq a$ holds if and only if $-a \leq x \leq a$. [*Note:* The term "if and only if" means that first one must show that $|x| \leq a$ implies $-a \leq x \leq a$ and then second one must show that $-a \leq x \leq a$ implies $|x| \leq a$.]

We leave the verification, which uses the procedure indicated in the preceding example, to the interested reader.

PROPERTY 5. Let a be any given real number. The inequality $|x| \geq a$ holds if and only if $x \leq -a$ or $x \geq a$. [*Note:* If $a \leq 0$, then $|x| \geq a$ for all real numbers x.] Figure 1.1.4 indicates geometrically the collection of all those numbers x such that

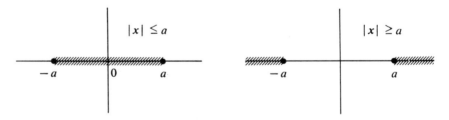

Figure 1.1.4

$|x| \leq a$ and $|x| \geq a$. The crosshatched region consists of all those numbers x where $|x| \geq a$. Before we proceed any further, let us solve some inequalities by finding the numbers x satisfying them.

▶ Example 6. Find all those real numbers x such that $|x + 2| \leq 4$.

Solution. Using Property 4, we see that the inequality $|x + 2| \leq 4$ is satisfied if and only if $-4 \leq x + 2 \leq 4$; that is, $-4 \leq x + 2$ and $x + 2 \leq 4$. From Rule 5

on inequalities, it is clear that $x + 2 \leq 4$ implies that $x \leq 2$ and $-4 \leq x + 2$ implies $-6 \leq x$. Therefore, those numbers x satisfying $|x + 2| \leq 4$ are $-6 \leq x \leq 2$.

▶ **Example 7.** Find all those numbers x such that $|x - 2| \geq 3$.

Solution. Using Property 5, we see that the inequality $|x - 2| \geq 3$ is satisfied if and only if $x - 2 \geq 3$ or $x - 2 \leq -3$. Now, it is clear (Why?) that $x - 2 \geq 3$ implies that $x \geq 5$ and $x - 2 \leq -3$ implies that $x \leq -1$. Thus those numbers x satisfying $|x - 2| \geq 3$ are $x \geq 5$ or $x \leq -1$. (See Figure 1.1.5.) The crosshatched

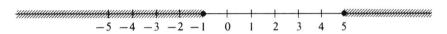

Figure 1.1.5

region in Figure 1.1.5 is the collection of those numbers x such that $x \leq -1$ or $x \geq 5$.

In our next example we solve a slightly different type of inequality. By a nonnegative number we mean a real number that is positive or zero.

▶ **Example 8.** Find all those numbers x such that $x(x + 5) \geq 0$.

Solution. Using the fact that the product of two nonnegative numbers is nonnegative if (1) both x and $x + 5$ are nonnegative or (2) both x and $x + 5$ are nonpositive, we have

from (1) $x \geq 0$ and $x + 5 \geq 0$, which yields $x \geq 0$;
or from (2) $x \leq 0$ and $x + 5 \leq 0$, which implies $x \leq -5$.

Thus, the solution is the set of all those numbers x such that $x \geq 0$ and $x \leq -5$. (See Figure 1.1.6.)

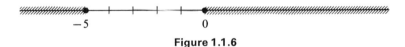

Figure 1.1.6

We conclude this section with a brief discussion of *sets*. We have frequently referred to a *collection* of numbers or *all* numbers satisfying certain conditions. It would be very convenient to have a notation for these concepts. For this purpose it is helpful to use the notation and terminology of set theory. In mathematics, we define a set simply as a collection of objects. For example, we speak of the set of real numbers, meaning the collection of all real numbers. The members of a set are called *elements* of the set and are said to *belong to* or to be *contained in* the set. In this text we are mainly interested in sets of real numbers.

Notation 1. We designate sets by capital letters $A, B, C, \ldots$ and elements of a set by small letters $a, b, c, \ldots$.

Notation 2. We write "$x \in A$" to mean x is an element of the set A.

We say that two sets A and B are equal and write $A = B$ if and only if they have the same elements.

Sets may be defined either by listing the elements in the set or by stating the specific properties possessed by each element of the set. For example, suppose A is the set of positive integers less than or equal to 5. We may either list the elements of A and write

$$A = \{1, 2, 3, 4, 5\},$$

where the brackets $\{ \ \}$ indicate the set, or we may write

$$A = \{x \mid x \text{ is a positive integer and } x \leq 5\},$$

which is read: A is the set of all those numbers x such that x is a positive integer and $x \leq 5$.

In general it is convenient to use the second method of defining a set, where we use $\mid$ to indicate the words "such that."

Notation 3. $A = \{x \mid x \text{ has property } P\}$. This notation means that A is the set of all those elements x such that x has property P. It is clear that if we write $a \in A$, then a has property P. We determine whether or not an element is a member of a specific set by checking to see if the element possesses the property P.

▶ **Example 9.** Write, using the second method of defining sets, the notation for the set of numbers whose absolute value is less than 5.

Solution. Let A designate the set. Then

$$A = \{x \mid |x| < 5\}.$$

In this example any element of A must have the property that its absolute value is less than 5. Note that 6 is not an element of A. We indicate this by writing $6 \notin A$.

▶ **Example 10.** Does the number $\frac{1}{2}$ belong to the set A, where

$$A = \{x \mid |x + 2| > 3\}?$$

Solution. Clearly $\frac{1}{2} \notin A$, since $|\frac{1}{2} + 2| < 3$. That is, $\frac{1}{2}$ does not possess the property necessary to make it an element of A.

Sometimes we want to speak of only a particular collection of elements of a set, rather than the set itself. For example, if $A = \{x \mid x \text{ is a real number}\}$, we may only want to discuss the set $B = \{x \mid x \text{ is a positive number}\}$. Note, every element of B is an element of A, but A consists of many elements not included in B. In this case we say that B is a *subset* of A and write $B \subset A$, read "the set B is contained in the set A." We use the word "proper" to indicate that A contains elements which are not in B. We can therefore make the following definition.

DEFINITION 3. *A set B is a subset of A, written $B \subset A$, if every element of B is an element of A.*

[*Note:* If $B \subset A$, and if every element of A is contained in B (that is, $A \subset B$), then $A = B$.]

There are certain operations on sets which will be useful in discussing the solutions to inequalities. These are the operations of set union (or addition) and set intersection.

DEFINITION 4. *The* union *of two sets A and B, written A ∪ B, is defined as follows:*

$$A \cup B = \{x \mid x \in A \text{ or } x \in B\}.$$

▶ Example 11. If $A = \{x \mid 0 \le x \le 5\}$ and $B = \{x \mid 1 \le x \le 6\}$, find $A \cup B$.

Solution. Observe that, by Definition 4, $A \cup B = \{x \mid 0 \le x \le 6\}$. This is clearly illustrated in Figure 1.1.7.

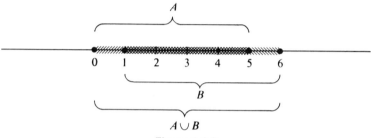

Figure 1.1.7

DEFINITION 5. *The* intersection *of two sets A and B, written A ∩ B, is defined as follows:*

$$A \cap B = \{x \mid x \in A \text{ and } x \in B\}.$$

▶ Example 12. Let A and B be defined as in Example 11. Find $A \cap B$.

Solution. By Definition 5,

$$A \cap B = \{x \mid 1 \le x \le 5\}.$$

The crosshatched interval in Figure 1.1.8 indicates $A \cap B$.

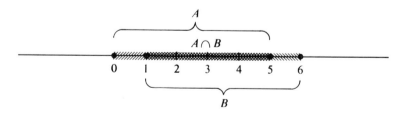

Figure 1.1.8

Certain subsets of the real numbers occur so often in our work that we shall give them special names. We first define these subsets and then illustrate them with examples on the real axis. These subsets are called *intervals*.

DEFINITION 6. *A closed interval on the real axis, written* $[a, b]$, *where a and b are any real numbers, is defined by*

$$[a, b] = \{x \mid a \leq x \leq b\}.$$

Geometrically, $[a, b]$ is illustrated in Figure 1.1.9.

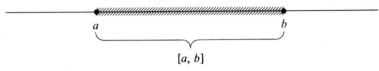

$$[a, b]$$

Figure 1.1.9

DEFINITION 7. *An open interval, written* (a, b), *where a, b are any real numbers, is defined by*

$$(a, b) = \{x \mid a < x < b\}.$$

That is, the numbers a and b are excluded from our set.

Figure 1.1.10 illustrates (a, b) on the real axis.

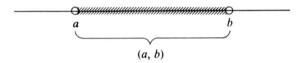

$$(a, b)$$

Figure 1.1.10

In obvious ways we can define

DEFINITION 8. $[a, b) = \{x \mid a \leq x < b\}.$

DEFINITION 9. $(a, b] = \{x \mid a < x \leq b\}.$

These sets are called half-open intervals. Other useful subsets of the real numbers are the sets

DEFINITION 10. $[a, \infty) = \{x \mid a \leq x\}.$

DEFINITION 11. $(-\infty, b] = \{x \mid x \leq b\}.$

The symbol ∞ (infinity) is a notational convenience, *not a number*, and we use it as we have done in Definitions 10 and 11. One could easily define (a, ∞) and $(-\infty, b)$.

We could now go back to our inequalities and interpret our solutions as subsets of the real numbers. These subsets are often referred to as *solution sets*. For example, recall from Example 6 that those numbers x satisfying the inequality $|x + 2| \leq 4$

were $-6 \leq x \leq 2$. We could now say that the solution set to this inequality is the closed interval $[-6, 2]$.

Let us consider several other examples.

▶ **Example 13.** Find the solution set of the inequality $|x - 4| \geq 5$.

Solution. Observe from Property 5 that $|x - 4| \geq 5$ implies that either $x - 4 \geq 5$ or $(x - 4) \leq -5$. Thus, either $x \geq 9$ or $x \leq -1$. The solution set S is given by

$$S = (-\infty, -1] \cup [9, \infty).$$

In Figure 1.1.11 the crosshatched region denotes S.

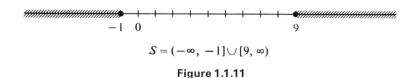

$$S = (-\infty, -1] \cup [9, \infty)$$

Figure 1.1.11

▶ **Example 14.** Find the solution set of the inequality $x(x - 2) \geq 0$.

Solution. Using the fact that the product of two numbers is positive if (1) both numbers x and $x - 2$ are positive or (2) both numbers x and $x - 2$ are negative, we have:

from (1), $x \geq 0$ and $x - 2 \geq 0$, which implies $x \geq 2$;
or from (2), $x \leq 0$ and $x - 2 \leq 0$, which implies $x \leq 0$.

Thus the solution set S is given by

$$A = (-\infty, 0] \cup [2, \infty).$$

There are other properties and operations with sets which could be discussed. These concepts will be introduced as they are needed.

Exercises

1. Suppose x and y are two numbers such that $x > y$. Which of the following inequalities do you know to be true? (Simply refer to one of the Rules 1–7 to justify your result.)

(a) $\frac{1}{5}x > \frac{1}{5}y$.

(b) $x + 5 > y + 5$.

(c) $-x > -y$.

(d) $-x + 5 < -y + 5$.

(e) $\dfrac{5}{x} < \dfrac{5}{y}$.

(f) $\dfrac{1}{x} + 5 < \dfrac{1}{y} - 5$.

(g) $-5x < -5y$.

2. For what real numbers x is

(a) $|x| = 3.$ (b) $|x - 2| = 3.$ (c) $|x + 2| = 4.$

(d) $2|x| = 4.$ (e) $|5x| = 2.$ (f) $|\frac{1}{2}x| = 5.$

(g) $|x - 2| = 2 - x.$ (h) $|2x| \leq 6.$ (i) $|\frac{1}{2}x| \leq 3.$

(j) $|x + 2| \leq 1.$ (k) $|x - 3| \geq 2.$ (l) $|\frac{1}{3}x - 9| \leq 7.$

3. Find all those real numbers x satisfying the following inequalities and indicate the appropriate intervals on the real line. Express the answer in terms of solution sets.

(a) $|2x| \leq 1.$ (b) $|x + 3| \leq 5.$ (c) $|x - 2| > 4.$

(d) $|x - 2| \leq 2 - x.$ (e) $|4 - 2x| \leq 1.$ (f) $|3 + 2x| \leq 7.$

4. Find all those real numbers x satisfying the following inequalities and indicate the appropriate intervals on the real line. Express your answers in terms of solution sets.

(a) $x(x + 3) \geq 0.$ (b) $x(3 - x) \geq 0.$

(c) $x(x + 3) \leq 0.$ (d) $x(3 - x) \leq 0.$

(e) $x(2x + 1) \leq 0.$ (f) $(x - 1)(2x + 3) \geq 0.$

(g) $(x - 1)/x \leq 0.$ [*Hint:* Consider two cases: (1) $x \geq 0$; (2) $x < 0.$]

(h) $x^2 + x \leq 6.$ [*Hint:* First rewrite the inequality as $x^2 + x - 6 \leq 0$ and then observe that $x^2 + x - 6 = (x + 3)(x - 2).$]

(i) $x^2 + 3x \geq -2.$ [*Hint:* Use the same technique as in (h).]

5. Explain why the following statements are true:

(a) $|x|$ is the larger of the numbers x and $-x.$

(b) If $-2 < x < 2$, then $|x| < 2.$

6. Assume that a used-car dealer budgets $400 for radio advertising and that a local radio station charges $80 for a one-minute commercial during the night call show. How many of these commercials can the dealer afford? [*Hint:* Let x denote the unknown number of commercials. Clearly, their cost is $80x$. Hence, you must solve the inequality $80x \leq 400.$]

7. An automatic elevator designed to carry a maximum load of 2,400 pounds is being used to lift pianos weighing 500 pounds each. What is the greatest number of pianos the elevator can lift?

8. Suppose that Mr. Brown, who has $10,000 in his savings account, wants to use some of his money to buy a certain kind of stock which sells for $80 a share, but does not want the balance of his savings account to go below $3,000. What is the greatest number of shares of stock that he can buy? (Assume he cannot buy a fraction of a stock.) [*Hint:* Let x denote the unknown number of shares. Clearly, $10,000 - 80x \geq 3,000.$]

9. Let A and B be any two sets such that $A = B$. Show that this is true if and only if $A \subset B$ and $B \subset A$. [*Hint:* You must first show that if $A = B$, then $A \subset B$ and $B \subset A$; and then show if $B \subset A$ and $A \subset B$, $A = B$. Use the definitions to do this.]

10. (a) Show that if $c > 0$, the closed interval $[-c, c]$ consists of all those real numbers x such that $|x| \leq c$.

    (b) Show that if $c > 0$, the closed interval $[a - c, a + c]$ consists of all those real numbers x such that $|x - a| \leq c$.

11. Show by examples that the following are in general *not* correct statements:

    (a) If $a < b$ and $c < d$, then $ac < bd$.

    (b) If $1/a > 1/b$ and $c > d$, then $c/a > d/b$.

12. (a) Let x be any real number. Show that $-|x| \leq x \leq |x|$.

    (b) Using part (a) and the rules for working with absolute values, show that the triangle inequality $|x + y| \leq |x| + |y|$ is valid.

*13. In this exercise we will show that $\sqrt{2}$ may not be written as the quotient of two integers. Answer all queries in parentheses.

Suppose we assume that it is possible to express $\sqrt{2}$ as the quotient of two integers, say $\sqrt{2} = m/n$, where m and n have no common divisors. If we square both sides of the equation $\sqrt{2} = m/n$, we obtain $2 = m^2/n^2$ and hence $2n^2 = m^2$. The number m^2 and thus the number m must be an even integer. (Why?) This implies that $m = 2p$, where p is any integer. The substitution of $m = 2p$ into the equation $2n^2 = m^2$ yields $n^2 = 2p^2$. (Why?) This last statement implies that n is also an even integer. (Why?) Since m and n are both even, they must have a common divisor greater than 1. This last statement contradicts the original assumption that m and n have no common divisor greater than 1. Thus our original assumption must have been false and hence $\sqrt{2}$ is not a rational number.

1.2 Coordinate Systems

Before proceeding any further, we need a precise way of measuring the distance between two points, say x_1 and x_2, on the real line. We define this concept as follows:

DEFINITION 1. *Given two points x_1, x_2 on the real line, the distance from x_1 to x_2 is $|x_1 - x_2|$.*

Thus if $x_1 = 1$ and $x_2 = 5$, there are 4 units separating them. In this case $|x_1 - x_2| = |1 - 5| = |-4| = 4$, which agrees with the more intuitive interpretation given above. Note the following facts:

1. The distance from x_1 to x_2 is positive or zero.
2. If the distance from x_1 to x_2 is zero, then $x_1 = x_2$.

3. Since $|(x_1 - x_2)| = |-(x_2 - x_1)| = |x_2 - x_1|$, the distance from x_1 to x_2 is the same as the distance from x_2 to x_1.
4. If $x_1 \neq x_2$, then the distance from x_1 to x_2 is greater than zero.

We have previously observed that once a unit of length is chosen we can represent numbers as points on a line and conversely. Let us now extend this procedure to the plane and to pairs of real numbers. Choose a point in the plane and call this point the origin, 0. Draw a horizontal line through 0 and construct a line perpendicular to it through 0. The horizontal line is called the x axis and the vertical line is called the y axis. Both of these axes are real lines with 0 at the origin.

The order of the numbers on the x and y axes can most easily be described by Figure 1.2.1. Observe that the two axes divide the plane into four quadrants. If we let P be any point in the plane, and draw perpendicular lines to the horizontal and vertical axes, we obtain two numbers a and b as indicated in Figure 1.2.1; a is called

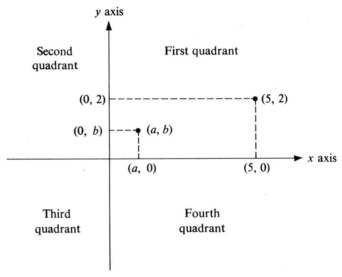

Figure 1.2.1

the x coordinate of P and b is called the y coordinate of P. We use the notation (a, b) to describe the point in the plane whose x coordinate is a and whose y coordinate is b. For example, to find the point $(5, 2)$, count 5 units along the x axis and then up 2 units. For a point in the first quadrant, both x and y coordinates are positive. What is the situation for the other quadrants? A little thought reveals the following:

1. Every pair (a, b) of real numbers corresponds to a point in the plane. (Just measure a units along the x axis—to the right if a is positive and to the left if a is negative—and b units up to determine P if b is positive. If b is negative, measure b units down.)
2. To every point P in the plane we can associate a pair (a, b). We simply drop perpendicular lines L_1 and L_2 to the x and y axes and determine the coordinates of P by noting that L_1 intersects the x axis at $(a, 0)$ and that L_2 intersects the y axis at $(0, b)$.

Thus the coordinates of P must be (a, b). (See Figure 1.2.2.) [Statements 1 and 2 say that there exists a one-to-one correspondence between points in the plane and pairs of real numbers.]

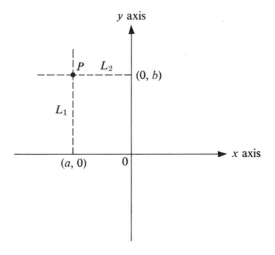

Figure 1.2.2

3. The order of the pair is *important*; thus $(-1, 5)$ is not the same point as $(5, -1)$. (See Figure 1.2.3.) For this reason the pair (a, b) is referred to as an ordered pair.

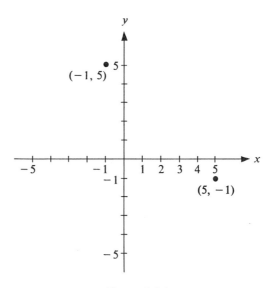

Figure 1.2.3

4. (a_1, b_1) and (a_2, b_2) represent the same point if and only if $a_1 = a_2$ and $b_1 = b_2$.

We now extend the concept of distance to points in the plane with the following definition.

DEFINITION 2. *The distance from $P_1 = (x_1, y_1)$ to $P_2 = (x_2, y_2)$, written $d[P_1, P_2]$, is defined by*

$$d[P_1, P_2] = \sqrt{(x_1 - x_2)^2 + (y_1 - y_2)^2}.$$

(This distance is just the length of the line segment $P_1 P_2$ in Figure 1.2.4.)

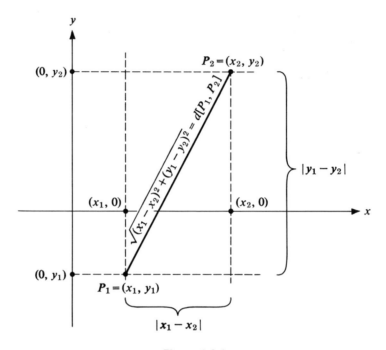

Figure 1.2.4

If we consider Figure 1.2.4, it becomes clear that Definition 2 follows easily from the Pythagorean theorem. Recall that $|a|^2 = a^2$.

▶ Example 1. Find the distance from $P_1 = (1, 2)$ to $P_2 = (4, 6)$.

Solution. $d[P_1, P_2] = \sqrt{(1 - 4)^2 + (3 - 6)^2} = \sqrt{18}.$

Note the following:

1. $d[P_1, P_2] \geq 0$.
2. $d[P_1, P_2] = 0$ if P_1 is the same point as P_2.
3. $d[P_1, P_2] \neq 0$ if $P_1 \neq P_2$.
4. $d[P_1, P_2] = \sqrt{(x_1 - x_2)^2 + (y_1 - y_2)^2}$
   $= \sqrt{(x_2 - x_1)^2 + (y_2 - y_1)^2} = d[P_2, P_1]$.

Actually it doesn't matter in which order you write the x's and the y's as long as you group the x's together and the y's together.

5. If P_1 and P_2 are points on the x axis, then our two definitions of distance agree. Thus, if $P_1 = (x_1, 0)$ and $P_2 = (x_2, 0)$, then

$$d[P_1, P_2] = \sqrt{(x_1 - x_2)^2 + (0 - 0)^2} = \sqrt{(x_1 - x_2)^2} = |x_1 - x_2|.$$

The same comment is valid for points on the y axis.

Exercises

1. Graph the following points:

   (a) $(-1, 9)$.   (b) $(5, 2)$.   (c) $(-1, -2)$.

   (d) $(-11, 6)$.   (e) $(0, 0)$.   (f) $(2, -1)$.

   (g) $(\frac{1}{2}, -\frac{2}{3})$.   (h) $(-\frac{2}{3}, -\frac{1}{2})$.   (i) $(-1, 0)$.

   (j) $(0, -1)$.

2. Find the distance from

   (a) $(1, 3)$ to $(5, 9)$.   (b) $(1, 1)$ to $(-3, 7)$.

   (c) $(-1, 0)$ to $(6, -2)$.   (d) $(-1, 4)$ to $(-1, -4)$.

   (e) $(19, -7)$ to $(0, 0)$.   (f) $(-2, -3)$ to $(-3, -4)$.

   (g) $(-1, 0)$ to $(0, -1)$.   (h) $(-\frac{1}{2}, -\frac{1}{2})$ to $(\frac{1}{2}, \frac{1}{2})$.

3. In the following exercises locate the indicated points on a graph:

   (a) All points whose y coordinates are 1.

   (b) All points whose x coordinates are -3.

   (c) The points for which $x > 0$. Note that this set consists of all points to the right of the y axis.

   (d) The points for which $y > 2$. This set consists of those points above the horizontal line through the point $(0, 2)$.

   (e) All points (x, y) for which $x > 0$ and $y > 2$. [*Hint:* Use (c) and (d).]

   (f) All points (x, y) for which $|x| < 2$.

   (g) All points (x, y) for which $x > 0$ and $y < 0$.

4. In what quadrants is

   (a) The ratio $y/x < 0$?   (b) The ratio $y/x \geq 0$?

5. Let L be the line segment joining (x_1, y_1) to (x_2, y_2). By means of the distance formula show that $\left(\dfrac{x_1 + x_2}{2}, \dfrac{y_1 + y_2}{2}\right)$ is the midpoint of L.

6. An IBM 704 computer is to be shipped from Los Angeles to Reno. It can be shipped directly by train at a cost of $5 per mile or by truck via San Francisco

at a cost of $3 per mile. Assuming that Los Angeles, Reno, and San Francisco are located as shown in Figure 1.2.5, which way is cheaper?

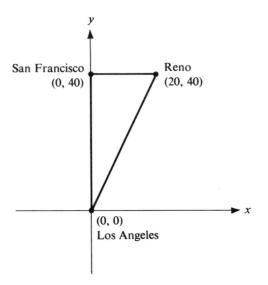

Figure 1.2.5

7. A vacuum cleaner salesman living in town A of Figure 1.2.6 must regularly visit customers in towns B and C. If all units are in miles, what is the total length of one round trip?

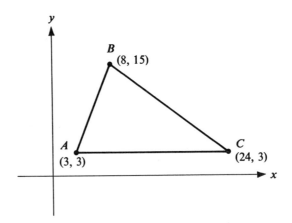

Figure 1.2.6

8. Let $C = \{(x, y) \mid$ distance from the point (x, y) to the origin is equal to $R\}$.

(a) Show that $x^2 + y^2 = R^2$. [Note that if $x^2 + y^2 = R^2$, then the distance from the point (x, y) to the origin is R; this equation represents the equation of a circle whose center is at the origin and having radius R.] Draw a picture of this circle in the rectangular coordinate system.

(b) Find the equation of the circle whose center is at the origin and whose radius is 5. Check by substitution whether or not the following points lie on this circle: $(2, 5)$, $(-1, 4)$, $(3, -4)$, $(-5, 0)$, $(0, -5)$.

9. After 8 hours the bacterial population in a certain culture is less than 400,000. If we let x represent the time and y represent the population, this statement may be expressed in the form for $x > 8$, $y < 400,000$. Shade in the appropriate region in the xy plane.

1.3 Straight Lines

Next to points, lines are the simplest and probably the most useful geometric objects we encounter. Consider the equation $y = 2x + 5$. If we plot a few points satisfied by the equation, we notice that they lie on a line. This is not an accident but an

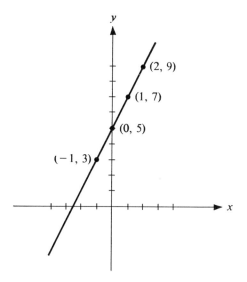

Figure 1.3.1

example of a general situation. To plot points we substitute values for x into the equation $y = 2x + 5$, and thereby arrive at the corresponding y values. In this case, for example, we find

x	2	1	0	-1
y	9	7	5	3

Thus the pairs $(2, 9)$, $(1, 7)$, $(0, 5)$, and $(-1, 3)$ satisfy the equation. The equation $y = mx + b$ describes a straight line for any pair of real numbers m and b. (Think of m and b as being fixed real numbers even though we do not know their values.)

There is one property of a line which should be obvious to anyone—its slope or steepness. What we need is a way of describing this concept of slope by a number. We do this as follows.

DEFINITION 1. *If (x_1, y_1) and (x_2, y_2) are two distinct points on the nonvertical line L, then the slope of L is*

$$\frac{y_2 - y_1}{x_2 - x_1}.$$

An alert reader may inquire whether a different choice of points on L might not give a different number for the slope. Consider Figure 1.3.2; a few elementary computa-

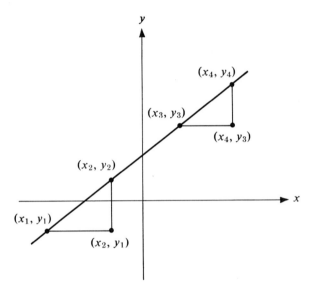

Figure 1.3.2

tions with similar triangles will show that the slope does not depend on the points chosen. One need only observe that

$$\frac{y_4 - y_3}{x_4 - x_3} = \frac{y_2 - y_1}{x_2 - x_1}.$$

If $x_1 = x_2$ (i.e., L is parallel to the y axis), we do not define the slope of L.

Observe that if the slope is positive, then the line rises to the right, and if the slope is negative, then the line sinks to the right. (See Figures 1.3.3 and 1.3.4.) It is not hard to see why. If

$$\frac{y_2 - y_1}{x_2 - x_1} = m > 0,$$

then

$$(y_2 - y_1) = m(x_2 - x_1) > 0 \quad \text{for } x_2 > x_1;$$

20

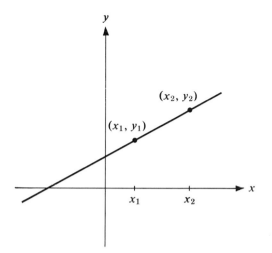

Figure 1.3.3

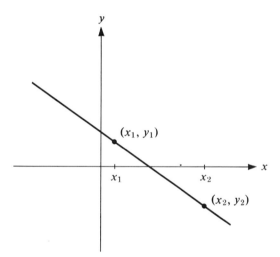

Figure 1.3.4

that is, $y_2 \geq y_1$, so the line is rising. A similar argument shows that if

$$\frac{y_2 - y_1}{x_2 - x_1} = m < 0,$$

then the line sinks to the right.

We shall discuss the formulas for the equations of various lines by considering the three problems that follow. It is important to note that the different forms for the equation of a line are really all equivalent.

▶ **Problem 1.** If a line L has slope m and passes through the point (x_1, y_1), what is the equation of L?

Solution.

$$(y - y_1) = m(x - x_1). \tag{1.3.1}$$

This is called the *point-slope form* of the equation of a line. It is a simple matter to derive Equation (1.3.1). Think of (x, y) as being a fixed point on L. Then

$$m = \frac{y - y_1}{x - x_1}$$

by the definition of slope and hence $(y - y_1) = m(x - x_1)$.

This equation may be reduced to the form

$$y = mx + b, \tag{1.3.2}$$

where $b = -mx_1 + y_1$. It is often convenient to use this form for the equation of a line. Suppose we wish to determine the equation of a line passing through $(1, 0)$ and having slope 1. Clearly, $m = 1$, so that $y = x + b$. We can determine b by using the fact that $y = 0$ when $x = 1$, and thus $b = -1$. The equation of the line is therefore $y = x - 1$.

▶ **Example 1.** Find the equation of the line with slope equal to $-\frac{1}{2}$ which passes through the point $(2, 1)$.

Solution. By Equation (1.3.1), $(y - 1) = -\frac{1}{2}(x - 2)$, or $y = -\frac{1}{2}x + 2$, which is in the form of Equation (1.3.2).

▶ **Problem 2.** Find the equation of the line L through the two (distinct) points (x_1, y_1) and (x_2, y_2).

Solution.

$$\frac{y - y_1}{x - x_1} = \frac{y_2 - y_1}{x_2 - x_1}. \tag{1.3.3}$$

We call this the *two-point form* of the equation for a straight line.

Since we are given two points (x_1, y_1) and (x_2, y_2) on the line, we can determine the slope of the line. Indeed,

$$m = \frac{y_2 - y_1}{x_2 - x_1}.$$

Thus, using formula (1.3.1), we see that

$$y - y_1 = \left(\frac{y_2 - y_1}{x_2 - x_1}\right)(x - x_1).$$

We could equally as well have written

$$y - y_2 = \left(\frac{y_2 - y_1}{x_2 - x_1}\right)(x - x_2).$$

The two equations represent the same line. However, we have chosen to write the equation as in (1.3.3) for reasons of symmetry.

▶ Example 2. Find the equation of the line through (1, 2) and (−3, −4).

Solution. By (1.3.2), we have

$$\frac{y-2}{x-1} = \frac{-4-2}{-3-1};$$

thus

$$\frac{y-2}{x-1} = \frac{3}{2},$$

whence

$$(y-2) = \tfrac{3}{2}(x-1),$$

and finally

$$y = \tfrac{3}{2}x + \tfrac{1}{2}.$$

The graph of this line is shown in Figure 1.3.5. Observe that two points are sufficient to graph a straight line.

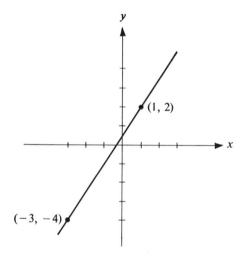

Figure 1.3.5

The different forms for the equation of a line are really equivalent. That is, one can convert from one to the other by simple algebraic manipulation. Example 2 illustrates this fact.

Lines which are parallel to the x or y axes are special cases, and the formulas do not always work in these cases. However, a little thought should resolve any difficulties in such situations.

▶ Example 3. The line parallel to the y axis and passing through the point (c, d) has as its equation $x = c$. This line has no slope. (Why?)

▶ Example 4. The line parallel to the x axis and passing through the point (a, b) has as its equation $y = b$. The slope of this line is zero. Note that using the point-slope form also yields the equation $y = b$.

An equation of the form $ax + by + c = 0$ also represents a line where a, b, c are real numbers such that both a and b are not zero. Conversely, *any* straight line has an equation of this form. We say that the equation $ax + by + c = 0$ is the most general form of the equation of a straight line. It is easy to verify that (1.3.1), (1.3.2), and (1.3.3) are all of this form. If $b \neq 0$, then

$$y = \left(-\frac{a}{b}\right)x - \frac{c}{d},$$

and the line has slope $(-a/b)$. If $b = 0$, then the line is parallel to the y axis and the equation reduces to $x = -c/a$.

▶ **Example 5.** Linear depreciation is one of several methods approved by the Internal Revenue Service for depreciating business property. If the original cost of the property is d dollars and it is depreciated linearly over N years, its value (undepreciated balance) y at the end of x years is given by

$$y = d - (d/N)x,$$

or equivalently

$$y = d(1 - x/N).$$

This is the equation of a straight line whose slope is given by $-d/N$. For instance, if football equipment worth \$2,000 is depreciated over 5 years, the undepreciated balance after x years is given by

$$y = 2{,}000(1 - x/5) = 2{,}000 - 400x.$$

After 5 years we obtain $y = 2{,}000 - 2{,}000 = 0$, which means that the equipment is completely depreciated.

▶ **Problem 3.** Given two lines $L_1: a_1x + b_1y + c_1 = 0$ and $L_2: a_2x + b_2y + c_2 = 0$, find their point of intersection. Notice first that the lines need not intersect, for they might be parallel. Secondly, if the two lines are the same, then every point is a point of intersection. Except for these special cases, two lines will intersect in exactly one point.

Suppose the point $P = (p, q)$ lies on the lines L_1 and L_2; then $a_1p + b_1q + c_1 = 0$ since P lies on L_1, and $a_2p + b_2q + c_2 = 0$ since P lies on L_2. Solving these equations for (p, q) we obtain the desired point P.

▶ **Example 6.** Find the point of intersection of the following lines: $2x + y + 1 = 0$ and $x + 2y + 1 = 0$.

Solution. Solving this system of equations we obtain $2x + 1 = 0$ and $3y + 1 = 0$, and thus $x = -\frac{1}{3}$, $y = -\frac{1}{3}$. The graphs of the two equations and their point of intersection are shown in Figure 1.3.6.

The following observations about lines are very useful. Let the distinct lines L_1 and L_2 have slopes m_1 and m_2, respectively. Then

1. L_1 and L_2 are parallel if and only if $m_1 = m_2$.
2. L_1 and L_2 are perpendicular if and only if $m_1 m_2 = -1$.

In Exercises 17 and 18, we outline the method of deriving these formulas.

▶ Example 7. Find the slope of all lines parallel to $2y = -3x + 2$.

Solution. Since this equation is equivalent to $y = -\frac{3}{2}x + 1$, the slope of all lines parallel to it must be $-\frac{3}{2}$.

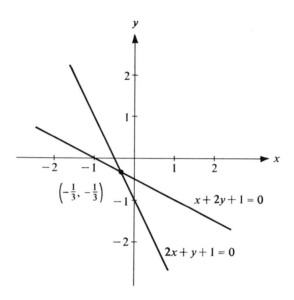

Figure 1.3.6

▶ Example 8. Find the equation of the line perpendicular to the line $y = 3x - 4$ and passing through $(-14, 9)$.

Solution. Let m be the slope of the line we are looking for. Then $3m = -1$ or $m = -\frac{1}{3}$. Thus the desired line has the equation $y - 9 = -\frac{1}{3}(x + 14)$, from the point-slope formula, (1.3.1).

Exercises

1. Find the slope of the line through the following pairs of points:

    (a) $(1, -1); (2, 6)$.
    (b) $(1, 0); (2, 1)$.

    (c) $(0, 0); (\frac{1}{9}, -1)$.
    (d) $(1, \frac{2}{5}); (4, -7)$.

    (e) $(3, -1); (2, 4)$.
    (f) $(4, -7); (3, 3)$.

    (g) $(2\frac{1}{3}, 6); (-1, 6)$.
    (h) $(\frac{19}{2}, -3); (0, 7)$.

    (i) $(0, 2); (1, 0)$.
    (j) $(-6, 3); (7, -\frac{2}{3})$.

2. Find an equation of the line passing through the pairs of points in Exercise 1.

3. (a) Find the equation of the line parallel to the y axis and 2 units to the right of it. What is its slope?

(b) Find the equation of the line parallel to the x axis and 5 units below it. What is its slope?

4. Find an equation for the line with slope m and passing through the indicated point:

(a) $5; (-1, 7)$.

(b) $2; (0, 1)$.

(c) $-8; (3, 2)$.

(d) $-\frac{1}{2}; (7, -2)$.

(e) $0; (0, 0)$.

(f) $0; (-1, 7)$.

(g) $\frac{2}{3}; (9, 1)$.

(h) $7; (11, -\frac{1}{3})$.

(i) $-2; (6, 5)$.

5. Find the slope of any line perpendicular to the line $y = -2x + 9$.

6. An apartment building worth \$300,000 was built in 1956. What is its value (for tax purposes) in 1975 if it is being depreciated linearly over 50 years?

7. In 1965 the Hoosier Cab Company purchased \$80,000 worth of new cabs. What is the value of these cabs in 1974, if they are being depreciated linearly over a period of 16 years?

8. (a) A machine purchased new for \$10,000 has a scrap value of \$5,000 after 10 years. If its value is depreciated linearly (from \$10,000 to \$5,000), find the equation of the line that enables us to determine its value after it has been in use any given number of years.

(b) What is its value after 3 years?

(c) Generalize the technique used in this problem so that it applies to a machine which is purchased at the original cost d, has the scrap value s after T years, and is depreciated linearly (from d to s) over a period of T years; that is, find a formula giving the value of the machine after x years. (This equation must involve as coefficients only the known quantities d, s, and T.)

9. Find an equation of the line through $(1, 2)$ and perpendicular to the line $y = 5x + 3$.

10. Find the equation of the line through $(7, 11)$ and parallel to the line $y = -2x + 7$.

11. Find the slope of the lines perpendicular to the following lines:

(a) $y = 3x - 4$.

(b) $4y + 5x - 2 = 0$.

(c) $x = 2y + 7$.

(d) $2x = 3y - 9$.

(e) $4x + 9y + 36 = 0$.

(f) $19x - 3y - 7 = 0$.

(g) $\frac{2}{3}x + \frac{3}{2}y - 1 = 0$.

12. Find the intersection of the following pairs of lines:

(a) $x - y = 1$,
    $2x + 4y = 3$.

(b) $x + \frac{1}{2}y - 9 = 0$,
    $y - x = 12$.

(c) $9x + 3y = 7$,
    $x + 3y = 6$.

(d) $6x + 9y = 4$,
    $5x - 4y = 3$.

(e) $x - y = 1$,
$\frac{1}{3}x + \frac{2}{5}y = 0$.

13. In studying the learning rate of a typical student, it was found that the size of his active vocabulary varied linearly with his age from the ages of 9 to 18. At the age of 9 his vocabulary was approximately 10,000 words while at the age of 18 his vocabulary was approximately 20,000 words. Find an equation relating the vocabulary size y and the student's age x. From the equation, find the size of the student's vocabulary at age 27 assuming the equation is still applicable then.

14. Find the equation of the line through the intersection of the two lines $x - y = 1$ and $2x - 3y = 7$ and perpendicular to the line through $(1, 2)$ and $(-3, 1)$.

15. Find the point or points on the line $y = 2x + 1$ whose distance from $(1, 0)$ is $\sqrt{5}$.

16. Find the shortest distance from the point $(1, 2)$ to the line $L: y - 2x = -3$. [*Hint:* Find the equation of the line perpendicular to L through $(1, 2)$. Find its intersection with L.]

17. Show that two lines are parallel if and only if their slopes are equal. [*Hint:* Let the equation of L_1 be $y = m_1 x + b_1$ and that of L_2 be $y = m_2 x + b_2$. Assume that two lines are parallel if and only if they coincide or have no points in common. If $m_1 = m_2$, show that one of the two possibilities must occur. If $m_1 \neq m_2$, show that there is precisely one point of intersection.]

18. Show that two lines are perpendicular if and only if the product of their slopes is -1. (See Figure 1.3.7.) [*Hint:* If $L_1: y = m_1 x + b_1$ and $L_2: y = m_2 x + b_2$, find their intersection (p, q). Solve for t and r. This is easy since $t = m_1 s + b_1$ and $r = m_2 s + b_2$. (Why?) Now use the Pythagorean theorem to check whether α is a right angle, given m_1 and m_2; conversely, given α a right angle, show that $m_1 m_2 = -1$.]

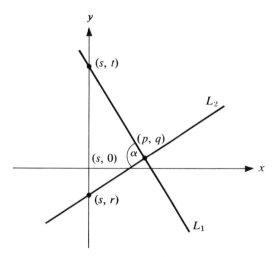

Figure 1.3.7

1.4 Functions. An Introduction

In our discussion of equations corresponding to straight lines we observed that once the real numbers m and b were fixed, we could determine y, where $y = mx + b$, by selecting the x's from the set of real numbers. For example, if we consider $y = 2x + 1$, it is clear that for each real number x we obtain one and only one real number y corresponding to that x. Notice that the entire set of y's we find in this way is exactly the set of real numbers. Thus, we may consider the equation $y = 2x + 1$ to be a rule which assigns to every real number x one and only one real number y. More precisely, this rule says: take twice x and add 1. This leads us to the idea of a function, one of the very basic concepts in all of mathematics.

We can define a function in abstract terms as follows.

DEFINITION 1. Let A and B be two given sets. A function is a rule which assigns to each element, a ∈ A, one and only one element b ∈ B.

The set A is called the *domain* of the function. We call the set of elements of B which correspond in this way to elements of A the *range* of the function. For most of our discussions, both A and B will be sets of real numbers. Although Definition 1 is really not a formal one, it will suffice for our purposes.

Before proceeding further we give some examples.

▶ Example 1. It is clear that a function was defined in the example in the first paragraph of this section. The domain is obviously the set of real numbers.

▶ Example 2. Let A be the set of tests of 250 students in this calculus course. To each test we assign a number between 0 and 100. Note that we have now defined a function. The domain is the set of 250 tests, rather than a set of numbers. The range consists of those numbers between 0 and 100 which are actual test scores. It is clear that to each test there corresponds one and only one grade. However, *many* tests may have the same grade. (That is, many elements of A may be assigned the *same* element of B—this is all right.)

▶ Example 3. Now we give an example that does *not* yield a function. Let A be a set of 10 specified books and let B be a set including all the authors of the books in A. Consider the rule that associates with each book in A its author or authors in B. Note we have not defined a function since a book in A may have two or more authors.

Before we proceed further in our discussion of functions we need some notation. Since we are dealing mainly with sets of real numbers, we concentrate on notation for this case. We denote functions by small letters, $f, g, h, \ldots$. It should be clear from the context when a letter denotes an element of a set and when it denotes a function. If we say that f is a function, we mean that f is the rule which assigns numbers to elements in its domain. If f is a function, then the number which f assigns to a number x in its domain is denoted by $f(x)$—read "f of x" or "f evaluated at x." Sometimes it is called the value of f at x. However, be careful to distinguish between f and $f(x)$. Observe that the symbol $f(x)$ makes sense only if $x \in$ domain of f; for other numbers x, the symbol $f(x)$ is *not* defined. The common procedure adopted for defining a function f is to indicate what $f(x)$ is for every number x in the domain of f. We illustrate this with some examples.

▶ **Example 4.** The function given in the example in the first paragraph of this section is defined by

$$f(x) = 2x + 1.$$

▶ **Example 5.** Let

$$f(x) = \begin{cases} x & \text{if } 0 \le x \le 1, \\ 2x & \text{if } 1 < x \le 2. \end{cases}$$

Note that the domain of f is $[0, 2]$. The rule is: to each $x \in [0, 1]$ assign the number x; to $x \in (1, 2]$ assign the number $2x$. It is clear that to each $x \in [0, 2]$ we have assigned one and only one real number. What is $f(\frac{1}{2})$? Since $\frac{1}{2} \in [0, 1]$, we see that $f(\frac{1}{2}) = \frac{1}{2}$. What is $f(\frac{3}{2})$? Since $\frac{3}{2} \in (1, 2]$, $f(\frac{3}{2}) = 3$. Observe that the range of f is $[0, 1] \cup (2, 4]$. Check this fact.

▶ **Example 6.** Let

$$f(x) = \sqrt{(1 - x^2)} \quad \text{for} \quad -1 \le x \le 1.$$

We have clearly specified the domain to be the interval $[-1, 1]$. The rule is obvious —for each $x \in [-1, 1]$ assign the number $\sqrt{(1 - x^2)}$. Note that the rule does not make sense for $x \notin [-1, 1]$. Indeed if we take $x = 2$, then $\sqrt{1 - 2^2} = \sqrt{-3}$, which is not a real number.

Examples 4–6 illustrate that a function need *not* be defined only by one formula. (See Example 5.) Notice that we carefully specified the domain of f in Example 5 but we did not include this specification in Definition 1. We adopt the following convention. *Unless the domain is explicitly stated, it is understood to consist of all numbers for which the definition makes any sense.* Thus, in Example 4, we understand that the domain is the set of all real numbers. In Example 6, if (without explicitly stating the domain) we had written $f(x) = \sqrt{(1 - x^2)}$, then our convention tells us that the domain must be a subset of $[-1, 1]$, since the definition would not make sense for other real numbers.

There are many useful ways to visualize functions. Two of these in particular will be of great aid to us, machine diagrams and graphs.

A function can be thought of as an input and output machine as follows (see Figure 1.4.1). If f is a function, then the domain of f is the set of all those numbers that are fed into our machine. (The machine accepts as input only these numbers.) Once the machine is set to apply the rule f, it produces the number $f(x)$ as its output. Our only restriction on this machine is that if the same number is put in on two different occasions then the two outputs must be the same. The range of the machine is the set of all outputs of the machine. As an illustration, we set up a machine for Example 5 (Figure 1.4.2). In this case the machine must be made to distinguish between the elements of $[0, 1]$ and the elements of $(1, 2]$.

The graph of a function plays a particularly important role in our applications. We define the graph of a function f as follows:

$$\text{graph of } f = \{(x, f(x)) \mid x \in \text{domain of } f\}.$$

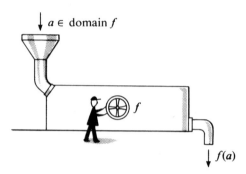

$a \in$ domain f

f

$f(a)$

Figure 1.4.1

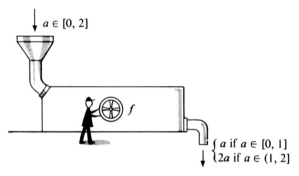

$a \in [0, 2]$

f

$\begin{cases} a \text{ if } a \in [0, 1] \\ 2a \text{ if } a \in (1, 2] \end{cases}$

Figure 1.4.2

Sometimes it is convenient when using the rectangular coordinate system to rewrite the definition as

$$\text{graph of } f = \{(x, y) \mid x \in \text{domain of } f \text{ and } y = f(x)\}.$$

We consider first the graph of the function $f(x) = 2x + 1$. The definition tells us that

$$\text{the graph of } f = \{(x, y) \mid x \text{ is a real number and } y = 2x + 1\}.$$

Now we plot this in the rectangular coordinate system. The graph of this function is clearly a straight line with slope 2, passing through the points $(0, 1)$ and $(-\frac{1}{2}, 0)$. (See Figure 1.4.3.) Since it is very cumbersome to use the set notation each time, we simply state that the graph of f is the straight line whose equation is $y = 2x + 1$.

As a second example, consider the graph of the function $f(x) = \sqrt{1 - x^2}$. Recall that the domain of $f = [-1, 1]$. By definition, the graph of $f = \{(x, y) \mid x \in [-1, 1]$ and $y = \sqrt{1 - x^2}\}$. Hence we set $y = \sqrt{1 - x^2}$ and observe that $y \geq 0$ (since $\sqrt{a}$ will *always* denote the positive root of a). Squaring both sides of this equation, we obtain $x^2 + y^2 = 1$. All points on the graph must be in the first two quadrants. It is not hard to see that $x^2 + y^2 = 1$ represents a circle of radius 1 with center at the origin. Let $P = (x, y)$ be a point in the plane whose distance from the origin is 1. Then the distance from P to 0 is given by $\sqrt{(x - 0)^2 + (y - 0)^2}$. Thus

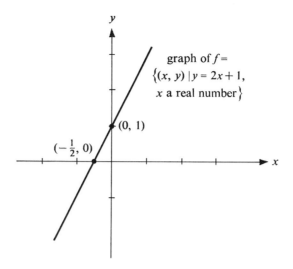

Figure 1.4.3

$\sqrt{(x-0)^2 + (y-0)^2} = 1$, and after a simplification we obtain $x^2 + y^2 = 1$. Hence the graph of our function is the upper semicircle illustrated in Figure 1.4.4.

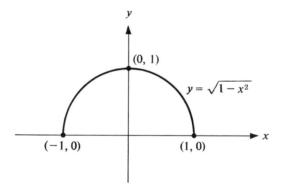

Figure 1.4.4

As a third example, we consider the function defined in Example 5:

$$f(x) = \begin{cases} x & \text{if } 0 \le x \le 1, \\ 2x & \text{if } 1 < x \le 2. \end{cases}$$

First set $y = f(x)$; that is, for $x \in [0, 1]$, $y = x$; and for $x \in (1, 2]$, set $y = 2x$. We observe that the graph is a straight line with slope 1 for $x \in [0, 1]$ and a straight line with slope 2 for $x \in (1, 2]$. (See Figure 1.4.5.) The open circle at $(1, 2)$ means that this point is not included in our graph. Observe the gap in the graph at $x = 1$. In Chapter 2, functions with gaps in their graphs will be discussed further.

For our last example, consider the function defined by $f(x) = |x|$. Note that the domain of this function is the set of all real numbers. If $x \ge 0$, then $f(x) = x$, and if $x < 0$, then $f(x) = -x$ by our definition of absolute value. Hence, we can see that the range of the function must be the set of all nonnegative real numbers. The

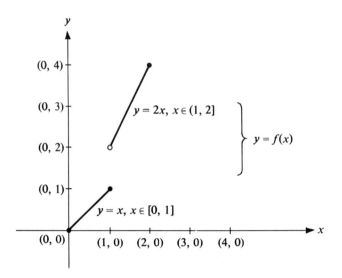

Figure 1.4.5

graph of f is shown in Figure 1.4.6. Observe that for $x > 0$, the graph of $f = \{(x, y) \mid y = x, x > 0\}$, and for $x < 0$, the graph of $f = \{(x, y) \mid y = -x, x < 0\}$.

Before proceeding further, we stop to make some simple observations concerning functions and their graphs.

Observe that in the notation for a function, $f(x)$, the x plays the role of a "dummy" in the sense that the function $f(x)$ is the same as $f(a)$, $f(b)$, etc. For example, if $f(x) = x^2 + 1$, then $f(a) = a^2 + 1$, $f(b) = b^2 + 1$, etc. The function is the *same* in all cases. Also, observe that if f is defined by the above formula, then

$$f(x + 2) = (x + 2)^2 + 1,$$

and

$$f(x - 2) = (x - 2)^2 + 1.$$

That is, no matter what we put in for x, we always come out with that thing squared

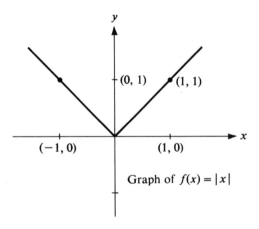

Graph of $f(x) = |x|$

Figure 1.4.6

plus one. Even if we put in a pink elephant we get (pink elephant)^2 + 1. Of course, this last remark assumes pink elephant is in the domain of f. It is essential to thoroughly comprehend this. Let us look at some examples.

▶ Example 7.   Let $f(x) = (x + 3)^3 + 1$. Find (1) $f(n + h)$; (2) $f(c)$; and (3) $f(\frac{1}{4}x)$.

Solution.   1. $f(n + h) = [(n + h) + 3]^3 + 1.$
2. $f(c) = (c + 3)^3 + 1.$
3. $f(\frac{1}{4}x) = (\frac{1}{4}x + 3)^3 + 1.$

With regard to the graph of f, notice that the domain of f is always found on the x axis while the range is the set of values given on the y axis. We illustrate this with the next example.

▶ Example 8.   Let

$$f(x) = \begin{cases} x, & \text{for} \quad x \in [0, 1), \\ 2, & \text{for} \quad x = 1, \\ 3x, & \text{for} \quad x \in (1, 2]. \end{cases}$$

The graph of $f = \{(x, y) \mid y = f(x), x \in [0, 2]\}$ is shown in Figure 1.4.7, with the domain and range of f as indicated in the figure.

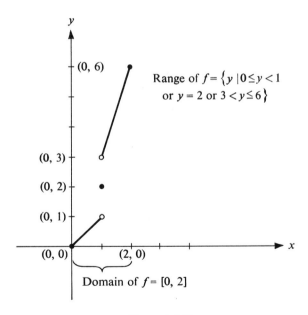

Figure 1.4.7

In Chapter 4, we shall carefully study the properties of graphs of functions and give simple procedures for accurately constructing them. However, it would be useful for us to have at our disposal certain elementary functions and their graphs. We list these functions and their graphs in the next section.

Exercises

1. Determine whether the following situations represent a function. If so, state the domains and ranges of the functions.

   (a) Let A be the set of all students (say 300) in a calculus class and suppose that B is the set of all grades (0–100) that the students receive on the final examination. Consider the rule that assigns to each student a number corresponding to his grade on the examination.

   (b) Suppose in (a) we consider the rule that assigns to numbers in B, students in A. Does this situation define a function? If so, state its range and domain.

   (c) Let A be a set of 20 authors and B the set of 30 books containing those written by the authors in A. Consider the rule that assigns to each author in A the book or books he has written.

   (d) Let A and B be the set of all real numbers. Consider the rule that assigns to each number n in A, the number n^5 in B.

   (e) Consider the rule that assigns to each mother her oldest child. Next, consider the rule that assigns to each mother her children.

   (f) The rule that associates to each person his social security number.

   (g) The rule that associates to each person the first letter of his first name.

   (h) The rule that associates with each letter of the alphabet all those people whose last name begins with that letter.

2. Let $f(x) = \begin{cases} x & \text{if } 0 \le x \le 1, \\ 3x & \text{if } 1 < x \le 3. \end{cases}$

   Find the domain and range of f. Show this explicitly on a graph; include the graph of the function. Find $f(\frac{5}{8}), f(\frac{1}{3}), f(2), f(\frac{7}{5}), f(1)$.

3. Let f be a function defined as follows:

   $$f(x) = \begin{cases} 1 & \text{if } -3 \le x < -1, \\ |x| & \text{if } -1 \le x < 0, \\ 1 & \text{if } \quad\quad x = 0, \\ x & \text{if } \quad 0 < x \le 3. \end{cases}$$

   (a) Carefully sketch a graph of f and clearly state the domain of f.

   (b) Find $f(\frac{2}{3}), f(2), f(-\frac{1}{2}), f(-\frac{2}{3}), f(-\frac{1}{3}), f(\frac{1}{3}), f(-2)$.

4. Consider the function

   $$f(x) = \begin{cases} \dfrac{x+1}{x-1} & \text{if } x \ne 1, \\ 1 & \text{if } x = 1. \end{cases}$$

   (a) What is the domain of f?

   (b) Determine each of the following: (i) $f(2)$, (ii) $f(-10)$, (iii) $f(a + 2)$, (iv) $f(a^2)$, (v) $f(1/a)$. (We assume a is a real number.)

5. Let $f(x) = (x + 1)^2$. Find (a) $f(n + h)$, (b) $f(c)$, (c) $f(\frac{1}{4}x)$.

6. If $f(x) = x^3$, find $f(\frac{1}{2}x)$, $f(x + 1)$, $f(2x)$, $f(x + \frac{1}{3})$.

7. Can a horizontal line be the graph of a function? Why? Is the line $x = 2$ the graph of a function? Why?

8. Graph the function $g(x) = \dfrac{1 - x}{1 - x}$. Is this the same function as $h(x) = 1$? Why not?

9. Let $f(x) = |x| - x$. What is $f(-1)$, $f(1)$, $f(-20)$? If $x \geq 0$, write another formula for $f(x)$. Do the same if $x < 0$.

10. Sketch a graph of the following functions and indicate at least two points on each graph:

    (a) $f(x) = 5x + \frac{1}{3}$.

    (b) $f(x) = \frac{1}{2}x + 1$.

    (c) $f(x) = |x| + x$.

    (d) $f(x) = \begin{cases} x & \text{if } x < 0, \\ 2 & \text{if } x = 0, \\ x & \text{if } x > 0. \end{cases}$

    (e) $f(x) = |x|/x$.

11. (a) Graph the following function and indicate its domain and range on the graph:
$$f(x) = \begin{cases} 2x & \text{if } x \in [0, 1), \\ 3 & \text{if } x = 1, \\ 5 - 3x & \text{if } x \in (1, \frac{5}{3}]. \end{cases}$$

    (b) If $0 \leq x \leq \frac{1}{2}$, what is (i) $f(x + 0.1)$, (ii) $f(x + 0.1) - f(x)$, (iii) $f(x + 1)$, (iv) $f(x + \frac{1}{2})$?

12. Can a circle be the graph of a function? Why? How about a vertical line?

13. Sketch the function whose graph goes through the point $(1, 1)$ and is

    (a) Perpendicular to the graph of the function $f(x) = 2x + 1$.

    (b) Parallel to the graph of the function $f(x) = 2x + 1$.

14. A book publisher agrees to pay an author royalties according to the following scheme: \$1.00 for the nth copy sold if $1 \leq n \leq 10,000$; \$1.50 for the nth copy if $10,000 < n \leq 20,000$; \$2.00 for the nth copy if $20,000 < n$. Let f be the function that assigns to the nth copy sold, the author's income for the first n books sold. Find formulas for $f(n)$ and graph the function f. Find $f(15,000)$, $f(20,000)$, $f(25,000)$, and $f(60,000)$.

15. The mathematics department's magnetic tape Selectric typewriter is supposed to be checked once a month. If this is not done (but the machine is checked at least once a year), the expected cost of repairs is \$10 plus ten times the square of the number of months the machine has gone without being checked. Express this relationship between the number of months the machine has gone without being checked and the expected cost of repairs by means of a formula, and use it to calculate the expected cost of repairs when the machine has not been

checked for (a) 2 months; (b) 4 months; (c) 6 months; (d) 8 months; and (e) 12 months.

16. The manager of a chain of bookstores has the following data on the supply s of certain paperbacks and their price p in dollars per cartons of books:

Price (dollars) p	4	6	8	10	17
Supply (hundred boxes) $s(p)$	0	12	21	25	27

(a) Display this information graphically.

(b) Check visually whether the function given by

$$s(p) = 30 - \frac{60}{p - 2}$$

can be used to approximate the above data by calculating $s(p)$ for each of the given values of p, plotting the corresponding points on the diagram constructed in part (a) and joining them by means of a smooth curve.

(c) Use the formula of part (b) to calculate $s(62)$ and $s(122)$ and discuss the practical significance of the results.

17. A farmer has 200 feet of fence with which he wishes to enclose three sides of a rectangular region bounded on the fourth side by a river. Express the area of the enclosure as a function of the length of one of its sides. [*Hint:* Let l denote one side and x another. Next, construct a diagram to illustrate the problem. Recall that the area of a rectangle is $l \cdot x$. Now, using the fact that the farmer has 200 feet of fence, express l in terms of x.]

1.5 Some Useful Functions. I

Before proceeding further, it would be advantageous to have at our disposal certain elementary functions and their graphs. We list these functions and their graphs.

(1) CONSTANT FUNCTIONS

Any function of the form $f(x) = c$, where c is any constant, is called a constant function. Its graph is shown in Figure 1.5.1.

(2) LINEAR FUNCTIONS

Any function of the form $f(x) = ax + b$, where a and b are constants, is called a linear function. The word linear is used because the graph of this function (see Section 1.3) is a straight line with slope a. Its graph is shown in Figure 1.5.2.

(3) POWER FUNCTIONS

Functions of the form $f(x) = x^n$, where n is a positive integer, are called power functions. We observe that if $n = 1$, the graph of f is a straight line with slope 1. For $n \geq 2$, the general shape of the graph depends upon whether n is an even or an

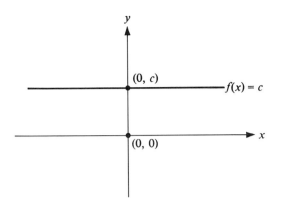

Graph of $f(x) = c$, where $c > 0$

Figure 1.5.1

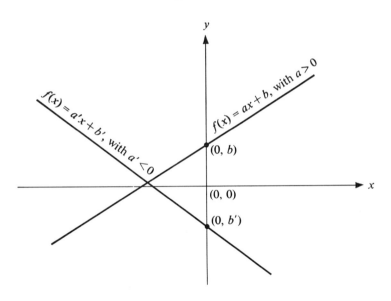

Figure 1.5.2

odd number. If n is even, the shape is similar to the graph of $f(x) = x^2$, shown in Figure 1.5.3. (The graph of this function is called a parabola.) If n is odd and > 3, then the graph is similar to $f(x) = x^3$ but flatter near zero and steeper outside of $[-1, 1]$ (Figure 1.5.4).

Notice that if $n = 2$ and we replace x by $-x$, we obtain

$$f(-x) = (-x)^2 = x^2 = f(x).$$

This means geometrically that the graph of f is *symmetric* about the y axis; that is, if (x, y) is on the graph, then so is $(-x, y)$. In fact, observe that this is true for any even integer $n \geq 2$. A function having the property that $f(-x) = f(x)$ is called an *even* function.

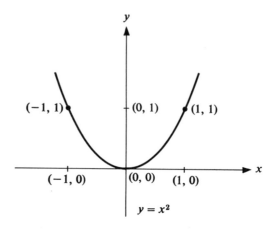

Figure 1.5.3

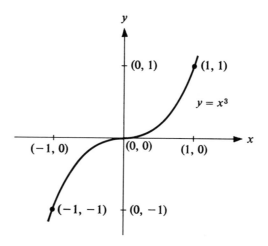

Figure 1.5.4

Also note that if $n = 3$ and we replace x by $-x$, we obtain

$$f(-x) = (-x)^3 = -x^3 = -f(x).$$

Geometrically, the graph of f is symmetric about the origin; that is, if (x, y) is on the graph, then so is $(-x, -y)$. Again, this is true for n odd and $n \geq 3$. A function possessing the property that $f(-x) = -f(x)$ is called an *odd* function. The notion of odd and even functions is introduced as a labor-saving device. If a function turns out to be odd or even, then once half of it is graphed the other half is obvious.

(4) POLYNOMIAL FUNCTIONS

A *polynomial function* is a function, defined for all real numbers x, of the form

$$f(x) = c_0 + c_1 x + c_2 x^2 + c_3 x^3 + \cdots + c_n x^n,$$

where $c_0, c_1, \ldots, c_n$ are given real numbers. Observe that constant functions, linear functions, and power functions are special cases of this function. For example, if all of the c's except c_n are zero and $c_n = 1$, we obtain the power function. We shall consider methods for graphing polynomials in Chapter 4.

(5) RATIONAL FUNCTIONS

A *rational function* is defined to be a function of the form

$$f(x) = \frac{g(x)}{h(x)},$$

where g and h are polynomial functions. Remember that we must exclude all those real numbers x for which $h(x) = 0$ from our domain. Some examples of rational functions follow.

▶ Example 1.

$$f(x) = \frac{x^2 + 3x + 1}{x^2 - 4}.$$

Thus, $\pm 2 \notin$ domain of f.

▶ Example 2.

$$f(x) = \frac{x^3 + 2x^2 + x + 1}{x^2 + 1}.$$

The domain of f includes all numbers, since $x^2 + 1 \neq 0$ for any real number x.

▶ Example 3.   A very important rational function is

$$f(x) = 1/x,$$

whose graph we draw in Figure 1.5.5. We note that $0 \notin$ domain of f. Any function of the form

$$g(x) = \frac{a}{x} + b,$$

where a and b are real numbers, is called a *hyperbolic* function. Since $f(-x) = -f(x)$ (Why?), we need only draw the graph in the first quadrant and by symmetry obtain the part of the graph in the third quadrant. It should be intuitively clear that as we choose larger and larger values of x the values $f(x)$ become smaller and smaller, that is, become closer to zero. Similarly for smaller and smaller positive values of x, the values of $f(x)$ become larger and larger.

▶ Example 4.   Hyperbolic curves play a dominant role in the study of the physiology of muscle and nerve.* A famous law due to Weiss is based on the finding that the intensity of an electric current required to just excite the tissue depends on the duration of current flow. In particular, Weiss discovered that the longer the duration of current flow the weaker will be the current required to excite. This relationship

---

* An interesting discussion of the role of hyperbolic curves in physiology can be found in J. G. Defares and I. N. Sneddon, *The Mathematics of Medicine and Biology* (North-Holland, Amsterdam, 1960), p. 71.

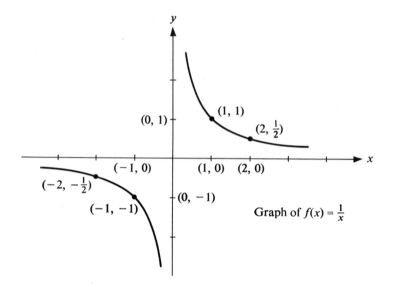

Figure 1.5.5

between current strength and duration was found experimentally to be a hyperbola, and could be represented by the following empirical formula:

$$i = \frac{a}{t} + b, \qquad t > 0, \tag{1.5.1}$$

where i is the current, t the time, and a, b are constants. The graph of the above relation for $a = b = 1$ is shown in Figure 1.5.6.

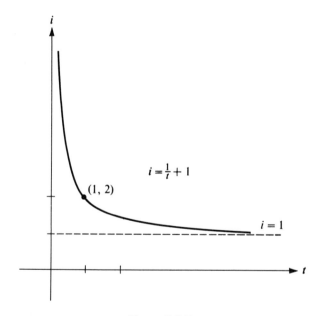

Figure 1.5.6

Exercises

1. (a) Sketch a graph of the following functions:
   (i) $f(x) = \frac{1}{2}x^2$,   (ii) $f(x) = 2x^2$,   (iii) $f(x) = \frac{1}{3}x^3$,
   (iv) $f(x) = 3x^3$,   (v) $f(x) = x^5$,   (vi) $f(x) = 3x^2$,
   (vii) $f(x) = 1/3x$.

   (b) In (a), which of the functions are odd functions? Even functions?

2. Suppose that in Example 4 of this section it has been discovered experimentally that at the end of four hours the current strength is 6 units of electricity, while at the end of eight hours the current strength is 8 units of electricity. Determine the relation between current strength and duration for this case. [*Hint:* In expression (1.5.1) determine the constants a and b.]

3. A psychology student believes that the score of a student on a given test worth 100 points can be expressed by the function

$$S(t) = \frac{I^2(50 + 5t - t^2)}{2,000},$$

   where I is the IQ of a given student and t is the number of hours spent studying for the test. Thus the score for a given individual is a function of the time spent studying. Find $S(t)$ for the following students with the given IQ's when $t = 0, 2, 10$:

Student	IQ
A	90
B	100
C	120
D	150

   Do you see anything wrong with this formula?

4. Taxis charge 25 cents for the first quarter mile of transportation and 5 cents for each additional quarter mile or part thereof. Graph this function for $0 \le x \le 4$, where x represents number of miles traveled.

5. A travel agency advertises all-expenses-paid trips to the Cotton Bowl for special groups. Transportation is by train. The agency charters one car on the train seating 50 passengers, and the charge is $100 plus an additional $5 for each empty seat. (Thus, if there are 4 empty seats each person pays $120, etc.) If there are x empty seats, how many passengers are there on the train, how much does each passenger have to pay, and what are the travel agency's total receipts? Plot the graph of the function that relates the travel agency's total receipts to number of empty seats and judge under what conditions its receipts will be greatest.

6. In connection with Exercise 5, how many passengers would have to go on the trip to give the travel agency total receipts of $3,000?

1.6 Inverse Functions

Before proceeding to give other useful examples, it is advantageous at this point to introduce a new concept involving functions. Recall from Section 1.4 the definition of a function: a rule which assigns to elements in one set of numbers *one* and *only* one number in another set. That is, if a is any number in the domain of f, then there is precisely one number b such that $f(a) = b$. Geometrically, this means that the vertical line through a meets the graph of the function f at just one point $(a, f(a))$. (See Figure 1.6.1.)

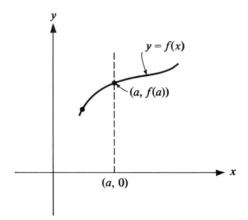

Figure 1.6.1

However, it is quite possible, given a point b in the range of values of f, to have two or more numbers $a_1, a_2, \ldots$, in the domain such that $f(a_1) = f(a_2) = f(a_3) = \cdots = b$. Geometrically, this means that the horizontal line through b may meet the graph of f at more than one point. (See Figure 1.6.2.)

There are, of course, some functions, such as $f(x) = 2x + 3, g(x) = x^3, h(x) = x^5$, for which this does not happen; that is, any horizontal line meets the graph of the

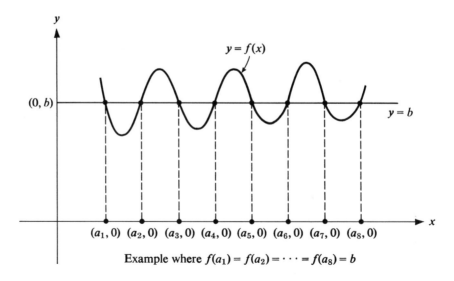

Figure 1.6.2

function in at most one point. Whenever this is the case, we can define a new function which is, roughly speaking, the reverse or inverse of the original function.

For the moment we concentrate our attention on functions which take on a given value *at most once*. That is, we are interested in the functions which take on every value in the range of the function exactly once. In this situation there is a simple way to define a new function g. If $f(2) = 6$, we set $g(6) = 2$. Thus if f takes P_1 to P_2, g takes P_2 back to P_1. This rule defines a function g whose domain is the range of f. More precisely, g takes every point in the range of f back into the domain of f. (See Figure 1.6.3.)

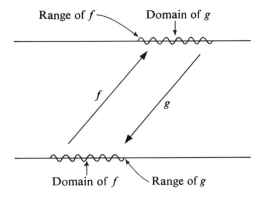

Range of f

Domain of g

f

g

Domain of f

Range of g

Figure 1.6.3

▶ Example 1. Let f be the function that assigns to every father his eldest son. Then the function g assigns to the eldest son, his father.

▶ Example 2. Let f be the function defined by $f(x) = 2x$. Observe that $f(1) = 2$, $f(2) = 4$, and $f(3) = 6$. According to the above discussion g should be the function such that $g(2) = 1$, $g(4) = 2$, and $g(6) = 3$. In fact, for any x, we want $g(2x) = x$. Thus, we would expect for any y, $g(y) = \frac{1}{2}y$, since if $y = 2x$, then $g(2x) = \frac{1}{2}(2x) = x$. In the following examples we shall establish a more systematic method for determining the function g.

We now give a mathematically precise definition of the foregoing concepts.

DEFINITION 1. A function f is 1–1 (read one-to-one) if $x_1 \neq x_2$ implies that $f(x_1) \neq f(x_2)$ for all x_1, x_2 in the domain of f. An equivalent formulation which is sometimes more convenient to check is: f is 1–1 if $f(x_1) = f(x_2)$ implies $x_1 = x_2$.

The last sentence in Definition 1 means that each horizontal line intersects the graph of f at most once.

▶ Example 3. Consider the function of Example 2, $f(x) = 2x$. Observe that, if we set $f(x_1) = f(x_2)$, we have $2x_1 = 2x_2$. Hence, $x_1 = x_2$ and the function f is one-to-one.

▶ Example 4. Consider the function $f(x) = x^2$. Observe that $f(x_1) = f(x_2)$ implies that $x_1^2 = x_2^2$ and hence $x_1 = \pm x_2$. Thus the condition in Definition 1 is not satisfied, and the function $f(x) = x^2$ is not one-to-one. If we had considered the function $f(x) = x^2$ for $x \geq 0$, then f would be a one-to-one function. (Why?)

DEFINITION 2. Let f be a one-to-one function and suppose that $y = f(x)$. Then the function g (which exists if f is one-to-one) defined by $g(y) = x$ for all x in the domain of f is called the inverse function of f. We denote g by f^{-1}. Thus, if $y = f(x)$, then $f^{-1}(y) = x$.

We might observe that if f is one-to-one the statements $y = f(x)$ and $x = f^{-1}(y)$ are identical. This will aid us in determining the inverse of various functions.

▶ Example 5. Let $y = 2x$. From Example 3, we see that $2x$ is a one-to-one function. To determine the *inverse* function, we need only solve for x in terms of y. Thus, $x = \frac{1}{2}y$ and $f^{-1}(y) = \frac{1}{2}y$. Observe that $f^{-1}[2x] = \frac{1}{2}(2x) = x$. That is, $f^{-1}[f(x)] = x$.

▶ Example 6. Let $y = f(x) = 3x - 2$. We wish to determine the inverse function f^{-1}. It is a simple matter to check that f is a one-to-one function. Let $x_1, x_2 \in$ domain of f. Then $f(x_1) = f(x_2)$ implies that $3x_1 - 2 = 3x_2 - 2$ and thus $3x_1 = 3x_2$ or, equivalently, $x_1 = x_2$. Hence, f is one-to-one. Now we can find f^{-1}. To proceed, we simply set $y = 3x - 2$ and solve for x in terms of y. Clearly, $x = (y + 2)/3$. Hence, $f^{-1}(y) = (y + 2)/3$ since $(y + 2)/3 = x$. Again note that

$$f^{-1}(3x - 2) = \frac{(3x - 2) + 2}{3} = \frac{3x}{3} = x.$$

If we have the graph of a function f, it is quite easy to obtain the graph of f^{-1}. Indeed, the graph of f^{-1} is just the graph of f rotated around the line $y = x$. (See Figure 1.6.4.) Let us see why this is the case. If we rotate the point (x_1, y_1) about

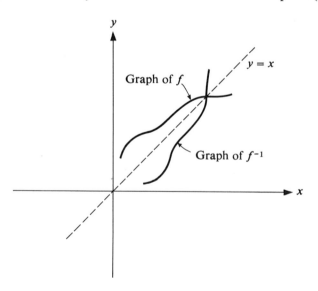

Figure 1.6.4

the line $y = x$, we obtain the point (y_1, x_1); we have simply interchanged the x and y coordinates. Let (a, b) be a point in the graph of f; i.e., $f(a) = b$. Then $f^{-1}(b) = a$ by definition of f^{-1}. If we rotate the point $(a, b) = (a, f(a))$ about the line $y = x$, we get the point $(b, a) = (b, f^{-1}(b))$. But the point $(b, f^{-1}(b))$ is a point in the graph of f^{-1}. This argument shows that if we rotate the graph of f about the line $y = x$, we end up in the graph of f^{-1}. It is not hard to see that we get *every* point in the graph of f^{-1} in this way. (Just reverse the roles of f and f^{-1}.)

Let us consider some examples.

▶ **Example 7.** Let $f(x) = 2x + 1$. Does f^{-1} exist? If so, find f^{-1} and construct its graph.

Solution. To show that f^{-1} exists, we must show that $f(x_1) = f(x_2)$ implies that $x_1 = x_2$. Suppose $f(x_1) = f(x_2)$. That is, $2x_1 + 1 = 2x_2 + 1$. Then clearly $x_1 = x_2$. To find f^{-1}, we must solve the equation $y = 2x + 1$ for x. We see that $x = (y - 1)/2$. Hence, $f^{-1}(y) = (y - 1)/2$. In order to construct its graph, we observe that if we first graph the function f, where $f(x) = 2x + 1$, and then rotate about $y = x$ (which amounts to interchanging the x and y coordinates) we obtain the graph of f^{-1}, where $f^{-1}(y) = (y - 1)/2$. (See Figure 1.6.5.)

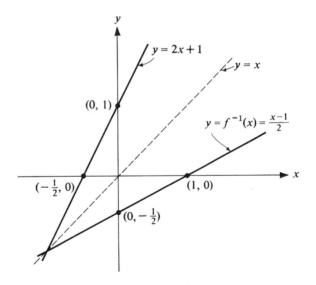

Figure 1.6.5

▶ **Example 8.** Recall from Example 4 that the function f defined by $f(x) = x^2$, for $x \geq 0$, is a one-to-one function. Hence an inverse f^{-1} exists. In order to construct its graph, we first graph f and then rotate about the line $y = x$ (see Figure 1.6.6). The inverse function f^{-1} is called the square root function, and we write $f^{-1}(x) = \sqrt{x}$. In the next section, we study the inverses of the power functions. Again, we remark that it is essential to specify that the domain is $x \geq 0$, since the function f where $f(x) = x^2$ for all real x is not one-to-one and hence does not have an inverse.

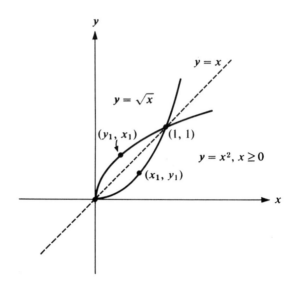

Figure 1.6.6

Exercises

1. Find the inverses of the following functions if they exist and draw the graphs of both f and f^{-1}. (In drawing the graph, show clearly how one can obtain the graph of f^{-1} from that of f.)

   (a) $f(x) = x + 2$.

   (b) $f(x) = 3x + 1$.

   (c) $f(x) = 1/x + 1, \quad x \neq 0$.

   (d) $f(x) = 2x^2$.

   (e) $f(x) = 2x^2, \quad x \geq 0$.

   (f) $f(x) = 3x - 2$.

   (g) $f(x) = |x|$.

   (h) $f(x) = |x|, \quad x \geq 0$.

   (i) $f(x) = 5x + \frac{1}{3}$.

2. (a) Let f be the function that assigns to each person his or her social security number. Does f^{-1} exist? If so, what is it?

   (b) Let f be the function that assigns numerical grades to 200 students in a calculus course. Does f^{-1} exist? If so, what is it?

3. Let f be a function defined by

$$f(x) = \frac{ax + b}{cx - a},$$

   where a, b, c are fixed nonzero numbers and $a^2 + bc \neq 0$. Show that f^{-1} exists and $f^{-1} = f$.

4. Show that the inverse of any function of the form $f(x) = ax + b$, where a and b are constants, always exists if $a \neq 0$. What is the graph of f^{-1}? Conclude that the inverse of any linear function is again a linear function.

1.7 Some Useful Functions. II

(6) ROOT FUNCTIONS

We have already seen in Example 8 of Section 1.6 that the function defined by $g(x) = x^2$, $x \geq 0$, has an inverse, which we write as $f(x) = x^{1/2}$, $x \geq 0$. It is not difficult to verify that the function $g(x) = x^n$, for $x \geq 0$ and n even, also has an inverse, which we denote by $f(x) = x^{1/n}$, $x \geq 0$. In like manner we observe that if $g(x) = x^n$ and n is odd, then g has an inverse given by $f(x) = x^{1/n}$. Note that if n is even, the domain of $f(x) = x^{1/n}$ is the set of all nonnegative real numbers, while if n is odd, the domain of f is the set of all real numbers. The function $f(x) = x^{1/n} = \sqrt[n]{x}$ is called a *root* function. The graphs of these functions are shown in Figures 1.7.1 and 1.7.2 for n even and odd, respectively.

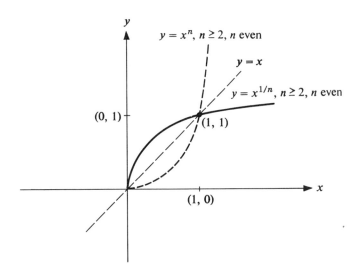

Figure 1.7.1

(7) EXPONENTIAL FUNCTIONS

Let a be a real number, where $a > 0$. We proceed to define the function $f(x) = a^x$, where x is any real number. This will take several steps. If you believe you can eliminate some of the steps, stop and think.

DEFINITION 1.   If a is any positive real number, then a^x is defined as follows:

1.   If x is a positive integer, say x = n,

$$a^x = a^n = \underbrace{a \cdot a \cdot a \cdots a}_{n \text{ times}}$$

(i.e., there are n factors of a).

2.   If x = 0, then $a^0 = 1$.

3.   If x is a negative integer, say x = $-n$, then $a^x = a^{-n} = 1/a^n$.

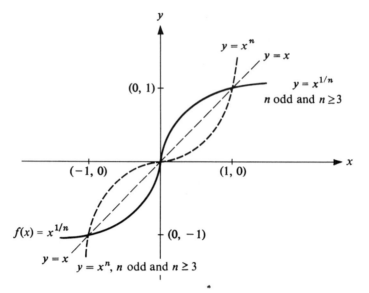

Figure 1.7.2

4. *If x is a rational number, say x = p/q, then $a^x = a^{p/q} = \sqrt[q]{a^p}$ (the qth root of a to the pth power).*

5. *After we have defined a^x for x rational, the graph of the function $f(x) = a^x$ for x rational would appear as in Figure 1.7.3. To define a^x on the irrational numbers we simply fill in the gaps (points) in the obvious way (i.e., we follow the pattern).*

It therefore follows that the function $f(x) = a^x$ has for its domain the set of real numbers. The graph of this function for two different choices of a is shown in Figure 1.7.4. Observe that as x becomes very large positively, $f(x)$ also becomes large. In like manner we observe that if $-x$ becomes large, then $f(x)$ approaches zero. We also see that a^x is always *positive*.

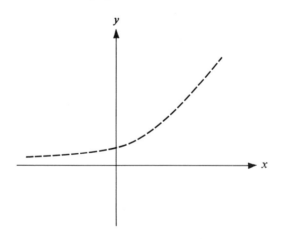

Figure 1.7.3

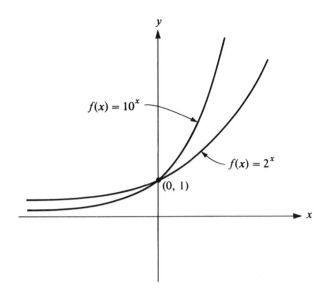

$f(x) = 10^x$

$f(x) = 2^x$

$(0, 1)$

Figure 1.7.4

In our future discussions we find that the irrational number $e = 2.71828\ldots$ appears very often as a choice for a. The function $f(x) = e^x$ is called *the exponential function* and is sometimes written $f(x) = \exp(x)$. This strange choice of a will be explained later.

From the definition of a^x, we can easily show that the following properties hold:

PROPERTY 1. $a^x \cdot a^y = a^{x+y}$.
[In function notation this states that $f(x) = a^x$ satisfies the equation $f(x)f(y) = f(x + y)$.]

PROPERTY 2. $(a^x)^y = a^{xy}$.

PROPERTY 3. $a^x/a^y = a^x \cdot a^{-y} = a^{x-y}$.

PROPERTY 4. If $a \neq 1$, then $a^x = a^y$ if and only if $x = y$.

These properties of exponential functions are very useful and should be remembered. Also observe the difference between a power and an exponential function—for a power function the *base* is variable and the exponent is fixed, while for an exponential function a is fixed and the exponent variable.

(8) LOGARITHM FUNCTIONS

Observe that Property 4 of exponential functions implies that if $a \neq 1$, then the function g, where $g(x) = a^x$, is one-to-one and hence g has an inverse. We denote this inverse function by $f(x) = \log_a x$, read "log of x to the base a." Thus we see from our discussion of inverses that the equation $\log_a x = y$ is equivalent to $a^y = x$. This is simpler than it sounds. Consider the following table:

$$
\begin{array}{ll}
2^0 = 1 & 2^6 = 64 \\
2^1 = 2 & 2^7 = 128 \\
2^2 = 4 & 2^8 = 256 \\
2^3 = 8 & 2^9 = 512 \\
2^4 = 16 & 2^{10} = 1024 \\
2^5 = 32 &
\end{array}
$$

What is $\log_2 64$? (That is, to what power do you have to raise 2 to get 64?) By glancing at the chart, it is easy to see that $2^6 = 64$ or $\log_2 64 = 6$.

What is $\log_2 512$? *Answer:* 9.

Using the information in the table we can construct a partial table of logarithms to the base 2.

$$
\begin{array}{ll}
\log_2 1 = 0 & \log_2 64 = ? \\
\log_2 2 = 1 & \log_2 128 = ? \\
\log_2 4 = 2 & \log_2 256 = ? \\
\log_2 8 = 3 & \log_2 512 = ? \\
\log_2 16 = 4 & \log_2 1024 = ?
\end{array}
$$

Of course, this table does not help us to find $\log_2 17$.

We now consider the function f, where $f(x) = \log_a x$, for $a \neq 1$ and $a > 0$. The domain of f is all positive real numbers, i.e., all x such that $x > 0$. We are mainly interested in the function $f(x) = \log_a x$ when $a > 1$. Its graph appears in Figure 1.7.5 for two choices of a. Note that since $\log_a x$ is the inverse of a^x, the graph of $\log_a x$ is obtained by reflecting the graph of $y = a^x$ about the line $y = x$. Thus if you can remember the graph of one of them you can easily obtain the other.

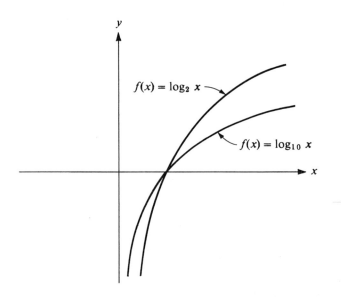

Figure 1.7.5

Again, from the standpoint of calculus, a very convenient base is $a = e = 2.71828\ldots$. We call the logarithm to the base e the *natural* logarithm, and write

$$f(x) = \log_e x = \log x = \ln x.$$

All of the foregoing notations are frequently used for natural logarithms. Notations vary between texts, tables, and professions. Engineers often use $\log x$ to denote $\log_{10} x$. Our convention is that we always mean the natural logarithm whenever we omit the base in our notation.

Properties similar to those for exponential functions hold for logarithmic functions. In fact, their derivation follows directly from the analogous properties for exponential functions. These properties are:

PROPERTY 1. $\log_a (xy) = \log_a x + \log_a y.$

PROPERTY 2. If r is any real number, then $\log_a x^r = r \log_a x$. In particular, $\log_a a^x = x \log_a a = x.$

PROPERTY 3. $\log_a (x/y) = \log_a x - \log_a y.$

PROPERTY 4. $\log_a x = \log_a y$ if and only if $x = y.$

We outline a method of obtaining Properties 1–4 for logarithmic functions in the exercises that follow.

A table of natural logarithms is found in Table A2 of the Appendix. We now consider some examples illustrating the preceding discussion on exponential and logarithm functions.

▶ Example 1. If $\log_4 x = 3$, determine x.

Solution. By definition of the logarithm, $x = 4^3 = 64.$

▶ Example 2. If $\log_x 3 = \frac{1}{3}$, determine x.

Solution. By definition, $x^{1/3} = 3$ and hence $x = 3^3 = 27.$

▶ Example 3. In 1924, J. S. Huxley [*Nature* **114**, 895 (1924)] offered the first quantitative analysis of differential growth. He used the formula

$$y = bx^k \tag{1}$$

to describe the relation of the growth of a part of an organ to the whole organism. In Equation (1), x and y denote the weights or lengths of parts of an organism and b and k are constants—b known as the initial growth index and k the equilibrium constant. Huxley referred to this equation as the simple allometry formula. The case $k = 1$ is described as isometry and k negative is described as enantiometry.

In the problem of most interest to biologists, one is given tables of the quantities x and y and from this must determine the appropriate values of b and k. By use of the properties of logarithms, we will solve this problem. First, observe that Equation (1) is equivalent (by taking natural logs of both sides) to

$$\log y = \log (bx^k).$$

By Property 1, $\log (bx^k) = \log b + \log x^k$, and by Property 2, $\log x^k = k \log x$. Hence, we see that

$$\log y = \log b + k \log x.$$

A table of values for x and y will permit us to determine b and k. Suppose that

x	1.6	1.9
y	1.1	1.4

Then,

$$\log 1.1 = \log b + k \log 1.6,$$

$$\log 1.4 = \log b + k \log 1.9.$$

From Table A2, $\log 1.1 = 0.0953$, $\log 1.6 = 0.4700$, $\log 1.4 = 0.3365$, and $\log 1.9 = 0.6419$. Subtracting the second equation from the first, we see that

$$[\log 1.1 - \log 1.4] = k[\log 1.6 - \log 1.9].$$

Hence, k is approximately 1.4. It then follows that

$$\log b = \log 1.1 - (1.4) \log 1.6$$

$$= -0.5627 \text{ (approximately)}.$$

Thus, $b = e^{-0.5627}$ and from Table A1 we see that b is approximately 0.58. Thus, the approximate relation between y and x is

$$y = 0.58 \, x^{1.4}.$$

▶ Example 4. A simple example of an exponential function arises in the study of *compound interest*. This is interest added to the principal at regular intervals of time; thereafter the interest itself earns interest. The interval of time between successive calculations of interest is called the conversion period. If the conversion period is one year, we say that the interest is compounded annually and if we borrow (or invest) P dollars compounded annually at the interest rate i, then the amount we owe (or have coming) at the end of n years is given by

$$A = P(1 + i)^n.$$

As an example, suppose that \$10,000 is invested at 10% compounded annually. We wish to find the value of this investment after 5 years. In this case $P = \$10,000$, $i = 0.10$, and $n = 5$. Thus the amount after 5 years is given by

$$A = 10,000 \, (1 + 0.10)^5 = 10,000 \, (1.10)^5.$$

To calculate $(1.10)^5$, we note that

$$(1.10)^5 = e^{\log (1.10)^5} = e^{5 \log (1.10)}.$$

From Table A2, $\log (1.10) \simeq (0.095)$, where the symbol $\simeq$ means approximately. Thus,

$$(1.10)^5 \simeq e^{5(0.095)} \simeq e^{0.475} \simeq 1.6$$

from Table A1. Hence, at the end of 5 years,

$$A \simeq \$16,000.$$

Exercises

1. Graph the following functions:

   (a) $f(x) = 4x^{1/2}$. [*Hint:* First graph $g(x) = \frac{1}{16}x^2$.]

   (b) $f(x) = 2x^{1/2}$.

   (c) $f(x) = 3x^{1/3}$. [*Hint:* First graph $g(x) = \frac{1}{27}x^3$.]

   (d) $f(x) = e^{2x}$.   (e) $f(x) = 10^x$.

   (f) $f(x) = \log_3 x$.   (g) $f(x) = \log_3 2x$.

2. (a) $2^2 \cdot 2^3 = 2^?$.   (b) $2^3 \cdot 3^3 = ?$.

   (c) $5^3 \cdot 5^{-4} = 5^?$.   (d) $7^4 \cdot 7^{1/3} = 7^?$

   (e) $(2^2)^5 = 2^?$.   (f) $6^2 = 2^? \cdot 3^?$.

   (g) $10^{15} \cdot 10^? = 10^{24}$.   (h) Does $3^{(2^4)} = (3^2)^4$?.

   (i) $5^4/5^3 = 5^?$.   (j) $10^{-5} \cdot 10^? = 10^{24}$.

   (k) $(8)^{1/3} = ?$.   (l) $(4^{1/2})^3 = (4^3)^{1/2} = ?$.

3. (a) $\log_3 9 = ?$. $9 = 3^x$ (b) $\log_3 27 = ?$.

   (c) $\log_3 243 = ?$.   (d) $\log_3 1 = ?$.

   (e) $\log_4 256 = ?$.   (f) $\log_4 16 = ?$.

   (g) $\log_{13} 169 = ?$.   (h) $\log_{10} 1000 = ?$.

   (i) $\log_a 450 = \log_a 15 + \log_a (?)$.

4. Determine x if

   (a) $\log_7 x = 3$.   (b) $\log_x 2 = \frac{1}{2}$.

   (c) $\log_{10} (10)^{(0.02)} = x$.   (d) $\log_8 x = -\frac{2}{3}$.

5. The atmospheric pressure at a height x is given by
$$y = q \cdot e^{-kx},$$
where q and k are constants. Solve for x in terms of y and the constants q and k.

   [*Hint:* Take the logarithm of both sides of the question.]

6. In the following expressions, change those in exponential form to logarithmic form and those in logarithmic form to exponential form:

   (a) $3^{-5} = \frac{1}{243}$.   (b) $81^{3/4} = 27$.

   Solution: $\log_3 (\frac{1}{243}) = -5$.

   (c) $(\frac{1}{2})^{-2} = 4$.   (d) $\log_8 (\frac{1}{512}) = -3$.

                                                Solution: $8^{-3} = \frac{1}{512}$.

   (e) $\log_{0.5} 16 = -4$.   (f) $\log_{(1/16)} (\frac{1}{8}) = \frac{3}{4}$.

   (g) $8^{1/3} = 2$.   (h) $\log_{1/3} 9 = -2$.

7. Sketch a graph of $f(x) = 2^x$ and $g(x) = x^2$ on the same axes, noting the differences.

8. This exercise refers to Example 4 of this section.

    (a) If $10,000 is invested at 10% compounded annually, find the value of this investment after 1 year and after 25 years.

    (b) If someone invests $3,000 at 6% compounded annually, what is the value of this investment after 3 years? 5 years? 20 years?

    (c) Suppose we invest $5,000 at 4% compounded quarterly. What amount do we have coming after 1 year? 3 years? 20 years? [*Hint:* $i = 0.04/4$; n must now represent the conversion period in terms of quarters. Thus, $n = 4$ if the period is 1 year. (There are 4 conversion periods in a given year.)]

9. (a) Consider 3^{3^3}. This is ambiguous without parentheses. Which is larger, $(3^3)^3$ or $3^{(3^3)}$?

    (b) What is the largest number you can write with four 2's? Make a quick guess as to the relative size of 22^{22}, 2^{222}, 2^a where $a = [2^{(2^2)}]$, 2222, $2^{(2^{2^2})}$. [*Note:* The last number is larger than 1 followed by one million zeros.]

10. Show how one can obtain the graph of the function $f(x) = \log_2 x$ from the graph of $g(x) = 2^x$. Sketch both graphs on the same axes. [*Hint:* Look at Figure 1.1.5.]

11. This exercise refers to Example 3 of this section. The values of mean cranium length and total face length in certain animals are given by the following table:

mean cranium length (cm)	0.2	0.4
mean total length (cm)	1.5	3.0

    It can be shown that the two quantities are related by an allometric law. Find the approximate values of the relevant constants and write out the formula describing the law.

12. The population size for a certain town can be expressed as

$$P = k10^{ax},$$

where P denotes the population of the town in the year x and k and a are constants. Suppose that figures for P in year x are as follows:

x	1910	1920
P	180,000	270,000

    Determine the constants a and k (approximately) and estimate the year in which the population will be 800,000. [*Hint:* In order to use Table A2 you will need to observe that $180,000 = 18 \times 10^4 = 2 \times 3^2 \times 10^4$ and that $\log 180,000 = \log 2 + 2 \log 3 + 4 \log 10$. Obviously the same type of reasoning must be used to take logs of other large numbers. Just estimate your results as best you can.)

13. Using Properties 1–4 for exponential functions, show that

    (a) If $f(x) = a^x$, then $f(x + h) - f(x) = a^x(a^h - 1)$.

    (b) If $f(x) = a^x$, then $f(x - 1)f(1) = f(x)$.

    (c) If $f(x) = a^x$, then $f(x + h)/f(h) = f(x)$.

14. Using Properties 1–4 for exponential functions, establish the corresponding Properties 1–4 for logarithm functions. [*Hint:* To prove Property 1, let $w = \log_a x$ and $v = \log_a y$; then, by definition, $x = a^w$ and $y = a^v$. Hence, $xy = a^w \cdot a^v = a^{w+v}$ (by Property 1 for exponential functions). Thus, $w + v = \log_a xy$ and hence $\log_a (xy) = \log_a x + \log_a y$. The other three properties are similarly derived.]

15. Using Properties 1–4 for logarithm functions, show that

    (a) If $f(x) = \log_a x$, then $f(x + h) - f(x) = \log_a (1 + h/x)$.

    (b) If $f(x) = \log_a x$, then $f(x) - f(x^2) = -f(x)$.

    (c) If $f(x) = \log_a x$, then $f(pq) - f(p/q) = 2f(q)$.

    (d) If $f(x) = \log_a x$, then $a^{f(x^2)} = x^2$.

1.8 Operations on Functions

It is possible to combine functions in several ways to arrive at new functions. We begin by adding two functions, f and g. We designate the new function by $[f + g]$ and define

$$[f + g](x) = f(x) + g(x).$$

The domain of $[f + g]$ is the set of real numbers common to both the domain of f and the domain of g. We employ the notation $[f + g]$ for emphasis here. In later sections we use the simpler notation $f + g$.

▶ Example 1. If $f(x) = x + 1$ and $g(x) = |x|$, then graph $[f + g](x)$.

Solution. First draw the graphs of f and g as in Figure 1.8.1 (a) and (b). The graph of $[f + g]$ can be drawn by inspecting the graphs of f and g. This method works in very simple cases. One could have drawn the graph by first simplifying the expression for $[f + g]$. More precisely,

$$[f + g](x) = x + |x| + 1.$$

If $x \geq 0$, $|x| = x$ and $[f + g](x) = 2x + 1$, while if $x < 0$, $|x| = -x$ and $[f + g](x) = 1$. Hence,

$$[f + g](x) = \begin{cases} 2x + 1 & \text{if } x \geq 0, \\ 1 & \text{if } x < 0. \end{cases}$$

This is precisely the graph shown in Figure 1.8.1 (c). Note that the domain of $[f + g]$ is the set of all real numbers, since that is the intersection of the domains of f and g.

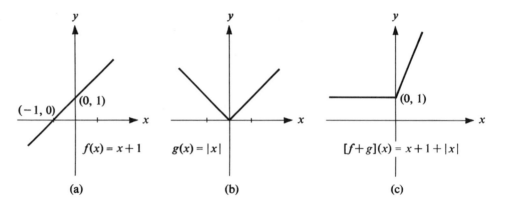

$f(x) = x + 1$ $g(x) = |x|$ $[f+g](x) = x+1+|x|$

(a)   (b)   (c)

Figure 1.8.1

The function obtained by subtracting g from f, that is, $f - g$ or

$$[f - g](x) = f(x) - g(x),$$

is similar to the addition case.

▶ **Example 2.** Let $f(x) = 2x + 1$ and $g(x) = 2^x$. Graph $[f - g](x)$.

    Solution.   Note that

$$[f - g](x) = f(x) - g(x) = 2x + 1 - 2^x.$$

Since the domain of both f and g is the set of all real numbers, so is the domain of $f - g$. (See Figure 1.8.2.)

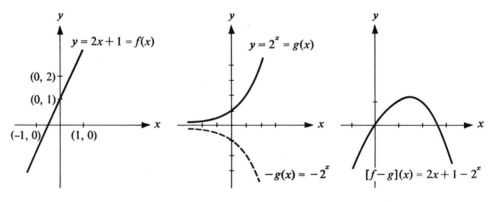

$y = 2x + 1 = f(x)$ $y = 2^x = g(x)$

$-g(x) = -2^x$ $[f-g](x) = 2x+1-2^x$

Figure 1.8.2

Observe that a polynomial can be considered to be the sum of several power functions.

Next, we consider the function obtained by multiplying the functions f and g. This function is designated by $f \cdot g$ or

$$[f \cdot g](x) = f(x) \cdot g(x).$$

The domain of the new function $f \cdot g$ is the set of real numbers common to the domain of f and of g. In many cases it is easy to see what $f \cdot g$ looks like by manipulating the product in terms of x. That is, if $f(x) = (1 - x^4)$ and $g(x) = 1/(1 + x^2)$, then

$$[f \cdot g](x) = (1 - x^4)/(1 + x^2)$$
$$= (1 - x^2)(1 + x^2)/(1 + x^2)$$
$$= (1 - x^2).$$

The graph of $[f \cdot g](x)$ is shown in Figure 1.8.3. The domain of f and g is the set

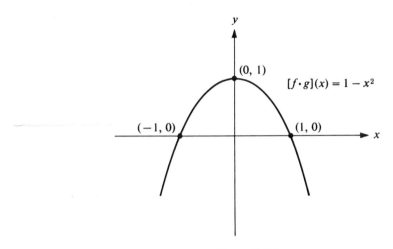

Figure 1.8.3

of all real numbers. Hence, so is the domain of $f \cdot g$. [Canceling the term $(1 + x^2)$ could lead to difficulties, but since the term is never zero, we are allowed to do it. In this text we will be casual about such cavalier treatment of rational functions but it should be mentioned from time to time.]

In some cases one can see fairly quickly what $f \cdot g$ is by graphing f and g.

▶ Example 3. Let

$$f(x) = \begin{cases} 2 & \text{for} \quad x \leq 1, \\ -2 & \text{for} \quad x > 1, \end{cases} \quad \text{and } g(x) = 2^x.$$

Graph the function $[f \cdot g](x)$.

Solution. We first note that

$$[f \cdot g](x) = f(x)g(x) = \begin{cases} 2^{x+1} & \text{for} \quad x \leq 1, \\ -2^{x+1} & \text{for} \quad x > 1. \end{cases}$$

Again, the domain of $f \cdot g$ is the set of all real numbers. The graphs of f, g, and $f \cdot g$ are shown in Figure 1.8.4.

Consider, now, the function obtained by dividing two functions. This is obviously very similar to multiplication. We designate it by f/g or

$$[f/g](x) = f(x)/g(x),$$

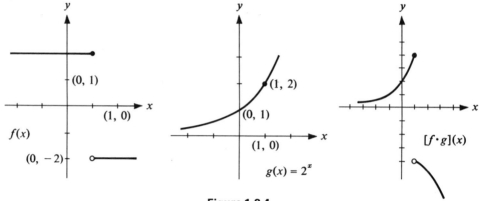

Figure 1.8.4

provided that $g(x) \neq 0$. In this case, the domain of f/g is the set of real numbers common to the domain of f, the domain of g, and the set of x's such that $g(x) \neq 0$.

▶ Example 4. Let $f(x) = x^2 + 1$ and $g(x) = x^2 - 1$. (See Figure 1.8.5.) Graph $[f/g](x)$.

Solution. Before we graph

$$[f/g](x) = (x^2 + 1)/(x^2 - 1),$$

let us make a few observations. First we rewrite it as

$$[f/g](x) = (x^2 + 1)/[(x - 1)(x + 1)].$$

Note that the denominator vanishes for $x = \pm 1$. Moreover, the expression

$$x^2 - 1 \quad \text{is} \quad \begin{cases} \text{positive} & \text{for} \quad x < -1, \\ \text{negative} & \text{for} \quad -1 < x < 1, \\ \text{positive} & \text{for} \quad x > 1, \end{cases}$$

while the numerator $x^2 + 1$ is always positive. Now, it is not too hard to see what

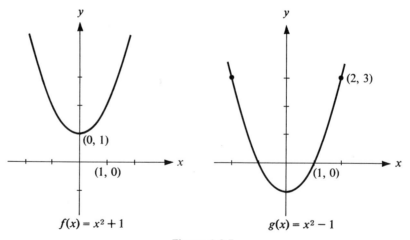

Figure 1.8.5

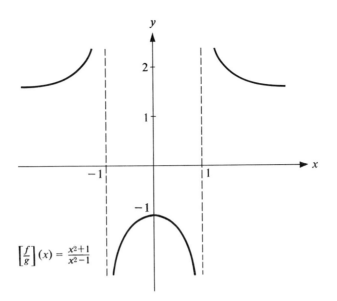

$$\left[\frac{f}{g}\right](x) = \frac{x^2+1}{x^2-1}$$

Figure 1.8.6

f/g look like. (See Figure 1.8.6.) The domain is the set of all real numbers except $x = \pm 1$.

We can also get a new function h from f and g by setting $h(x) = g(f(x))$. In this case h is a function of a function, and we say that h is the composition of g and f and write $h = [g \circ f]$ or

$$h(x) = [g \circ f](x) = g[f(x)].$$

The domain of $g \circ f$ is

$\{x \mid x$ is in the domain of f and $f(x)$ is in the domain of $g\}$.

Consider the picture of the rule for h given in Figure 1.8.7. Thus, if a is fed into the first machine, $f(a)$ is the output. Then, $f(a)$ is fed into the second machine, whose output is $g(f(a))$.

Let us consider several examples.

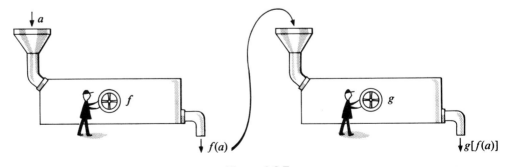

Figure 1.8.7

▶ Example 5. Let $f(x) = x^2 + 2$ and $g(x) = 1 - x$. Then

$$[g \circ f](x) = g[f(x)]$$
$$= 1 - f(x)$$
$$= 1 - (x^2 + 2)$$
$$= -(1 + x^2)$$

or

$$g[f(x)] = g(x^2 + 2)$$
$$= 1 - (x^2 + 2)$$
$$= -(1 + x^2).$$

The domain of $g \circ f$ is clearly the set of all real numbers. The graphs of f, g, and $g \circ f$ are shown in Figure 1.8.8.

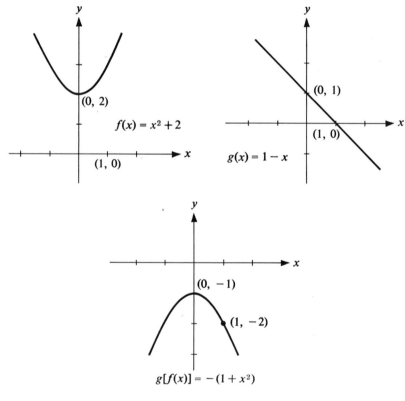

Figure 1.8.8

▶ Example 6. Let

$$f(x) = \begin{cases} x & \text{if } 0 \le x \le 1, \\ 2 & \text{if } x > 1, \end{cases} \quad \text{and} \quad g(x) = x^2.$$

Then

$$[g \circ f](x) = g[f(x)] = \begin{cases} g(x) & \text{if } 0 \le x \le 1, \\ g(2) & \text{if } x > 1; \end{cases}$$

therefore

$$[g \circ f](x) = \begin{cases} x^2 & \text{if } 0 \le x \le 1, \\ 4 & \text{if } x > 1. \end{cases}$$

The domain of $g \circ f$ is $\{x \mid x \ge 0\}$. The graphs of f, g, and $g \circ f$ are shown in Figure 1.8.9.

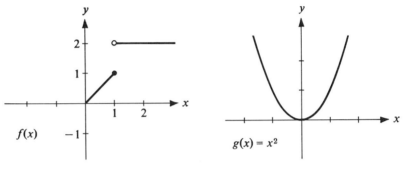

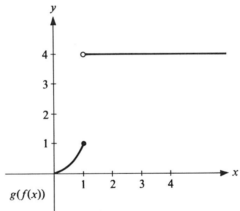

Figure 1.8.9

Note that

$$f[g(x)] = \begin{cases} g(x) & \text{if } 0 \le g(x) \le 1 \\ 2 & \text{if } g(x) > 1 \end{cases} = \begin{cases} x^2 & \text{for } x^2 \le 1, \\ 2 & \text{for } x^2 > 1. \end{cases}$$

Thus

$$f[g(x)] = \begin{cases} x^2 & \text{if } -1 \le x \le 1, \\ 2 & \text{for all other } x. \end{cases}$$

The domain of $f \circ g$ is the set of all real numbers. (See Figure 1.8.10.) Hence, we see that, in general, $g[f(x)] \ne f[g(x)]$.

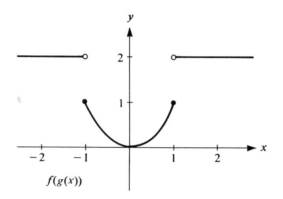

$f(g(x))$

Figure 1.8.10

At this point we might remark that if f is a one-to-one function (see Section 1.6), then its inverse could have been defined as that function g such that

$$[f \circ g](x) = [g \circ f](x) = x.$$

In fact, a further glance at Definition 2 of Section 1.6 reveals that this is precisely the definition we have adopted.

Exercises

1. Let $f(x) = 4x$ and $g(x) = x + 1$. Graph the following:

   (a) $[f + g](x)$.   (b) $[f \cdot g](x)$.   (c) $[f \circ g](x)$.

   What are the domains of the functions in (a), (b), and (c)?

2. Let $f(x) = 1$ and $g(x) = x$. Graph f/g and g/f.

3. Let f and g be given functions.

   (a) Is $[f + g] = [g + f]$? Why?

   (b) Is $[f \cdot g] = [g \cdot f]$? Why?

   (c) Is $[f/g] = [g/f]$? Why? (Consider Exercise 2.)

4. Let $f(x) = 3^x$ and $g(x) = 1 + x$. Graph $(g \circ f)(x)$. Is $[g \circ f](x) = [f \circ g](x)$?

5. Let $f(x) = \log x$ and $g(x) = e^x$. Graph $[g \circ f](x)$ and $[f \circ g](x)$. (Note of caution: $g[f(x)]$ takes on no negative values.)

6. Write the following functions as the composition of two functions.

   ▶ **Example.** $h(x) = e^{x^2 + 9}$.

   Solution. $h(x) = g[f(x)]$ when $f(x) = x^2 + 9$ and $g(x) = e^x$.

   (a) $h(x) = \log(x^2 + 7)$.   (b) $h(x) = \sqrt{\dfrac{x + 1}{x - 1}}$.

(c) $h(x) = \log(x + e^{-x} + 1)$.

(d) $h(x) = e^{(1 + x + x^2 + x^3)}$.

(e) $h(x) = \left[\dfrac{\log(x^2 + 1)}{x^4 + 1}\right]^{1/3}$.

(f) $h(x) = e^{|x|}$.

(g) $h(x) = 2^{(3x)}$.

(h) $h(x) = 3^{(x+4)^3}$.

(i) $h(x) = \log(\log x)$ for $x > 1$.

(j) $h(x) = (\log|x + 1|)^2$.

(k) $h(x) = e^{\sqrt{1+x}}$ for $x \geq 0$.

(l) $h(x) = \left(\dfrac{x + 1}{x - 1}\right)^{1/3}$.

(m) $h(x) = 2^{x^2 + 2}$.

(n) $h(x) = e^{(x^2 + 1)^3}$.

7. The domain of f consists of all numbers x such that $-2 \leq x \leq 2$ and

$$f(x) = \begin{cases} -2 & \text{if } x \text{ is an integer,} \\ 2 & \text{if } x \text{ is not an integer.} \end{cases}$$

(a) Draw the graph of f.

(b) What is $f(2)$? $f(\tfrac{1}{2})$?

(c) Let g be a function with the same domain as f, defined by $g(x) = [f \circ f](x)$. Draw the graph of g.

8. Let $f(x) = x^2$ and $g(x) = x^{1/2}$. Is $[f \circ g](x) = [g \circ f](x)$? Why?

9. Let f be a function whose domain is the set of real numbers.

(a) Let $g(x) = 7 + f(x)$. Geometrically, how is the graph of g obtained from the graph of f?

(b) Let $h(x) = f(x - 3)$. How is the graph of h obtained from the graph of f? Can you get the graph of h by leaving f fixed and moving the coordinate axis? Write h as the composition of f and something else.

10. (a) If $f(x) = \log_2 x$ and $g(x) = 2^x$, find $[f \circ g](x)$ and $[g \circ f](x)$. What is the domain of $[g \circ f](x)$?

(b) Sketch the graph of $f \circ g$ and $g \circ f$.

11. Give an explicit example of a function (whose domain is the set of real numbers) such that

$$f(x) < 2 \text{ for } x < 0, \quad f(x) > 3 \text{ for } x > 0, \text{ and } |f(0)| < \tfrac{1}{2}.$$

12. Let $f(x) = |x| - 1$ and $g(x) = |x|$.

(a) Graph $g[f(x)]$. How many corners does the curve have? How can you get a graph with five corners?

(b) Graph $f[g(x)]$. Note that $f[g(x)] = f(x)$. But $g[f(x)] \neq f(x)$ or anything like it. This would reinforce the suspicion that $g \circ f = f \circ g$ is not very likely if f and g are chosen at random.

*1.9   Trigonometric Functions

In order to facilitate the study of the trigonometric functions, we first discuss angle measurement in radians. Let us assume we have a circle of radius 1 and a fixed angle α. (See Figure 1.9.1.) We have a tape measure but no protractor. How can

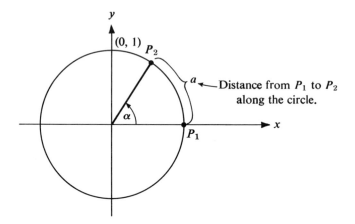

Figure 1.9.1

we measure the angle α? The obvious answer is to measure the distance from P_1 to P_2 along the circle. Call this distance a. Then by definition the angle α has radian measure a. Since a circle is 360 degrees, it is clear that 360 degrees $= 2\pi$ radians. If we started with a circle of radius r, then the angle α would have radian measure a/r. That is, the length of the subtended arc is simply ar. We now return to the trigonometric functions.

We define the *sine* and *cosine* functions as follows: Consider a circle of radius 1 with its center at the origin of the rectangular coordinate system. (See Figure 1.9.2.)

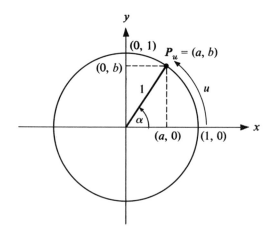

Figure 1.9.2

---

* Sections with asterisks may be omitted without loss of continuity.

For any *nonnegative u*, let $P_u = (a, b)$ be the point obtained by rotating the point $P = (1, 0)$ a distance of u radians around the circle in a counterclockwise direction. Note that since the length of the circumference of the circle is 2π units, if $u > 2\pi$, then P will be rotated about the circle more than once. We define

$$\sin u = b \quad \text{and} \quad \cos u = a.$$

We immediately see from the Pythagorean theorem that $a^2 + b^2 = 1$; hence for any u, $\sin^2 u + \cos^2 u = 1$.

We may define the sine and cosine functions for u negative by taking P_u to be the point obtained by rotating P through $|u|$ units in a clockwise direction. (See Figure 1.9.3.) From Figure 1.9.3, we note that if P_u is in the fourth quadrant, then $b < 0$ and

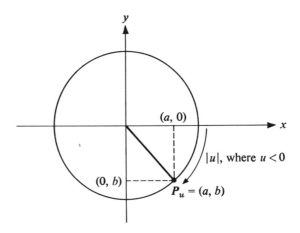

Figure 1.9.3

$a > 0$; that is, $\cos u > 0$ and $\sin u < 0$. In like manner, if P_u is in the second quadrant, $\cos u < 0$, $\sin u > 0$. If P_u lies in the third quadrant, then both $\sin u$ and $\cos u$ are negative. Using the fact that $u = 2\pi$ measures the length of the circumference of the circle, we observe that the points $(1, 0)$, $(0, 1)$, $(-1, 0)$, $(0, -1)$ correspond to $u = 0$, $\frac{1}{2}\pi$, π, and $\frac{3}{2}\pi$, respectively. Thus, we can state the following:

PROPERTY 1. $\sin 0 = 0$, $\sin \frac{1}{2}\pi = 1$, $\sin \pi = 0$, $\sin \frac{3}{2}\pi = -1$, $\cos 0 = 1$, $\cos \frac{1}{2}\pi = \cos \frac{3}{2}\pi = 0$, $\cos \pi = -1$.

PROPERTY 2. If $0 \leq u \leq \frac{1}{2}\pi$, then $\cos u \geq 0$ and $\sin u \geq 0$.

PROPERTY 3. If $\frac{1}{2}\pi \leq u \leq \pi$, then $\cos u \leq 0$ and $\sin u \geq 0$.

PROPERTY 4. If $\pi \leq u \leq \frac{3}{2}\pi$, then $\cos u \leq 0$ and $\sin u \leq 0$.

PROPERTY 5. If $\frac{3}{2}\pi \leq u \leq 2\pi$, then $\cos u \geq 0$ and $\sin u \leq 0$.

PROPERTY 6. The largest value of both $\cos u$ and $\sin u$ for $0 \leq u \leq 2\pi$ is $+1$ and the smallest value is -1. (That is, $|\cos u| \leq 1$ and $|\sin u| \leq 1$.)

PROPERTY 7. Also observe that $\sin (u + 2\pi) = \sin u$ and $\cos (u + 2\pi) = \cos u$. (This is easily deduced from our definition of the sine and cosine.)

Property 7 states that the sine and cosine functions are *periodic*, with period 2π. In general, we say that a function f is periodic with period p if p is the smallest positive number such that $f(x + p) = f(x)$. (We assume p is such that $x + p \in$ domain of f.)

From the information given in Properties 1–7, we can easily construct the graphs of the functions $f(x) = \sin x$ and $g(x) = \cos x$. The x may be thought of as the angle measured in radians. Their graphs are shown in Figures 1.9.4 and 1.9.5.

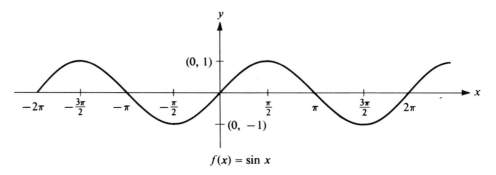

$f(x) = \sin x$

Figure 1.9.4

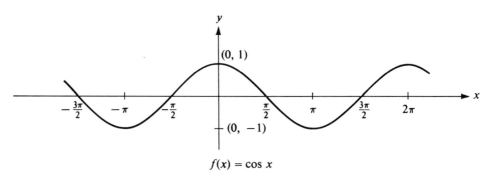

$f(x) = \cos x$

Figure 1.9.5

The domain of both the cosine and sine functions is the set of real numbers, while the range of these functions is $[-1, 1]$. From the definitions of the sine and cosine, it is easy to see that the cosine function is an even function, $[\cos (-x) = \cos x]$, and the sine function is an odd function, $[\sin (-x) = -\sin x]$. [Figures 1.9.2 and 1.9.3 show that $\sin u = b$ and $\sin (-u) = -b$, while $\cos u = a$ and $\cos (-u) = a$.]

The remaining trigonometric functions can all be defined in terms of the cosine and sine functions.

DEFINITION 1. *The tangent is defined by*

$$\tan x = \frac{\sin x}{\cos x}.$$

DEFINITION 2. *The secant is defined by*

$$\sec x = \frac{1}{\cos x}.$$

DEFINITION 3. *The cotangent is defined by*

$$\cot x = \frac{\cos x}{\sin x}.$$

DEFINITION 4. *The cosecant is defined by*

$$\csc x = \frac{1}{\sin x}.$$

Other properties of the trigonometric functions will be discussed as we need them.

Exercises

1. Using Figures 1.9.4 and 1.9.5, sketch the graph of the following functions:

   (a) $f(x) = 2 \sin x.$ (b) $f(x) = -\sin x.$

   (c) $f(x) = |\sin x|.$ (d) $f(x) = \sin |x|.$

   (e) $f(x) = 3 \cos x.$ (f) $f(x) = -\cos x.$

   (g) $f(x) = |\cos x|.$ (h) $f(x) = \cos |x|.$

2. Using the definitions of the sine and cosine functions, derive the following identities:

   (a) $\sin (t + \tfrac{1}{2}\pi) = \cos t.$ (b) $\cos (t + \tfrac{1}{2}\pi) = -\sin t.$

   [*Hint:* Let $0 < t < \tfrac{1}{2}\pi$; then $y = \sin (t + \tfrac{1}{2}\pi)$ in Figure 1.9.6. Clearly,

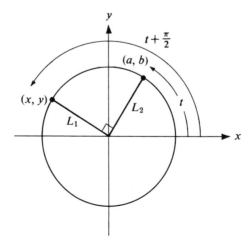

Figure 1.9.6

$y = \sin(t + \frac{1}{2}\pi)$, $x = \cos(t + \frac{1}{2}\pi)$, $a = \cos t$ and $b = \sin t$. We must show that $y = a$. Since the lines L_1 and L_2 are perpendicular,

$$\frac{b}{a}\left(\frac{y}{x}\right) = -1.$$

(Why?) Hence $b^2y^2 = a^2x^2$, and thus $(1 - a^2)y^2 = a^2(1 - y^2)$. Hence $y = a$. (Why?) A similar type of derivation yields (b).]

3. Find all those numbers x satisfying the following equations:

   (a) $\sin x = \cos x$.   (b) $\cos x = 0$.   (c) $\sin x = 0$.

4. Let $f(x) = \log x$ and $g(x) = \cos x$. Graph $f + g$.

5. Let $f(x) = \cos x$ and $g(x) = 1 + x$. Graph $g[f(x)]$. What is $[f \circ g](x)$?

6. If $f(x) = e^x + \sin x$, and $g(x) = \log x$, find $[f \circ g](x)$ and $[g \circ f](x)$.

7. If $f(x) = x^2$ and $g(x) = \sin x$, graph $g[f(x)]$. What is $[f \circ g](x)$?

8. Using the definition of $\tan \theta$ and the identity $\sin^2 \theta + \cos^2 \theta = 1$, derive the following identities:

   (a) $\cot \theta = \dfrac{\cos \theta}{\sin \theta}$.   (b) $1 + \tan^2 \theta = \sec^2 \theta$.

   (c) $1 + \cot^2 \theta = \csc^2 \theta$.

1.10 Chapter 1 Summary

1. Let a and b be two real numbers. We say that $a \geq b$ if $a - b \geq 0$. In like manner, $a \leq b$ if $b - a \geq 0$.

2. Let a, b be any two real numbers. Then

   (a) $[a, b] = \{x \mid a \leq x \leq b\}$,

   (b) $(a, b) = \{x \mid a < x < b\}$,

   (c) $[a, b) = \{x \mid a \leq x < b\}$,

   (d) $(a, b] = \{x \mid a < x \leq b\}$.

3. (a) If a is any real number, we define the absolute value of a, written $|a|$, as follows:

$$|a| = \begin{cases} a & \text{if } a \geq 0, \\ -a & \text{if } a < 0. \end{cases}$$

   (b) If b is any real number, then $|x| \leq b$ if and only if $x \in [-b, b]$. (See Figure 1.10.1.)

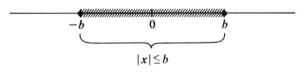

$$|x| \leq b$$

Figure 1.10.1

(c)   If b is any real number, then $|x| \geq b$ if and only if $x \in [-\infty, -b] \cup [b, \infty]$. (See Figure 1.10.2.)

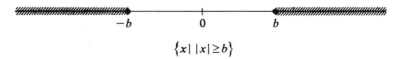

$$\{x \mid |x| \geq b\}$$

Figure 1.10.2

4.   If $P_1 = (x_1, y_1)$ and $P_2 = (x_2, y_2)$ are two points in the coordinate plane, then the distance from P_1 to P_2, denoted by $d[P_1, P_2]$, is

$$d[P_1, P_2] = \sqrt{(x_1 - x_2)^2 + (y_1 - y_2)^2}.$$

In particular, if $y_1 = y_2 = 0$, then

$$d[P_1, P_2] = |x_1 - x_2|.$$

5.   Let (x_1, y_1) and (x_2, y_2) be any two distinct points on a line L. The slope m of the line L is given by

$$m = \frac{y_2 - y_1}{x_2 - x_1}.$$

6.   Forms of the equation for a straight line

(a)   *Point-slope formula:* If (x_1, y_1) is any point on a line L, with slope m, the equation of the line is given by

$$y - y_1 = m(x - x_1).$$

(b)   *Slope-intercept formula:* If L is a line with slope m and passes through the point $(0, b)$, then its equation is given by

$$y = mx + b.$$

(c)   *Two-point formula:* If (x_1, y_1) and (x_2, y_2) are any two points on a line L, its equation is given by

$$y - y_1 = \left(\frac{y_2 - y_1}{x_2 - x_1}\right)(x - x_1).$$

(d)   If a, b, and c are any triple of real numbers, the equation

$$ax + by + c = 0 \qquad (1.10.1)$$

represents the most general form of an equation for a straight line. That is, the equation of any straight line may be written as (1.10.1). Equation (1.10.1) always represents a line with slope $-a/b$ when $b \neq 0$.

7.   Let m_1 and m_2 be the respective slopes of lines L_1 and L_2. Then

(a) L_1 is parallel to L_2 if and only if $m_1 = m_2$.

(b) L_1 is perpendicular to L_2 if and only if $m_1 m_2 = -1$.

8. Let A and B be any two sets of real numbers. A function is a rule that assigns to each number $x \in A$ one and only one number $y \in B$. The set A is called the domain of the function. If f denotes the function, we write $y = f(x)$. Unless the domain is explicitly given, it is understood to consist of all numbers for which the definition makes any sense. The graph of a function is

$$\{(x, y) \mid x \in \text{domain of } f \text{ and } y = f(x)\}.$$

9. If f is a one-to-one function, then its inverse function g exists and satisfies the relations $f[g(y)] = g[f(y)] = y$.

10. Useful functions

    (a) *Constant functions:* $f(x) = c$, where c is a given real number.

    (b) *Linear functions:* $f(x) = ax + b$, where a and b are given real numbers.

    (c) *Power functions:* $f(x) = x^n$, where n is a positive integer.

    (d) *Root functions:* $f(x) = x^{1/n}$, where n is a positive integer. Note that $y = x^{1/n}$ if and only if $x = y^n$. This function is the inverse of the power function.

    (e) *Polynomial functions:* $f(x) = a_0 + a_1 x + a_2 x^2 + \cdots + a_n x^n$, where the a_i's, $i = 1, 2, \ldots, n$, are given real numbers.

    (f) *Rational functions:* $f(x) = p(x)/q(x)$, where $p(x)$ and $q(x)$ are polynomials.

    (g) *Exponential functions:* $f(x) = a^x$, where $a > 0$. The most useful function of this type in calculus is $f(x) = e^x$, where $e = 2.71828 \ldots$. The domain of the exponential function is the set of all real numbers.

    (h) *Logarithm functions:* $f(x) = \log_a x$, where $a > 0$. This is the logarithm of x to the base a. Note that $y = \log_a x$ if and only if $x = a^y$. The domain of the logarithm function is the set of all those real numbers $x > 0$. If $a = e$, we say that $\log x$ is the natural logarithm of x and write $\log x = \log_e x = \ln x$. Log functions are the inverses of exponential functions.

The graphs of the logarithm and exponential functions are shown in Figure 1.10.3.

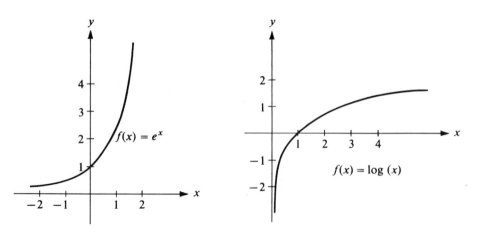

Figure 1.10.3

70

11. Operations on functions

    (a) $[f + g](x) = f(x) + g(x)$.

    (b) $[f \cdot g](x) = f(x)g(x)$.

    (c) $[f/g](x) = f(x)/g(x)$, provided that $g(x) \neq 0$.

    (d) $[g \circ f](x) = g[f(x)]$.

Review Exercises

1. (a) Find all those real numbers x satisfying the following inequalities:

    (i) $|x + 3| \leq 4$.   (ii) $|x + \sqrt{2}| > 1$.   (iii) $|x - 5| \leq 2$.

    (iv) $\dfrac{1}{x - 2} > 3$.   (v) $\dfrac{1}{|x - 2|} > 3$.

    (b) Determine the real numbers x for which $x(x - 1)(x + 2) < 0$. In both (a) and (b), illustrate your results on the coordinate axes.

2. (a) Find the point on the x axis equidistant from $(1, 2)$ and $(2, -3)$.

    (b) A line through the point $(4, -1)$ has slope $\frac{1}{2}$. At what point does the line cross the y axis?

3. What does each statement imply about a, b, c in the line whose equation is $ax + by + c = 0$?

    (a) The slope is $\frac{3}{2}$.

    (b) The line goes through the points $(4, 0)$ and $(0, 3)$.

    (c) The line goes through the origin.

    (d) The line goes through $(1, 1)$.

    (e) The line is parallel to the x axis.

    (f) The line is perpendicular to the x axis.

    (g) The line is parallel to the line whose equation is $3y = 2x - 4$.

    (h) The line is perpendicular to the line whose equation is $2x - 5y = 7$.

    (i) The line is identical with $y = 3x - 4$.

4. Consider the two lines L_1 and L_2, whose equations are given by

$$L_1: x - y = 1, \qquad L_2: 2x + y = 2.$$

    (a) Graph the two lines and find their point of intersection.

    (b) Find the equation of the line through the point of intersection of the above lines and perpendicular to the line $-\frac{1}{2}x + 2y = \frac{1}{4}$.

    (c) Sketch the graph of the line found in (b).

5. Find the area of the triangle with sides $x - 2 = 0$, $x - y = 0$, and $3x - y - 8 = 0$.

6. Suppose that the total cost of two different lawnmowers A and B is $600. If it is known that the total cost of 3 of the A mowers and 7 of the B mowers is no more than $3,000, what is the minimum cost of mower A? [*Hint:* Let x be the cost of mower A. Then the cost of mower B is $600 - x$. Now, solve $3x + 7(600 - x) \leq 3,000$.]

7. Two milk trucks must make deliveries over a distance totaling 200 miles. Truck A gets 10 miles per gallon of fuel and truck B gets 8 miles per gallon. If the total fuel consumption is to be no more than 22 gallons, what is the minimum number of miles truck A must travel?

8. Let
$$f(x) = \begin{cases} 2^x & \text{for } 0 \leq x < 1, \\ 1 & \text{for } x = 1, \\ 3 - x & \text{for } 1 < x \leq 3. \end{cases}$$

(a) Sketch a graph of f, clearly indicating the domain and range of the function.

(b) If a is a number such that $\frac{1}{2} \leq a \leq 1$, what is $f(a + \frac{1}{2})$?

9. Sketch a graph of $y = \log(1/x)$ and $y = \log|x|$.

10. (a) Find x if $\log x + \log(5 - x) = \log 2 + \log 3$.

(b) Find x if $e^{\log x} = 100$.

(c) Find x if $e^{2x} = 4$.

(d) Find x if $\dfrac{e^x + 1}{e^x - 1} = 3$.

11. If $f(x) = x^2$ and $g(x) = \sqrt{-x}$, state the domain of f and domain of g. What is $(f \circ g)(x)$? $(g \circ f(x))$? Graph $f \circ g$. What is the domain of $(f \circ g)$?

12. Same as Exercise 11 for $f(x) = -x^2$, $g(x) = \log x$.

13. Let $f(x) = x^2$ and $g(x) = \sqrt{1 - x^2}$.

(a) What is $[f + g](x)$? What is the domain of $f + g$?

(b) What is $[f \cdot g](x)$? What is the domain of $f \cdot g$?

(c) What is $[f \circ g](x)$? What is $[g \circ f](x)$? What is the domain of $f \circ g$? What is the domain of $g \circ f$? Sketch a graph of $f \circ g$.

14. (a) Show how one can obtain the graph of $f(x) = \log_3 x$ from the graph of $g(x) = 3^x$. Draw the graphs of both functions on the same set of axes.

(b) What is the value of y if $y = 2^{\log_2 34}$?

15. (a) Suppose a and b are positive constants. Let the point $(0, a)$ represent an off-shore rock in the ocean. Let the x axis denote the shoreline, with $(b, 0)$ a lifeguard station. The mile is the unit of distance. A man swims from the rock to the point $(x, 0)$ and then runs to the lifeguard station. If he

swims s miles per hour and runs r miles per hour, and if T is the total time required for the trip, express T as a function of x. Consider only values of x such that $0 \le x \le b$.

(b) Choose $a = \frac{3}{4}$, $b = 1$, $s = 2$, $r = 6$, and construct a rough graph of the function using $x = 0, \frac{1}{4}, \frac{1}{2}, \frac{3}{4}, 1$.

16. A Swiss bank pays interest compounded quarterly at a 4% interest rate. What deposit should be made to yield $2,000 at the end of six years? (Refer to Exercise 8 in Section 1.7.)

17. The rate at which a certain chemical salt dissolves in a liquid obeys a law nearly equal to the law $y = 2^{-x}$, where y is the rate of solution in grams per second and x is the time in seconds. Make a sketch of the graph of the function defined by $y = 2^{-x}$ and give the domain and range of the function.

18. (a) If $f(x) = x^2 + x$, compute the value of

$$\frac{f(2 + h) - f(2)}{h}, \qquad \text{where} \quad h > 0.$$

(b) If $f(x) = |x|$, compute

$$\frac{f(1 + h) - f(1)}{h}, \qquad \text{where} \quad h > 0.$$

Compute $f(h)/h$, where $h > 0$ and $h < 0$.

19. Show that if $f(x) = (x + 3)/(2x - 1)$, then $f \circ f$ gives the identity function g, defined by $g(x) = x$.

Limits and Continuity*

2.1 Introduction

After the concept of function, probably the most important concept is that of a limit. We first illustrate the use of the word "limit" and some related words as they occur in situations already known to the reader.

Suppose a rubber ball is dropped from the top of a two-story building. Assume that, on each successive bounce, it bounces half as high as it did on the previous bounce. Thus it bounces 8 feet high on the first bounce, 4 feet high on the second bounce, and so on, as recorded in the table.

Bounce	1	2	3	4	5	6
Height	8	4	2	1	$\frac{1}{2}$	$\frac{1}{4}$

It is reasonable to say that the ball is approaching the ground or, equivalently, that the distance from the ball to the ground is approaching zero. In this simple example, we have already encountered a rudimentary notion of limit. It is a special instance of the following question: If f is a function and a is any given number, then what happens to $f(x)$ as x gets closer and closer to a? Alternatively we could ask, "What is the limit of $f(x)$ as x approaches a?" Thus, we are in some sense equating the word "limit" with the words "closer and closer." In the next two sections, we attempt to give a satisfactory explanation that answers this question.

2.2 A Geometric View of Limits

Imagine a fly walking along a wall as in Figure 2.2.1. We watch the fly up to the time he reaches the heavy black line and then try to predict at what point he will cross it.

---

* A note to the instructor concerning the topic of limits appears in the Preface.

For example, if the fly took the path in Figure 2.2.2 or Figure 2.2.3, then we would predict that he would cross the heavy black line at the arrow. However, consider the path of the fly in Figure 2.2.4. His path moves back and forth between the dotted lines infinitely often. Here it is not possible to predict just where he will cross the heavy black line. Indeed, it appears he may cross it at any or every point in the crosshatched region.

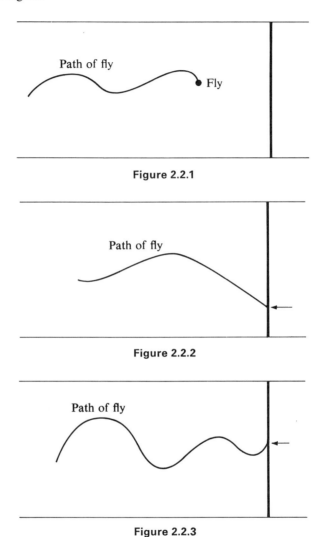

Figure 2.2.1

Figure 2.2.2

Figure 2.2.3

At this point in the discussion it is convenient to replace the fly by his path. Thus, instead of the fly, we have a curve $y = f(x)$. Rather than ask where the fly crosses the line $x = a$ (the heavy black line), we ask What is the limit of $f(x)$ as x approaches a from the left? (Notice the fly was walking from left to right. He was always to the left of the heavy black line $x = a$.) We write

$$\lim_{x \to a^-} f(x) = L$$

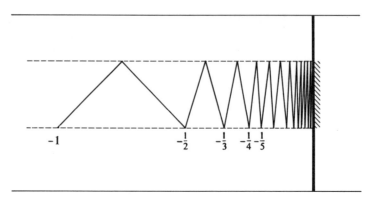

Figure 2.2.4

if, as the fly approaches the line $x = a$ from the left, it is possible to predict that he will cross the line at the point (a, L). (The symbol a^- indicates that x is approaching a from the left.) In this case we say that the *left-hand limit* exists. We ignore completely the value of f at a; that is, $f(a)$ has nothing to do with the question of limits. (See Figure 2.2.5.)

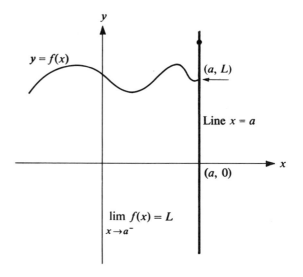

Figure 2.2.5

Let us now consider the analogous situation where the fly approaches from the right. Again we ask if it is possible to predict where the fly will cross the heavy black line. In Figure 2.2.6(a) he will clearly cross the heavy black line at the arrow. Note that the fly is now approaching the black line from the right. In this case we say the right-hand limit exists. Again, we replace the path of the fly by a curve $y = f(x)$ and replace the heavy black line by the line $x = a$. If we can predict where the fly will cross the line $x = a$, we say $\lim_{x \to a^+} f(x)$ exists. If the fly will cross at the point (a, R), we say that $\lim_{x \to a^+} f(x) = R$. (The symbol a^+ indicates that x is approaching a from the right.) [See Figure 2.2.6(b).]

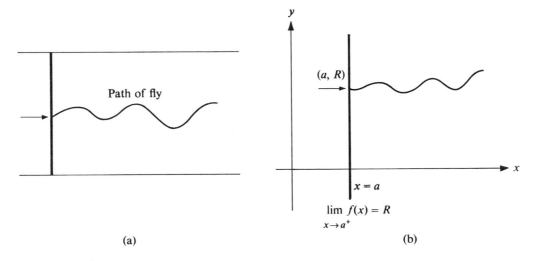

Figure 2.2.6

The following rule is extremely important.

HEURISTIC RULE. We are given a curve $y = f(x)$. We say

$$\lim_{x \to a} f(x) = L;$$

that is, the limit of f as x approaches a exists and is equal to L if the left-hand limit at a and the right-hand limit at a both exist and are equal to L.

The statement that $\lim_{x \to a} f(x)$ exists can be assigned the following geometric interpretation. Given the curve $y = f(x)$, consider it to be the path of two flies, one fly approaching the line $x = a$ from the left and one from the right. If you can predict where each fly will cross the line $x = a$ and if both flies will cross at the same point (that is, hit head on), then $\lim_{x \to a} f(x)$ exists. Otherwise it does not.

We now present several examples to clarify this rule. To reinforce or comment on the unimportance of $f(a)$ in the examples, we do not bother to graph f at a.

▶ **Example 1.** In Figure 2.2.7 it is clear that the fly approaching from the left will cross at the dot; the fly approaching from the right will also cross at the dot. Thus both the right- and left-hand limits of f at a exist. Moreover, the crossing point is the same in both cases. (The flies will coincide.) Thus, since the right- and left-hand limits exist and are equal, we conclude that $\lim_{x \to a} f(x)$ exists.

▶ **Example 2.** In Figure 2.2.8, both the right- and left-hand limits exist, but they are different. Thus, $\lim_{x \to a} f(x)$ does *not* exist.

▶ **Example 3.** In Figure 2.2.9, the left-hand limit exists but the right-hand limit does not exist. Hence, $\lim_{x \to a} f(x)$ does *not* exist.

▶ **Example 4.** In Figure 2.2.10 both the right- and left-hand limits exist and they are equal. Thus, $\lim_{x \to a} f(x)$ exists.

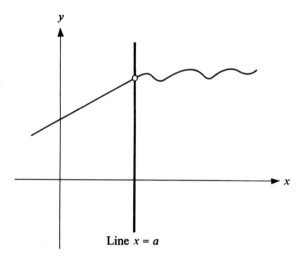

Line $x = a$

Figure 2.2.7

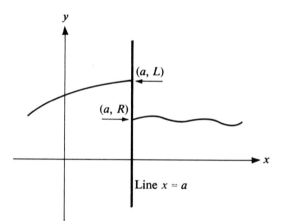

Line $x = a$

Figure 2.2.8

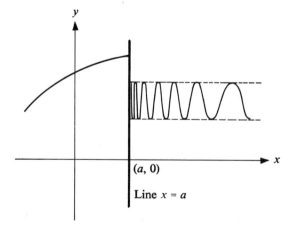

Line $x = a$

Figure 2.2.9

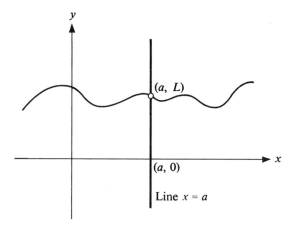

Figure 2.2.10

▶ Example 5. In Figure 2.2.11, the left-hand limit exists but the right-hand limit does not exist. Thus, $\lim_{x \to a} f(x)$ does *not* exist. One might argue that the fly approaching from the right will cross the line $x = a$ at $+\infty$. Geometrically this is

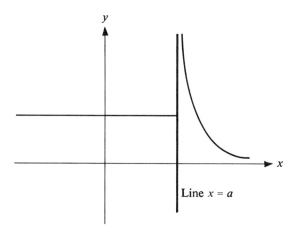

Figure 2.2.11

correct. However, we do not want to work with infinities and so we will say that the right-hand limit (or left as the case may be) does not exist in such situations. (Infinite limits is a subtle, complex topic, replete with pitfalls, which in the long run is not that useful.)

▶ Example 6. In Figure 2.2.12, neither the right-hand limit nor the left-hand limit exists. Thus, $\lim_{x \to a} f(x)$ does *not* exist.

Observe that we have ignored completely the value of the function at a in all of the above examples. The value $f(a)$ has no effect on the existence or nonexistence of a limit.

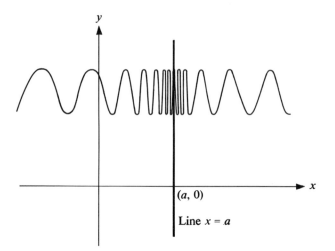

Figure 2.2.12

▶ Example 7. In Figure 2.2.13, we see that the left-hand limit and the right-hand limit exist and that they are equal. Thus, $\lim_{x \to a} f(x)$ exists. In fact, we see in the next section that we can assign the limit a numerical value, namely $\lim_{x \to a} f(x) = 2$. However, $f(a) = 3$. Again we see that the value of f at a has *nothing* to do with the existence of a limit.

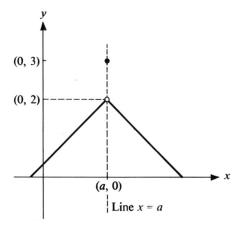

Figure 2.2.13

In most of the foregoing examples we have not included any numerical values. In simple examples such as these, one can decide whether or not the limit exists by simply examining the picture. Numbers are not necessary. In the next section we study limits by more algebraic techniques.

Exercises

1. Decide for each function in Figure 2.2.14 whether the left-hand limit, the right-hand limit, and the $\lim_{x \to a} f(x)$ exist.

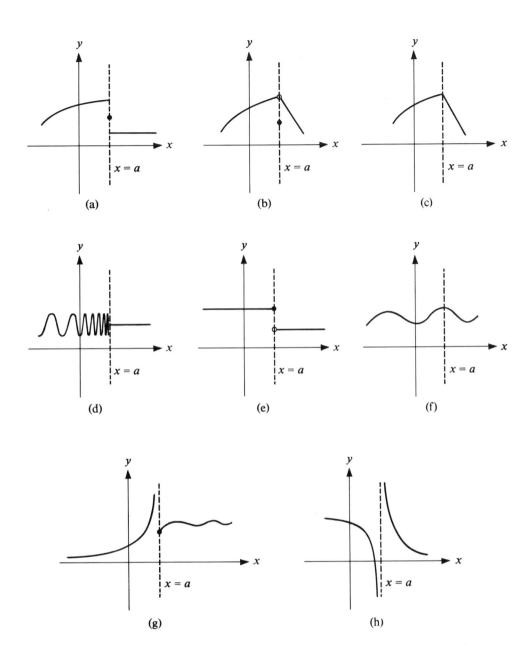

Figure 2.2.14

2. Graph the following functions and decide from the graph whether the right-hand limit, the left-hand limit, and $\lim_{x \to a} f(x)$ exist for a as specified in each part:

(a) $f(x) = \begin{cases} x, & x < 0, \\ 0, & x \geq 0, \end{cases}$ $a = 0$.

(b) $f(x) = \begin{cases} -x, & x \leq 0, \\ 1, & x > 0, \end{cases}$ $a = 0$.

(c) $f(x) = \begin{cases} x, & x < 0, \\ 1, & x \geq 0, \end{cases} \quad a = 0.$

(d) $f(x) = \begin{cases} x + 1, & x \leq 2, \\ 3, & x > 2, \end{cases} \quad a = 1.$

(e) $f(x) = \begin{cases} x + 1, & x < 2, \\ 3, & x \geq 2, \end{cases} \quad a = 2.$

(f) $f(x) = \begin{cases} x + 1, & x < 2, \\ 5, & x = 2, \quad a = 2. \\ 3, & x > 2, \end{cases}$

3. Graph the following functions and decide from the graph whether the right-hand limit, the left-hand limit, and the $\lim_{x \to a} f(x)$ exist for a as specified in each part:

(a) $f(x) = \begin{cases} 2x - 1 & \text{for} \quad x \leq 3, \\ -x + 1 & \text{for} \quad x > 3, \end{cases} \quad a = 3.$

(b) $f(x) = \begin{cases} 9 - x & \text{for} \quad x < 4, \\ 2x - 3 & \text{for} \quad x \geq 4, \end{cases} \quad a = 4.$

(c) $f(x) = \begin{cases} x + 1 & \text{for} \quad x < 3, \\ 2 & \text{for} \quad x = 3, \quad a = 3. \\ -2x + 10 & \text{for} \quad x > 3, \end{cases}$

(d) $f(x) = \begin{cases} 3x - 4 & \text{for} \quad x < 1, \\ 0 & \text{for} \quad x = 1, \quad a = 1. \\ x + 1 & \text{for} \quad x > 1, \end{cases}$

(e) $f(x) = \begin{cases} x^2 & \text{for} \quad x < 0, \\ 5 & \text{for} \quad x = 0, \quad a = 0. \\ x & \text{for} \quad x > 0, \end{cases}$

(f) $f(x) = \begin{cases} -3x + 4 & \text{for} \quad x < 2, \\ 1037 & \text{for} \quad x = 2, \quad a = 2. \\ x - 4 & \text{for} \quad x > 2, \end{cases}$

(g) $f(x) = \begin{cases} x + 11 & \text{for} \quad x < -3, \\ 8 & \text{for} \quad x = -3, \quad a = -3. \\ 3 + 2x & \text{for} \quad x > -3, \end{cases}$

4. Suppose f is defined as follows:

$$f(x) = \begin{cases} |x| & \text{for} \quad -1 \leq x \leq 0, \\ 2 & \text{for} \quad 0 < x \leq 1. \end{cases}$$

(a) Draw the graph of f.

(b) Do $\lim_{x \to 0^-} f(x)$ and $\lim_{x \to 0^+} f(x)$ exist?

(c) Does $\lim_{x \to 0} f(x)$ exist? Why?

5. Suppose f is defined as follows:

$$f(x) = \begin{cases} |x + 2| & \text{for} \quad -2 \leq x \leq 0, \\ 3 & \text{for} \quad 0 < x \leq 4. \end{cases}$$

(a) Draw the graph of f.

(b) Do $\lim_{x \to 0^+} f(x)$ and $\lim_{x \to 0^-} f(x)$ exist?

(c) Does $\lim_{x \to 0} f(x)$ exist?

(d) Evaluate $\lim_{x \to 0^+} f(x)$ and $\lim_{x \to 0^-} f(x)$.

6. Let

$$g(x) = \begin{cases} x^2 & \text{for} \quad x < 0, \\ x & \text{for} \quad x \geq 0. \end{cases}$$

(a) Graph the function g.

(b) Do $\lim_{x \to 0^+} g(x)$ and $\lim_{x \to 0^-} g(x)$ exist? If so, try to evaluate them.

(c) Does $\lim_{x \to 0} g(x)$ exist?

(d) Does (i) $\lim_{x \to 0^-} \dfrac{g(x) - g(0)}{x}$ exist?

Does (ii) $\lim_{x \to 0^+} \dfrac{g(x) - g(0)}{x}$ exist?

$$\left[Hint: \text{ First construct the graph of } \frac{g(x) - g(0)}{x}. \right]$$

(e) Does $\lim_{x \to 0} \dfrac{g(x) - g(0)}{x}$ exist? Why?

2.3 Limits of Functions

In the discussion that follows, we rely heavily on intuition. Most of this section is not intended to be precise and cannot be used in a rigorous way. Let us begin by restricting our attention to the function defined by

$$f(x) = x^2, \qquad x \in [0, 3].$$

It is clear that $f(2) = 4$. However, we might ask what the value of f is when x takes on values slightly larger than 2 and slightly smaller than 2. That is, we are no longer concerned with the value of f at 2, but rather the values of f when x gets close to 2 through values larger than 2 and for values smaller than 2. Observe that

$$f(2.2) = 4.84,$$

$$f(2.1) = 4.41,$$

$$f(2.01) = 4.0401,$$

$$f(2.001) = 4.004001, \text{ etc.}$$

Thus, we see that as x gets closer to 2 through values of $x > 2$, it appears that $f(x)$ gets closer and closer to 4. Comparing this with our geometrical discussion in

Section 2.2, we see that this is equivalent to stating that the right-hand limit $\lim_{x \to 2^+} f(x) = 4$. In like manner, we see that

$$f(1.8) = 3.24,$$
$$f(1.9) = 3.61,$$
$$f(1.99) = 3.9601,$$
$$f(1.999) = 3.996001, \text{ etc.}$$

Again, we observe that as x gets closer to 2 through values of $x < 2$, $f(x)$ gets closer and closer to 4. This statement simply says that the left-hand limit $\lim_{x \to 2^-} f(x) = 4$. Thus, as x tends to 2, $f(x)$ tends to 4. Note that the value of $f(x)$ for $x = 2$ plays *no* role. This is roughly what we mean by saying that the limit of $f(x)$ as x tends to 2 is 4. We abbreviate the preceding statement as follows:

$$\lim_{x \to 2} x^2 = 4.$$

You are reminded of the fact that in order to find $\lim_{x \to 2} x^2$, we had to observe values of x^2 for both $x < 2$ and $x > 2$ (that is, we look at $\lim_{x \to 2^+} x^2$ and $\lim_{x \to 2^-} x^2$)—not caring at all about $x = 2$. Let us investigate what this means geometrically. In Figure 2.3.1, one clearly sees that

$$\lim_{x \to 2^+} f(x) = \lim_{x \to 2^-} f(x) = 4.$$

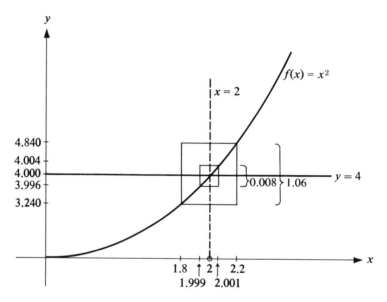

Figure 2.3.1

If in Figure 2.3.1 we restrict our attention to the interval (1.8, 2.2), the graph of f lies in a rectangle of height 1.06, while for $x \in$ (1.999, 2.001) the graph of f lies in a rectangle of height 0.008000. That is, the heights of the rectangles get closer and closer to 0. All the rectangles contain the line $y = 4$ (and thus the heights as $x \to 2$ cluster about the line $y = 4$), indicating geometrically that $\lim_{x \to 2} f(x) = 4$. This is certainly not a precise formulation of the concept of a limit, but should give us some feeling for the meaning of a limit. Note that the point (2, 0) is to be excluded from our intervals. [This is the reason for the circle about (2, 0).]

Next we consider the function f defined by

$$f(x) = \frac{x - 2}{2|x - 2|}.$$

First, observe that $2 \notin$ domain of f. We wish to investigate this function for values of x near 2; in fact, we ask whether $\lim_{x \to 2} f(x)$ exists. First observe that if $x > 2$, then $|x - 2| = x - 2$ and hence

$$f(x) = \frac{x - 2}{2(x - 2)} = \frac{1}{2};$$

while if $x < 2$, $|x - 2| = -(x - 2)$ and

$$f(x) = -\frac{x - 2}{2(x - 2)} = -\frac{1}{2}.$$

Thus, it is clear that as x approaches 2 through values of $x > 2$, $f(x)$ approaches $+\frac{1}{2}$, and as x approaches 2 through values of $x < 2$, $f(x)$ approaches $-\frac{1}{2}$. Hence,

$$\lim_{x \to 2^+} f(x) = \frac{1}{2} \quad \text{and} \quad \lim_{x \to 2^-} f(x) = -\frac{1}{2}.$$

This is clearly illustrated in Figure 2.3.2. In this case, we must say that $\lim_{x \to 2} f(x)$ does not exist since the left- and the right-hand limits are not equal. [*Note:* The fact that $2 \notin$ domain of f did not enter into our discussion.] The graph of f is shown in Figure 2.3.2.

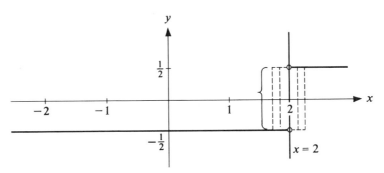

Figure 2.3.2

If in Figure 2.3.2 we restrict our attention to various intervals about (2, 0) (excluding that point), the graph of f always lies in a rectangle of height 1, and in *no* smaller rectangle. (In fact, this rectangle has $y = \frac{1}{2}$ and $y = -\frac{1}{2}$ as two of its sides.) Thus, in this case the heights *never* approach zero. This is another geometric argument that $\lim_{x \to 2} f(x)$ does not exist.

Let us now consider the function defined by

$$f(x) = \begin{cases} \dfrac{2x^2 - x - 3}{x + 1} & \text{for} \quad x \neq -1, \\ 2 & \text{for} \quad x = -1. \end{cases}$$

We wish to investigate the behavior of $f(x)$ as x approaches -1. (That is, we are really asking if we can determine $\lim_{x \to -1} f(x)$ heuristically.) Again we first look at values of x close to -1 with $x > -1$. Clearly,

$$f(0) = -3,$$
$$f(-0.9) = -4.8,$$
$$f(-0.95) = -4.90,$$
$$f(-0.99) = -4.98,$$
$$f(-0.999) = -4.998, \text{ etc.}$$

Thus, as x approaches -1 for $x > -1$, we see that $f(x)$ approaches -5. We can then say that $\lim_{x \to -1^+} f(x) = -5$. In like manner we observe that

$$f(-2) = -7,$$
$$f(-1.5) = -6,$$
$$f(-1.1) = -5.2,$$
$$f(-1.01) = -5.02,$$
$$f(-1.001) = -5.001, \text{ etc.}$$

Hence, it appears that as x approaches -1 for $x < -1$, $f(x)$ approaches -5 and thus $\lim_{x \to -1^-} f(x) = -5$. Since the right- and left-hand limits are equal, we see that $\lim_{x \to -1} f(x) = -5$. Note that $f(-1) = 2$ did not enter into our discussion at all. We see in fact that $\lim_{x \to -1} f(x) = -5$ even though $f(-1) = 2$. The graph of f is shown in Figure 2.3.3.

The circle indicates that the point $(-1, f(-1))$ is excluded from the straight line. If we choose intervals about $(-1, 0)$, excluding that point, we can see that the graph of f lies in rectangles whose heights get closer and closer to zero. In fact, they seem

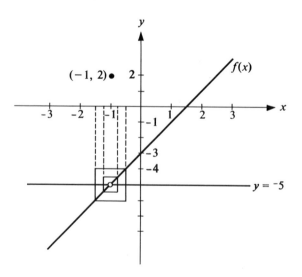

Figure 2.3.3

to cluster about the line $y = -5$. Once again, we have an indication that $\lim_{x \to -1} f(x) = -5$.

Our arguments for the existence of a limit are fairly awkward to go through each time. As you can see, we rely very heavily on our intuition. In the preceding example, a convenient use of algebra leads us to the conclusion much more quickly. We note that

$$f(x) = \frac{2x^2 - x - 3}{x + 1} = \frac{(2x - 3)(x + 1)}{x + 1}.$$

If $x \neq -1$, then

$$f(x) = 2x - 3$$

and one can easily see that as $x \to -1$ for $x > -1$, then $f(x) \to -5$ [that is, $\lim_{x \to -1^+} f(x) = -5$] and as $x \to -1$, for $x < -1$, then $f(x) \to -5$ [that is, $\lim_{x \to -1^-} f(x) = -5$]. Hence, we conclude that $\lim_{x \to -1} f(x) = -5$. We emphasize once again the fact that the value of f at $x = -1$ did not play a role in our discussion.

Let us consider one more example. Suppose we define a function f as follows:

$$f(x) = \begin{cases} 2x & \text{for} \quad 0 \leq x < 1, \\ 4 & \text{for} \quad x = 1, \\ 5 - 3x & \text{for} \quad 1 < x \leq 2. \end{cases}$$

We wish to determine whether or not $\lim_{x \to 1} f(x)$ exists. (We have not really defined what is meant by "$\lim_{x \to 1} f(x)$ exists"—we rely solely on our previous intuitive discussion.) First, we investigate the behavior of $f(x)$ for values of x close to but greater than 1. Observe that

$$f(1.2) = 1.4,$$
$$f(1.1) = 1.7,$$
$$f(1.01) = 1.97,$$
$$f(1.001) = 1.997, \text{ etc.}$$

It appears that as $x \to 1$, $x > 1$, $f(x) \to 2$. Now we see that

$$f(0.8) = 1.6,$$
$$f(0.9) = 1.8,$$
$$f(0.99) = 1.98,$$
$$f(0.999) = 1.998, \text{ etc.}$$

We also note that as $x \to 1$, $x < 1$, $f(x) \to 2$. Thus,

$$\lim_{x \to 1^+} f(x) = \lim_{x \to 1^-} f(x) = 2.$$

Hence, we may conclude that $\lim_{x \to 1} f(x) = 2$. Again, note the fact that $f(1) = 4$ plays no role in the determination of $\lim_{x \to 1} f(x)$. The graph of f is shown in Figure 2.3.4.

We could again give an intuitive geometrical argument similar to those of the preceding examples to add further verification that $\lim_{x \to 1} f(x) = 2$, but we leave that discussion to the reader. (It would be worthwhile for you to go through this argument.)

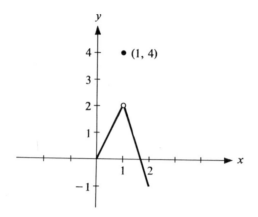

Figure 2.3.4

Before proceeding to a workable (but *not* formal) definition of a limit, we again illustrate a case where the limit does not exist. Consider the function

$$f(x) = 1/x.$$

We wish to investigate $\lim_{x \to 0} f(x)$. First, we observe values of $f(x)$ for x near 0 and $x > 0$. Note that

$$f(0.1) = 10,$$
$$f(0.01) = 100,$$
$$f(0.001) = 1000,$$
$$f(0.0001) = 10^4, \text{ etc.}$$

That is, we observe that as $x \to 0$, $x > 0$, $f(x)$ becomes larger and larger and does not seem to approach any particular number. [We say that as $x \to 0$, for $x > 0$, $f(x) \to \infty$.] Now, observe that

$$f(-0.1) = -10,$$
$$f(-0.01) = -100,$$
$$f(-0.001) = -1000,$$
$$f(-0.0001) = -10^4, \text{ etc.}$$

It appears that as $x \to 0$, $x < 0$, $|f(x)|$ becomes larger and larger and does not appear to approach any particular number. [We could say that as $x \to 0$, for $x < 0$, $f(x) \to -\infty$.] Since, as $x \to 0$, $|f(x)|$ gets larger and larger and does not approach a fixed number, we conclude that $\lim_{x \to 0} f(x)$ does not exist. The graph of f is shown in Figure 2.3.5.

Since $f(x)$ becomes arbitrarily large as we approach 0 from the right (and $-f(x)$ becomes arbitrarily large as we approach 0 from the left); for x near 0, the graph of f is not contained in any rectangle of height say 1, or even 10,000,000 for that matter. This observation also implies that $\lim_{x \to 0} f(x)$ does not exist.

We conclude this section with one more negative result for the existence of a limit. Consider the function f defined by

$$f(x) = \begin{cases} x & \text{for } 0 \le x \le 1, \\ 3 - x & \text{for } 1 < x \le 4. \end{cases}$$

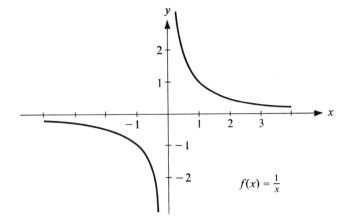

Figure 2.3.5

We wish to determine whether or not $\lim_{x \to 1} f(x)$ exists. Looking at values of $f(x)$ for x close to 1 and $x < 1$, we see that

$$f(0.9) = 0.9,$$

$$f(0.99) = 0.99,$$

$$f(0.999) = 0.999, \text{ etc.}$$

Thus we intuitively see that for $x < 1$, as $x \to 1$, then $f(x) \to 1$. Now for x close to 1 and $x > 1$, we note that

$$f(1.1) = 1.9,$$

$$f(1.01) = 1.99,$$

$$f(1.001) = 1.999, \text{ etc.}$$

Thus for $x > 1$, as $x \to 1$, we see that $f(x) \to 2$. Hence, $\lim_{x \to 1^-} f(x) = 1$ and $\lim_{x \to 1^+} f(x) = 2$. We say that $\lim_{x \to 1} f(x)$ does *not* exist since $f(x)$ approaches two different numbers as we let x approach 1. In order that a limit may exist, both numbers must be the same. The graph of f is shown in Figure 2.3.6. Again, a geometric argument similar to that given in the example on page 85 could be given to add further justification to our conclusion.

The problem of giving a precise meaning to $\lim_{x \to a} f(x) = L$ is not an easy one. The calculus reached a reasonable stage of development at the time of Newton and Leibniz (about 1700), but a satisfactory definition of limit was not arrived at until after the year 1800. Nevertheless an immense amount of mathematics was produced in the eighteenth century. Part of the difficulty seems to be a psychological fluke. The natural (but wrong) approach to the problem is to let the x values get close to a and then claim the $f(x)$ values to be close to L. Slightly reworded, this definition takes the form "as x gets closer and closer to a, $f(x)$ gets closer and closer to L." Unfortunately this is not mathematically precise. The correct approach is to take an open interval about L and claim that $f(x)$ lies in this interval for the x's in some interval about a, but $x \neq a$. If this can be done for any open interval about L, then we say $\lim_{x \to a} f(x) = L$.

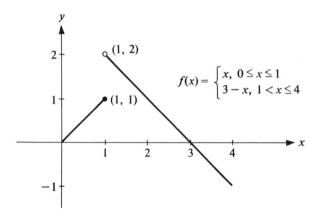

$$f(x) = \begin{cases} x, & 0 \le x \le 1 \\ 3 - x, & 1 < x \le 4 \end{cases}$$

Figure 2.3.6

Thus the trick, which took 100 years to discover, is to start *not* with a small interval about *a*, but with a small interval about *L*. This procedure is examined in more detail in the appendix to this chapter.

The definition we give is clearly *not* a formal one but only one that will aid us in determining the existence of a limit.

PROVISIONAL OR WORKING DEFINITION OF A LIMIT. We say that $\lim_{x \to a} f(x)$ *exists and is equal to a number L if the numbers f(x) remain arbitrarily close to L when we take x sufficiently close to a, for values of x greater than a and values of x less than a, but x ≠ a. If the numbers f(x) remain arbitrarily close to L when x > a is sufficiently close to a, we say that the right-hand limit exists and write* $\lim_{x \to a^+} f(x) = L$. *In like manner we define the left-hand limit,* $\lim_{x \to a^-} f(x) = L$.

Observe certain facts about the definition:

1. If $f(x)$ approaches two different numbers, say L_1 and L_2, depending on whether we take *x* close to *a* for values of $x > a$ or $x < a$, then we say that $\lim_{x \to a} f(x)$ does *not* exist.

2. The fact that *f* is or is not defined at *a* does not play any role in the determination of the existence of a limit.

Exercises

In the following exercises, use arguments similar to those given in Section 2.3 to determine the existence of a limit. Construct a graph of each of the functions.

1. Let $f(x) = x - 1$. Does $\lim_{x \to 2} f(x)$ exist? If so, what is it?

2. Let $f(x) = x + 1$. Does $\lim_{x \to -1} f(x)$ exist? If so, what is it?

3. If $f(x) = x + 2$, does $\lim_{x \to -1} f(x)$ exist? If so, what is it?

4. If $f(x) = x^2 + 2$, does $\lim_{x \to 1} f(x)$ exist? If so, what is it?

5. Let $f(x) = 2x^2$. Does $\lim_{x \to -1} f(x)$ exist? If so, what is it?

6. Let $f(x) = \sqrt{x - 1}$. Does $\lim_{x \to -2} f(x)$ exist? If so, what is it?

7. Let $f(x) = \sqrt{x - 1}$. Does $\lim_{x \to 3} f(x)$ exist? If so, what is it?

8. Let $f(x) = \dfrac{2x^2 - x}{2x - 1}$. Does $\lim_{x \to 1/2} f(x)$ exist? If so, what is it? [*Hint:* Use

   an algebraic manipulation similar to that used in the example on page 87.]

9. Let $f(x) = \dfrac{2x - 2x^2}{x - 1}$. Does $\lim_{x \to 1} f(x)$ exist? If so, what is the limit? [*Hint:*

   Use an algebraic manipulation similar to that used in the example on page 87.]

10. Let $f(x) = |x|/2x$. Does $\lim_{x \to 0} f(x)$ exist? If so, what is the limit?

11. Let

$$f(x) = \frac{2(|x| - 1)}{x - 1}.$$

   Does $\lim_{x \to 1^+} f(x)$ exist? Does $\lim_{x \to 1^-} f(x)$ exist? Does $\lim_{x \to 1} f(x)$ exist? Why?

12. Consider the following graphs, representing functions $f(x)$, $g(x)$, $h(x)$, and $s(x)$, where $f(1) = 1$, $g(1) = 2$, $h(2) = 3$, and s is *not* defined at $x = \frac{1}{2}$. From the

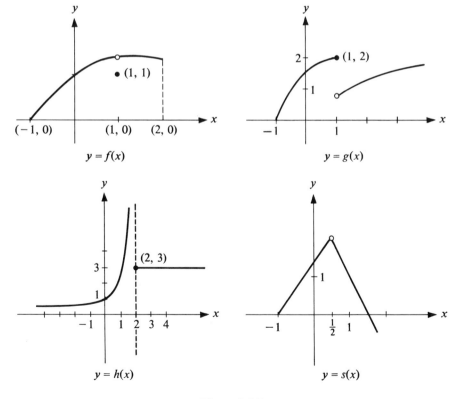

Figure 2.3.7

graphs in Figure 2.3.7 (using intuitive arguments) do you expect that the following limits exist?

(a) $\lim\limits_{x \to 1} f(x)$.

(b) $\lim\limits_{x \to 1} g(x)$.

(c) $\lim\limits_{x \to 2} h(x)$.

(d) $\lim\limits_{x \to 1/2} s(x)$.

13. Let

$$f(x) = \begin{cases} x & \text{for} & -1 \le x < 0, \\ 1 & \text{for} & x = 0, \\ 2 & \text{for} & 0 < x \le 1. \end{cases}$$

Does $\lim_{x \to 0} f(x)$ exist? If so, what is the limit?

14. Let

$$f(x) = \begin{cases} x & \text{for} & -1 \le x < 0, \\ 1 & \text{for} & x = 0, \\ 2x & \text{for} & 0 < x \le 1. \end{cases}$$

Does $\lim_{x \to 0} f(x)$ exist? If so, what is the limit?

15. If $f(x) = 1/(x - 1)$, does $\lim_{x \to 1} f(x)$ exist? Graph this function.

16. Let $f(x) = x^2$.

(a) Calculate

$$\frac{f(x) - f(2)}{x - 2} \quad \text{for} \quad x \ne 2.$$

(b) Does

$$\lim_{x \to 2} \frac{f(x) - f(2)}{x - 2}$$

exist? If so, what is it?

17. Let $f(x) = |x|$.

(a) Find

$$\lim_{x \to 0^+} \frac{f(x) - f(0)}{x}, \qquad \lim_{x \to 0^-} \frac{f(x) - f(0)}{x}.$$

(b) Does $\lim\limits_{x \to 0} \dfrac{f(x) - f(0)}{x}$ exist? Why?

2.4 Properties of Limits

The methods for evaluating limits in Section 2.3 often prove to be very cumbersome and awkward. It would be advantageous for us to develop some general properties of limits in order to make our calculations easier. We do not have at our disposal the necessary machinery to derive these properties rigorously. The formal definition given in the appendix to this chapter would allow us to rigorously derive the properties listed below. Although we will not use these properties very often in our future work, the reader should at least understand the statements of them.

PROPERTY 1. If $f(x) = c$, where c is a constant, then $\lim_{x \to a} f(x) = c$ for any a. The graph of $f(x) = c$ is given in Figure 2.4.1.

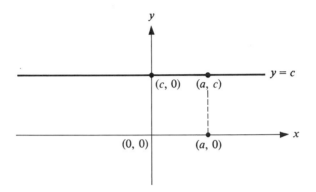

Figure 2.4.1

PROPERTY 2. If $f(x) = x$, then $\lim_{x \to a} f(x) = a$.

You can easily see the validity of this statement from the graph shown in Figure 2.4.2.

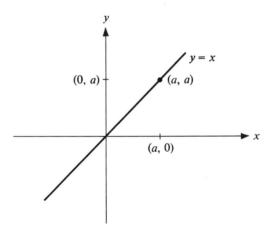

Figure 2.4.2

PROPERTY 3. Let $f(x) = x^n$ for any positive number n; then $\lim_{x \to a} f(x) = a^n$.

For example, if $f(x) = x^3$, then $\lim_{x \to 2} f(x) = 2^3 = 8$; if $f(x) = x^{2/3}$, then $\lim_{x \to 3} f(x) = 2^{2/3}$; if $f(x) = x^{99}$, then $\lim_{x \to 10} f(x) = 10^{99}$. In order to verify this property for any n, we must resort to the formal definition of a limit. (Figure 2.4.3 shows $n = 3$ and $a = 2$.)

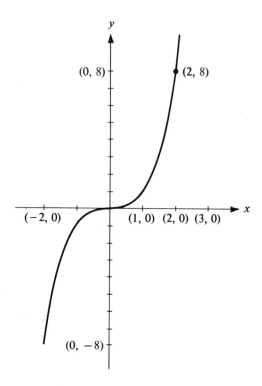

Figure 2.4.3

PROPERTY 4. Suppose that f and g are defined for x close to a number a (a need *not* be in the domain of f and g), and assume that $\lim_{x \to a} f(x)$ exists and $\lim_{x \to a} g(x)$ exists. Then

$$\lim_{x \to a} (f + g)(x) = \lim_{x \to a} f(x) + \lim_{x \to a} g(x).$$

In other words, Property 4 states that the limit of the sum of two functions is the sum of the limits.

Let us now look at some examples illustrating Properties 1–4.

▶ **Example 1.** Find $\lim_{x \to 1} (1 + x^2)$.

Solution. By Property 4,

$$\lim_{x \to 1} (1 + x^2) = \lim_{x \to 1} 1 + \lim_{x \to 1} x^2,$$

and by Property 1, $\lim_{x \to 1} 1 = 1$. Property 3 tells us that

$$\lim_{x \to 1} x^2 = 1^2 = 1.$$

Hence,

$$\lim_{x \to 1} (1 + x^2) = 1 + 1 = 2.$$

▶ Example 2. Find $\lim_{x\to 4} (x^{1/2} + x^{3/2})$.

Solution. By Property 4,

$$\lim_{x\to 4} (x^{1/2} + x^{3/2}) = \lim_{x\to 4} x^{1/2} + \lim_{x\to 4} x^{3/2} = 4^{1/2} + 4^{3/2} = 10.$$

A similar property holds for the difference of two functions. That is,

$$\lim_{x\to a} (f - h)(x) = \lim_{x\to a} f(x) - \lim_{x\to a} h(x)$$

provided the appropriate limits exist. This result *follows* from Property 4 by letting $g(x) = -h(x)$.

PROPERTY 5. Let f and g be two functions defined near a, such that $\lim_{x\to a} f(x)$ and $\lim_{x\to a} g(x)$ exist. Then

$$\lim_{x\to a} f(x)g(x) = \lim_{x\to a} f(x) \lim_{x\to a} g(x).$$

In other words, the limit of the product is the product of the limits.

Consider the following example. Let us find $\lim_{x\to 3} 2x^3$. By Property 5,

$$\lim_{x\to 3} 2x^3 = \left(\lim_{x\to 3} 2\right)\left(\lim_{x\to 3} x^3\right) = 2 \cdot 3^3 = 54.$$

The last equality follows from Properties 1 and 3.

We might also note that Property 3 really follows from repeated application of Properties 2 and 5. Namely,

$$\lim_{x\to a} \underbrace{x \cdot x \cdot x \cdots x}_{n \text{ times}} = \underbrace{\left(\lim_{x\to a} x\right)\left(\lim_{x\to a} x\right)\cdots\left(\lim_{x\to a} x\right)}_{n \text{ times}} = \underbrace{a \cdot a \cdot a \cdots a}_{n \text{ times}} = a^n.$$

PROPERTY 6. Let f and g be two functions defined near a such that $\lim_{x\to a} f(x)$ and $\lim_{x\to a} g(x)$ exist and such that $\lim_{x\to a} g(x) \neq 0$. Then $\lim_{x\to a} f(x)/g(x)$ exists and

$$\lim_{x\to a} \frac{f(x)}{g(x)} = \frac{\lim_{x\to a} f(x)}{\lim_{x\to a} g(x)}.$$

In other words, the limit of the quotient of two functions is the quotient of the limits provided that the limit of the function in the denominator is not equal to zero.

▶ Example 3. Let f be defined as follows:

$$f(x) = \frac{x^2 + 1}{3x + 2}.$$

We wish to see if $\lim_{x\to -1} f(x)$ exists. From Properties 3 and 4 we see that

$$\lim_{x\to -1} (x^2 + 1) = 2 \quad \text{and} \quad \lim_{x\to -1} (3x + 2) = -1.$$

Hence, we may use Property 6 and obtain

$$\lim_{x \to -1} \frac{x^2 + 1}{3x + 2} = \frac{\lim_{x \to -1} (x^2 + 1)}{\lim_{x \to -1} (3x + 2)} = \frac{2}{-1} = -2.$$

▶ **Example 4.** Suppose that

$$f(x) = \frac{(x^2 - 1)(x + 2)}{(x - 1)(x - 2)}.$$

Does $\lim_{x \to 1} f(x)$ exist? First, we observe that

$$\lim_{x \to 1} (x^2 - 1)(x + 2) = \lim_{x \to 1} (x^2 - 1) \lim_{x \to 1} (x + 2) = 0 \cdot 3 = 0,$$

by Properties 3, 4, and 5. Using Properties 4 and 5, we find that

$$\lim_{x \to 1} (x - 1)(x - 2) = 0 \cdot (-1) = 0.$$

Hence we may *not* apply Property 6, since the limit of the function in the denominator is zero. This does *not* mean that the limit does not exist, just that Property 6 does *not* apply. However, a convenient algebraic manipulation does yield a limit. First note that for $x \neq 1$,

$$f(x) = \frac{(x - 1)(x + 1)(x + 2)}{(x - 1)(x - 2)} = \frac{(x + 1)(x + 2)}{x - 2}.$$

Thus, since we are only concerned with numbers near 1 and do not care about $x = 1$, we see that

$$\lim_{x \to 1} f(x) = \lim_{x \to 1} \frac{(x + 1)(x + 2)}{x - 2} = \frac{\lim_{x \to 1} (x + 1)(x + 2)}{\lim_{x \to 1} (x - 2)}$$

$$= \frac{\lim_{x \to 1} (x + 1) \lim_{x \to 1} (x + 2)}{\lim_{x \to 1} (x - 2)} = \frac{2 \cdot 3}{-1} = -6.$$

Here we have used Properties 3, 4, 5, and 6.

This last example is quite instructive. If it turns out that the denominator tends to zero in the limit, then an attempt should be made to see if there is an algebraic manipulation that eliminates the troublesome expression, as in Example 4.

There are two other properties of limits involving inequalities that are extremely useful in the sequel. We state these properties here, but their application appears in the next two chapters.

PROPERTY 7. Let f and g be defined near a but not necessarily at a. Suppose that for every x in the domain of f and g, $f(x) \leq g(x)$ and $\lim_{x \to a} f(x)$ and $\lim_{x \to a} g(x)$ exist. Then,

$$\lim_{x \to a} f(x) \leq \lim_{x \to a} g(x).$$

▶ **Example 5.** Suppose $g(x) = x$ and $f(x) = x^2$, both having domain $[0, 1]$. Then $f(x) \leq g(x)$, for $x \in [0, 1]$. In particular,

$$\lim_{x \to 1/2} f(x) = \frac{1}{4} \leq \lim_{x \to 1/2} g(x) = \frac{1}{2}.$$

PROPERTY 8. Suppose the assumptions given in Property 7 hold and in addition $\lim_{x \to a} f(x) = \lim_{x \to a} g(x)$. Let h be another function having the same domain as f and g and such that $f(x) \le h(x) \le g(x)$. Then $\lim_{x \to a} h(x)$ exists and

$$\lim_{x \to a} h(x) = \lim_{x \to a} g(x) = \lim_{x \to a} f(x).$$

This property is often referred to as the *squeezing* or *pinching* process.

We now proceed to further illustrate Properties 1–8 with further examples.

▶ **Example 6.** Suppose we let $h(x)$ be a function defined on $[0, 2]$ such that $|h(x)| \le (x - 1)^2$ for $x \in [0, 2]$. We wish to determine $\lim_{x \to 1} h(x)$. Now, since $|h(x)| \le (x - 1)^2$, we know that $-(x - 1)^2 \le h(x) \le (x - 1)^2$. Observe that

$$\lim_{x \to 1} (x - 1)^2 = \lim_{x \to 1} \left[-(x - 1)^2 \right] = 0.$$

If we let

$$f(x) = -(x - 1)^2 \quad \text{and} \quad g(x) = (x - 1)^2,$$

then applying Property 8,

$$\lim_{x \to 1} \left[-(x - 1)^2 \right] \le \lim_{x \to 1} h(x) \le \lim_{x \to 1} (x - 1)^2.$$

Thus, $\lim_{x \to 1} h(x) = 0$.

At this point, we remark that all of the preceding eight properties hold for right- and left-hand limits. For example, if

$$\lim_{x \to a^+} f(x) = L_1 \quad \text{and} \quad \lim_{x \to a^+} g(x) = L_2,$$

then

$$\lim_{x \to a^+} \left[f(x) + g(x) \right] = L_1 + L_2,$$

etc.

▶ **Example 7.** Let $f(x) = x^2$. Find

$$\lim_{x \to 3} \frac{f(x) - f(3)}{x - 3}.$$

Solution. Since $f(x) - f(3) = x^2 - 9$, we may write

$$\lim_{x \to 3} \frac{f(x) - f(3)}{x - 3} = \lim_{x \to 3} \frac{x^2 - 9}{x - 3}$$

$$= \lim_{x \to 3} \frac{(x - 3)(x + 3)}{x - 3}$$

$$= \lim_{x \to 3} (x + 3) = 6.$$

The reader should justify each of the steps. Compare this example with Example 8.

▶ **Example 8.** Let $f(x) = 1/(2x + 1)$. Evaluate

$$\lim_{x \to 2} \frac{f(x) - f(2)}{x - 2}.$$

Solution. First observe that

$$f(x) - f(2) = \frac{1}{2x + 1} - \frac{1}{5} = \frac{5 - (2x + 1)}{5(2x + 1)} = \frac{2(2 - x)}{5(2x + 1)}$$

and

$$\frac{f(x) - f(2)}{x - 2} = \frac{2(2 - x)}{5(2x + 1)(x - 2)}$$

$$= -\frac{2(x - 2)}{5(2x + 1)(x - 2)}$$

$$= \frac{2}{5(2x + 1)},$$

since $x \neq 2$. Thus,

$$\lim_{x \to 2} \frac{f(x) - f(2)}{x - 2} = \lim_{x \to 2} \left[-\frac{2}{5(2x + 1)} \right] = -\frac{2}{25},$$

by Properties 4 and 6.

Exercises

In Exercises 1–3 state which of the properties 1–6 you use in order to evaluate the indicated limits.

1. Find

    (a) $\lim_{x \to 2} (2x + 1)$.

    (b) $\lim_{x \to 1} (x^2 - 1)$.

    (c) $\lim_{x \to 5} (5x^2 + 2x + 1)$.

    (d) $\lim_{x \to 3} \dfrac{(x^2 + 3x - 1)}{x + 2}$.

    (e) $\lim_{x \to 1} (x^2 + 2x + 1)(x^3 - 1)$.

    (f) $\lim_{x \to -1} (x - 1)(x + 5)(x - 3)$.

2. Find

    (a) $\lim_{x \to 1} (5x^2 - 9x + 3)$.

    (b) $\lim_{x \to 3} \left(x^3 - \dfrac{1}{x} \right)$.

    (c) $\lim_{x \to -2} (x + 5)(x^2 - 9)$.

    (d) $\lim_{x \to 4} \left(x^{1/2} - \dfrac{4}{x^2} \right)$.

    (e) $\lim_{x \to 25} (3x^{1/2} + x - 4)$.

    (f) $\lim_{x \to -1} (x^{35} - x^2 + 11)$.

    (g) $\lim_{x \to 12} \dfrac{x + 3}{x - 7}$.

    (h) $\lim_{x \to 1} \dfrac{x^2 + 1}{2x^2 - 1}$.

(i) $\displaystyle\lim_{x\to 1} \frac{x-1}{x+1}$.

(j) $\displaystyle\lim_{x\to 1} \frac{x^2-x}{x-1}$.

(k) $\displaystyle\lim_{x\to 1} \frac{x^3-x}{x-1}$.

(l) $\displaystyle\lim_{x\to 1} \frac{(x^2-1)^2}{x-1}$.

3. Find

   (a) $\displaystyle\lim_{x\to 2} \frac{x^2-x-6}{x-3}$.

   (b) $\displaystyle\lim_{x\to 3} \frac{x^2-x-6}{x-3}$.

   (c) $\displaystyle\lim_{x\to 1} \frac{x^2+2x-3}{x^2+x-2}$.

   (d) $\displaystyle\lim_{x\to -2} \frac{3(x^2-4)}{5x(x+2)}$.

4. Evaluate

   (a) $\displaystyle\lim_{x\to 1} \frac{f(x)-f(1)}{x-1}$, if $f(x) = x$.

   (b) $\displaystyle\lim_{x\to 2} \frac{f(x)-f(2)}{x-2}$, if $f(x) = x^2$.

   [*Hint:* First factor the quantity $f(x) - f(2)$ and then recall the example on page 87.]

5. If $f(x) = x + 1$, evaluate $\displaystyle\lim_{x\to a} \frac{f(x)-f(a)}{x-a}$.

6. Find

   (a) $\displaystyle\lim_{x\to 2} \frac{f(x)-f(2)}{x-2}$ if $f(x) = \dfrac{1}{x^2}$.

   (b) $\displaystyle\lim_{x\to a} \frac{f(x)-f(a)}{x-a}$ if $f(x) = \dfrac{1}{x^2}$ and $a \in \operatorname{dom} f$.

7. (a) Find $\displaystyle\lim_{x\to 1} \frac{|x-2|}{x-2}$.

   (b) Does

$$\lim_{x\to 2} \frac{|x-2|}{x-2}$$

   exist? Why? (Recall the discussion in Section 2.3.) Sketch a graph of the function $f(x) = |x-2|/(x-2)$.

8. Is

$$\lim_{x\to 1} \frac{x^2-x}{x-1},$$

   the same as $\lim_{x\to 1} x$? Why? Draw a graph of the function

$$f(x) = \frac{x^2-x}{x-1}.$$

*9. Let n be a positive integer. Evaluate the following limit:

$$\lim_{x \to a} \frac{x^n - a^n}{x - a}.$$

[*Hint:* From elementary algebra,

$$x^n - a^n = (x - a)(x^{n-1} + x^{n-2}a + x^{n-3}a^2 + \cdots + a^{n-1}).]$$

*10. Let

$$f(x) = \begin{cases} 1 & \text{if } x \text{ is an integer,} \\ 0 & \text{if } x \text{ is } not \text{ an integer.} \end{cases}$$

Intuitively, do you think that $\lim_{x \to 2} f(x)$ exists? If so, what is its value?

*11. Evaluate the following limits:

(a) $\displaystyle \lim_{x \to 0} \frac{ax + b}{cx + d}$, where a, b, c, d are given real numbers, and $d \neq 0$.

(b) $\displaystyle \lim_{x \to 0} \frac{ax^2 + bx + c}{dx^2 + ex + f}$, where a, b, c, d, e, f are given real numbers, and $f \neq 0$.

*12. Suppose that f is a function defined on $[a, b]$ such that $|f(x)| \leq M|x - c|$, where $c \in (a, b)$. Show that $\lim_{x \to c} f(x) = 0$.

2.5 Continuity

In Section 2.2 and 2.3 we observed several examples of functions where $\lim_{x \to a} f(x)$ exists but $\lim_{x \to a} f(x) \neq f(a)$. We kept emphasizing that the limit is completely independent of the value of f at a—in fact, f need not even be defined at a. (See the example on page 87.) The various possibilities are shown in Figure 2.5.1.

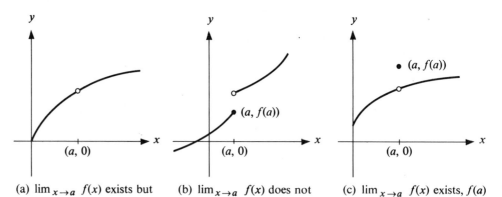

(a) $\lim_{x \to a} f(x)$ exists but $f(a)$ is not defined.

(b) $\lim_{x \to a} f(x)$ does not exist, but $f(a)$ is defined.

(c) $\lim_{x \to a} f(x)$ exists, $f(a)$ defined, but $\lim_{x \to a} f(x) = f(a)$.

Figure 2.5.1

---

* Exercises marked with an asterisk are more difficult and/or deal with subjects that may be omitted without loss of continuity.

We like to think of the behavior of the functions in Figure 2.5.1 near a as *abnormal* and give a name to functions that do not exhibit such peculiarities. The term used is "continuous." Intuitively, we say that a function is continuous if its graph contains no jumps, breaks, or wild oscillations; that is, if we can draw its graph without removing our pencil from the paper. This description would enable you to judge, by looking at its graph, whether or not a function is continuous. There can be pitfalls to this approach and thus we need a more precise definition of continuity.

DEFINITION. A function f is continuous at a point a if (1) $a \in$ domain of f and (2) $\lim_{x \to a} f(x)$ exists and $\lim_{x \to a} f(x) = f(a)$. We say that f is continuous over an interval if f is continuous at every point in the interval.

The definition of continuity involves $\lim_{x \to a} f(x)$, for which we have only given a provisional definition. In all the cases we will consider, and for all reasonable functions, the provisional definition is sufficiently discerning for us to decide whether the function in question is or is not continuous.

Let us now look again at the examples given in Section 2.3 to determine whether or not those functions are continuous.

▶ **Example 1.** Let $f(x) = x^2$. Is f continuous at $x = 2$?

Solution. It is clear from Section 2.3 that $2 \in$ domain of f and $\lim_{x \to 2} x^2 = 4 = f(2)$. Hence f is continuous at $x = 2$. We can observe from Figure 2.5.2 that f is continuous at every point in its domain.

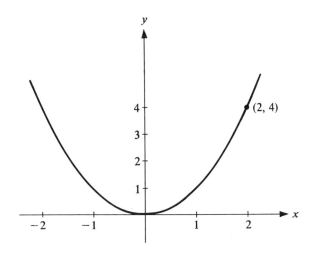

Figure 2.5.2

▶ **Example 2.** Let $f(x) = \dfrac{x - 2}{2|x - 2|}$. Is f continuous at $x = 2$?

Solution. Since $2 \notin$ domain of f, the first condition for continuity is violated and hence f is not continuous at $x = 2$. From the graph of f in Figure 2.5.3, we see that f is continuous at every other point in its domain. This is an example of a *jump* discontinuity: at $x = 2$ we have a jump in the graph.

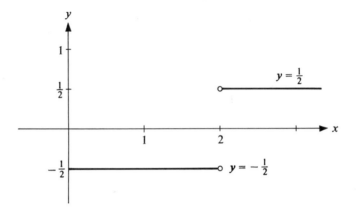

Figure 2.5.3

▶ Example 3. Let

$$f(x) = \begin{cases} 3 - x & \text{for } x \neq -1, \\ 1 & \text{for } x = -1. \end{cases}$$

Is f continuous at $x = -1$?

Solution. Note that

$$\lim_{x \to -1^-} f(x) = 2 = \lim_{x \to -1^+} f(x).$$

Hence $\lim_{x \to -1} f(x) = 2$. However, $f(-1) = 1 \neq 2$, and thus the second condition is violated. ($-1 \in$ domain of f so the first condition is satisfied.) Thus, f is not continuous at $x = -1$.

A natural question to ask is: Can we redefine f at $x = -1$ so that f is continuous at that point? The answer is clearly yes. Define $f(-1) = 2$ rather than 1. Then

$$\lim_{x \to -1} f(x) = 2 = f(-1)$$

and hence we have continuity.

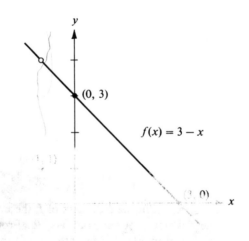

$(0, 3)$

$f(x) = 3 - x$

Note that (as can be seen from Figure 2.5.4) f is continuous at every other point in its domain.

▶ **Example 4.** Let

$$f(x) = \begin{cases} 2x & \text{for } 0 \le x < 1, \\ 4 & \text{for } x = 1, \\ 5 - 3x & \text{for } 1 < x \le 2. \end{cases}$$

Is f continuous at $x = 1$? If not, can we redefine f at 1 in order to make f continuous at that point?

Solution. The graph of f is shown in Figure 2.5.5. Clearly $1 \in$ domain of f. We know also that

$$\lim_{x \to 1^-} f(x) = 2 = \lim_{x \to 1^+} f(x).$$

Hence $\lim_{x \to 1} f(x) = 2$. However, $f(1) \ne 2$; hence f is not continuous at $x = 1$. If we define $f(1) = 2$, then f is continuous at $x = 1$. Again for every other $x \in [0, 2]$, it is clear that f is continuous.

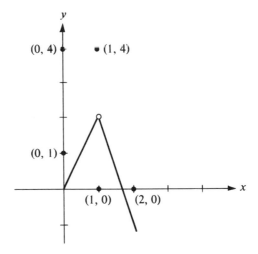

Figure 2.5.5

▶ **Example 5.** Let

$$f(x) = \begin{cases} x & \text{for } 0 \le x \le 1, \\ 3 - x & \text{for } 1 < x \le 4. \end{cases}$$

Is f continuous at $x = 1$? If not, is it possible to redefine f at $x = 1$ in order to make f continuous at that point?

Solution. The graph of f is shown in Figure 2.5.6. Clearly $1 \in$ domain of f. Now, $\lim_{x \to 1^-} f(x) = 1$ while $\lim_{x \to 1^+} f(x) = 2$. Hence $\lim_{x \to 1} f(x)$ does not exist. Thus, f is *not* continuous at that point. Since $\lim_{x \to 1^-} f(x) \ne \lim_{x \to 1^+} f(x)$, we *cannot* redefine f at 1 in order to make f continuous.

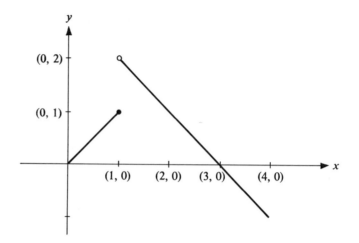

Figure 2.5.6

▶ **Example 6.** Let f be the function represented in Figure 2.5.7. Is f continuous at 0? [Note that $f(0) = 0$.]

Solution. It should be clear from the picture that neither $\lim_{x \to 0^+} f(x)$ nor $\lim_{x \to 0^-} f(x)$ exists. Thus f cannot be continuous at 0. A discontinuity of this kind is called an oscillatory discontinuity.

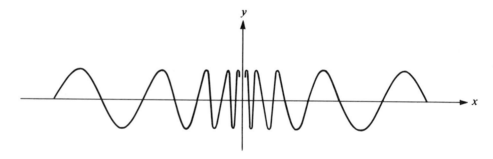

Figure 2.5.7

The following general comment should prove useful. If one knows what a continuous function is, the notion of limit can be handled in the following way. The $\lim_{x \to a} f(x) = L$ if the function is continuous at a when we set $f(a) = L$. (This definition allows the possibility that f is originally not defined at a or is defined at a but $f(a)$ was different from L originally.) For example, if $f(x) = x^3 + 1$, then f is continuous at $x = 1$ and $f(1) = 2$. Hence, $\lim_{x \to 1} f(x) = 2$.

There are certain properties of continuous functions that follow immediately from Properties 1–6 in Section 2.4. The reader can easily verify this. We simply list those properties.

PROPERTY 1. If $f(x) = c$, where c is a constant, then f is continuous at every point in its domain.

PROPERTY 2. If $f(x) = x^n$, for any positive number n, then f is continuous at every point in its domain.

PROPERTY 3. If f and g are continuous at any a, where $a \in$ domain of f and $a \in$ domain of g, then $f + g$ and $f - g$ are continuous at a. (That is, the sum and difference of any two continuous functions are again continuous.)

PROPERTY 4. If f and g are continuous at any point $a \in$ domain of f and $a \in$ domain of g, then $f \cdot g$ is continuous at a.

PROPERTY 5. If f and g are continuous at a and $g(a) \neq 0$, then f/g is continuous at a.

Note that Properties 3 and 4 are valid for 3, 4, or more functions.

We now give a supply of continuous functions that will be useful for our future work.

▶ **Example 7.** (*Polynomials*) A function $f(x) = a_0 + a_1 x + a_2 x^2 + \cdots + a_n x^n$, where $a_0, a_1, \ldots, a_n$ are constants, is called a polynomial of degree n. We assert that if b is any point in the domain of f, then f is continuous at b. This result follows from Properties 2, 3, and 4 above. It is necessary to verify that $\lim_{x \to b} f(x) = f(b)$. However,

$$\lim_{x \to b} f(x) = \lim_{x \to b} (a_0 + a_1 x + \cdots + a_n x^n) = a_0 + a_1 b + \cdots + a_n b^n = f(b).$$

▶ **Example 8.** (*Rational Functions*) Let $f(x) = p(x)/q(x)$, where $p(x)$ and $q(x)$ are polynomials. If $q(a) \neq 0$, then f is continuous at a. (This is just a direct consequence of Property 5.)

▶ **Example 9.** (*Exponential Function*) Recall that in Chapter 1 we discussed the function given by $f(x) = e^x$. The graph of this function is shown in Figure 2.5.8. It is intuitively clear from Figure 2.5.8 that f is continuous at every point in its domain (otherwise there would be gaps in the graph of f). The continuity of e^x could be rigorously derived by proving $\lim_{x \to a} e^x = e^a$, but we omit the details and rely only on our intuition.

▶ **Example 10.** (*Logarithmic Function*) Let $f(x) = \log x$ (where we understand the base to be e). The graph of f is shown in Figure 2.5.9. It should be intuitively clear from Figure 2.5.9 that f is continuous at every nonnegative point except $x = 0$. However, 0 is not in the domain of f; hence the first condition for continuity is violated. Thus, if $a > 0$, $\lim_{x \to a} \log x = \log a$. Again, we rely only on our intuition for this function. A rigorous proof is beyond the scope of the book.

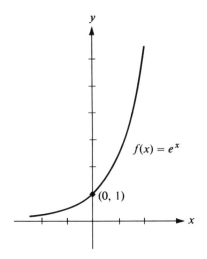

Figure 2.5.8

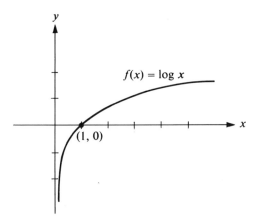

Figure 2.5.9

Exercises

1. Graph the function $f(x) = x + 1$ for $-4 \le x \le 3$. Is f continuous for $-4 \le x \le 3$?

2. Suppose we consider the function $f(x)$, defined by

$$f(x) = \begin{cases} x + 1, & -2 \le x < 1, \quad 1 < x \le 3; \\ 5, & x = 1. \end{cases}$$

Determine from the graph of the function whether or not f is continuous on $-2 \le x \le 3$.

3. Define f as follows:

$$f(x) = \begin{cases} x & \text{for} \quad -1 \le x < 0, \\ 1 & \text{for} \quad x = 0, \\ 2x & \text{for} \quad 0 < x \le 1. \end{cases}$$

(a) Draw a graph of f.

(b) Is f continuous at $x = -\frac{1}{2}$? Why? What about $x = +\frac{1}{2}$?

(c) Is f continuous at $x = 0$? Why? If not, can you redefine f at 0 in order to make f continuous at that point?

4. Define f as follows:

$$f(x) = \begin{cases} -2 & \text{for} & -2 \le x < -1, \\ -1 & \text{for} & -1 \le x < 0, \\ 0 & \text{for} & 0 \le x < 1, \\ 1 & \text{for} & 1 \le x \le 2. \end{cases}$$

(a) Draw a graph of f.

(b) At what points in its domain is f discontinuous? Why? Is it possible to redefine f at these points in order to make it continuous there?

5. (a) Let $f(x) = 1 - 3x^2$. Is f continuous at $x = 1$? Why?

(b) Let $f(x) = (1 - 3x^2)^{1/2}$. Is f continuous at $x = 1$? Why?

6. Draw a graph of the function defined by $f(x) = 5e^x$. From its graph, what can you say about the continuity of the function. Do the same for $f(x) = 5e^{-x}$.

7. Let

$$f(x) = \begin{cases} x & \text{for} & 0 \le x < 1, \\ 7 & \text{for} & x = 1, \\ 1 & \text{for} & x > 1. \end{cases}$$

Draw a graph of f. Is f continuous at $x = 3$? Is f continuous at $x = 1$? If not, can you redefine f at 1 in order to make f continuous at that point?

8. The domain of definition of each of the following functions is to be $[-2, +2]$. Specify a function and draw its graph if:

(a) It has a limit at each point and is continuous at each point.

(b) It has a limit at each point but is not continuous at $x = 0$ and $x = 1$.

(c) It has no limit when $x = 0$ and is not continuous at $x = 0, 1, 2$.

9. Let f be defined by

$$f(x) = \begin{cases} 1/x & \text{for} & x \ne 0, \\ 0 & \text{for} & x = 0. \end{cases}$$

Is f continuous at $x = 0$? Why?

10. Draw a graph of the function f given by $f(x) = e^{3x}$. From its graph, what can you say about the continuity of f? Do the same for $f(x) = e^{-3x}$.

11. Draw a graph of the function $f(x) = \log(x - 1)$, for $x > 1$. What can you infer about the continuity of f? If $f(x) = \log(x - b)$, where $x > b$ (b a constant), where would you expect f to be continuous?

12. A discount in freight rates is often offered on a large shipment. Consider the cost function

$$c(x) = \begin{cases} 0.50x & \text{for} & 0 < x \le 100, \\ 0.45x & \text{for} & 100 < x \le 500, \\ 0.42x & \text{for} & 500 < x, \end{cases}$$

where x is the number of pounds shipped and $c(x)$ is the cost.

(a) Find $\lim_{x\to 50} c(x)$, $\lim_{x\to 100} c(x)$, and $\lim_{x\to 500} c(x)$.

(b) Find those points where c is *not* continuous.

13. Assume the postage rate for airmail letters is 13¢ per ounce (and each fractional part of an ounce) up to 9 ounces, and assume that from then on it is $1.17 up to one pound. Sketch a graph showing the airmail postage of a letter weighing any amount up to one pound. At what points is this graph discontinuous?

14. Let

$$f(x) = \begin{cases} 1 & \text{if} & x < -1, \\ |x| & \text{if} & -1 < x < 1, \\ \frac{1}{2} & \text{if} & 1 \le x \le 2, \\ \frac{1}{4}x & \text{if} & 2 < x. \end{cases}$$

(a) Graph the function carefully.

(b) List all points at which f is *not* continuous and give reasons why.

2.6   Chapter 2 Summary

1. Heuristic definition of limit

(a) We say that the limit as x approaches a from the right exists and is L_1 if the numbers $f(x)$ remain arbitrarily close to L_1 when we take x sufficiently close to a (but $x \ne a$) for values of $x > a$. We write $\lim_{x\to a^+} f(x) = L_1$.

(b) The limit as x approaches a from the left exists and is L_2 if the numbers $f(x)$ remain arbitrarily close to L_2 when we take x sufficiently close to a (but $x \ne a$) for values of $x < a$. We write $\lim_{x\to a^-} f(x) = L_2$.

(c) We say that $\lim_{x\to a} f(x)$ exists and is L if and only if $\lim_{x\to a^+} f(x)$ and $\lim_{x\to a^-} f(x)$ both exist and are *both* equal to L.

2. Continuity

(a) A function f is continuous at a if
 (i) $a \in$ domain of f.
 (ii) $\lim_{x\to a} f(x)$ exists and $\lim_{x\to a} f(x) = f(a)$.

(b) A continuous function (one that is continuous at every point in its domain) has a smooth graph, no jumps or sustained oscillations.

3. Examples of continuous functions

(a) polynomials

(b) rational functions

(c) the exponential function

(d) the logarithm function

Review Exercises

1. In Figures 2.6.1–2.6.7, decide by inspection whether (a) the functions have a right-hand limit at a; (b) the functions have a left-hand limit at a; (c) $\lim_{x\to a} f(x)$ exists; (d) f is continuous at a.

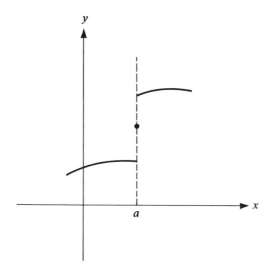

Figure 2.6.1

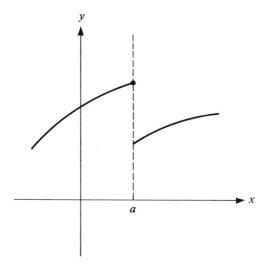

Figure 2.6.2

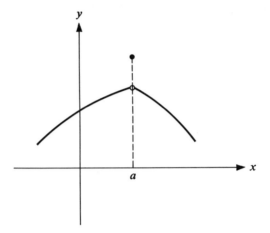

Figure 2.6.3

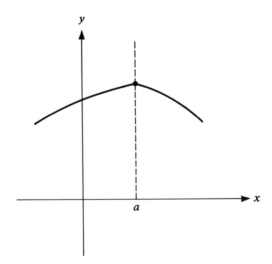

Figure 2.6.4

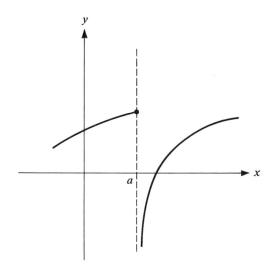

Figure 2.6.5

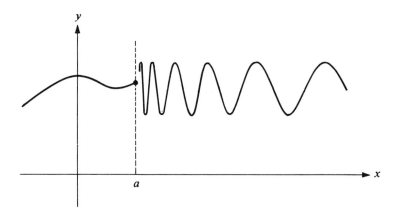

Figure 2.6.6

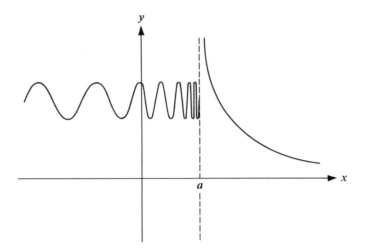

Figure 2.6.7

2. Graph the following function and decide whether $\lim_{x \to 3} f(x)$ exists:

$$f(x) = \begin{cases} 2x - 4 & \text{for} \quad x < 3, \\ 3 & \text{for} \quad x = 3, \\ 5 - x & \text{for} \quad x > 3. \end{cases}$$

Is f continuous at 3?

3. Graph the function $f(x) = x^2 - |x|$. Is this function continuous at $x = 0$?

4. Define the function f as follows:

$$f(x) = \begin{cases} |x| - x & \text{for} \quad x \neq 0, \\ 2 & \text{for} \quad x = 0. \end{cases}$$

(a) Graph f.

(b) Evaluate $\lim_{x \to 0^+} f(x)$; $\lim_{x \to 0^-} f(x)$; $\lim_{x \to 0} f(x)$.

(c) Is f continuous at $x = 0$? If not, can f be redefined at 0 in order to make the function continuous there?

5. Consider the function defined by

$$f(x) = \begin{cases} 0 & \text{if} \quad x < 0, \\ x^2 & \text{if} \quad 0 \leq x < 1, \\ 1 & \text{if} \quad x \geq 1. \end{cases}$$

(a) Sketch a graph of f.

(b) Evaluate $\lim_{x \to 0} f(x)$; $\lim_{x \to 1} f(x)$.

(c) Is f continuous at $x = 0$?

6. Evaluate the following limits, if they exist:

(a) $\displaystyle \lim_{x \to 1} \frac{x^2 - x}{x - 1}$.

(b) $\displaystyle \lim_{x \to 3} \frac{|x - 3|}{x - 3}$.

7. Find

(a) $\lim\limits_{x \to 1^-} \dfrac{(1 + x + x^2)|x - 1|}{x - 1}$.

(b) $\lim\limits_{x \to 1^+} \dfrac{(1 + x + x^2)|x - 1|}{x - 1}$.

8. Let

$$f(x) = \begin{cases} 0 & \text{if} & x < -1, \\ x^2 & \text{if} & -1 \le x \le 1, \\ 1 & \text{if} & 1 < x < 2, \\ \tfrac{1}{2}x & \text{if} & 2 < x. \end{cases}$$

(a) Graph the function carefully.

(b) List all points at which f is *not* continuous and the reasons for this.

9. Let

$$f(x) = \begin{cases} \dfrac{x^2 - 9}{x - 3} & \text{if} \quad x \ne 3, \\ 3 & \text{if} \quad x = 3. \end{cases}$$

(a) Graph $f(x)$.

(b) Is f continuous at $x = 3$? Why? If not, can you redefine f at $x = 3$ in order to make it continuous at that point?

10. Evaluate the following limits, if they exist:

(a) $\lim\limits_{x \to 2} \dfrac{x^2 - x - 2}{x(x - 2)}$.

(b) $\lim\limits_{x \to -1} \dfrac{(x - 1)^2 - 4}{x + 1}$.

(c) $\lim\limits_{x \to 1} \dfrac{2x^2 - x - 1}{x(x - 1)}$.

(d) $\lim\limits_{x \to 2} \dfrac{|x - 2|}{x - 2}$.

(e) $\lim\limits_{x \to 1} \dfrac{x^2 - x}{x^2 - 1}$.

11. If $f(x) = x^2 - 1$, evaluate $\lim\limits_{x \to 1} \dfrac{f(x) - f(1)}{x - 1}$.

12. Suppose that $f(x) = 1/(x + 1)$. Evaluate the following limit:

$$\lim\limits_{x \to 2} \dfrac{f(x) - f(2)}{x - 2}.$$

13. (a) Draw a figure showing the graph of a function f such that for $x < 2$ the graph is part of a straight line; for $x > 2$ the same is true (though not the same straight line as when $x < 2$); and $f(2) = 1$, $\lim_{x \to 2} f(x) = 0$. Is f continuous?

(b) The same problem as (a) except that $f(2) = -1$, $\lim_{x \to 2} f(x) = -1$.

*14. A continuous function $y = f(x)$ is known to be negative at $x = 0$ and positive at $x = 1$. Why is it true that the equation $f(x) = 0$ has at least one root between $x = 0$ and $x = 1$? [That is, there exists at least one point $a \in (0, 1)$ such that $f(a) = 0$.] Illustrate with a sketch. (Intuitive answer only.)

113

2.7 Definition of a Limit

As we pointed out in the first part of this chapter, it is essential in any careful study of calculus to make the terms "closer and closer" and "approaches" precise. If we return to the provisional definition of a limit in Section 2.3, we see that we could make this more precise by noting that the statement "$f(x)$ remains arbitrarily close to L when x is sufficiently close to a" means that the distance between $f(x)$ and L is made small or equivalently $|f(x) - L|$ is small whenever $|x - a|$ is sufficiently small and $x \neq a$. This leads us to adopt the following definition of a limit.

DEFINITION 1. *The function f approaches the limit L at a, written* $\lim_{x \to a} f(x) = L$ *if for every number* $r > 0$ *there is some* $d > 0$ *such that* $|f(x) - L| < r$ *for* $0 < |x - a| < d$.

Observe that in general the choice of d depends upon the previous choice of r. In particular, we do *not* require that the same number d works for *all* r, but rather that for *each* r there exists a number d which works for it.

Let us now look at a geometrical interpretation of the definition. We do this in various steps. We begin with a number L and a number a on the y and x axes, respectively. (Figure 2.7.1.) Next we select a number $r > 0$ such that $L \in (L - r, L + r)$. (Figure 2.7.2.) We must then find a number $d > 0$ such that $x \in (a - d, a + d)$ and $x \neq a$ (Figure 2.7.3) implies that $|f(x) - L| < r$ or $L - r < f(x) < L + r$.

It should be evident to you that the definition of the limit requires that for each $r > 0$, we must find a $d > 0$ such that the graph of the function lying about the interval $(a - d, a + d)$ is contained in a rectangle of width $2d$ and height $2r$. (Figure 2.7.4.) The circle in the graph indicates that we do not care whether or not f is defined at a.

This definition offers us a precise way of verifying that the limit is a certain number. We have some intuitive idea from our previous discussion of what the limit is and then we verify that this is indeed the case via the definition. However, if we want to show that a limit does not exist, we must negate the statement in the definition. That is,

if it is *not* true that for every $r > 0$ there is some $d > 0$ such that $0 < |x - a| < d$ implies $|f(x) - L| < r$;

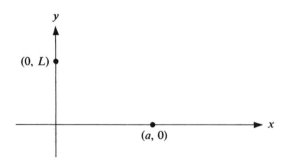

Figure 2.7.1

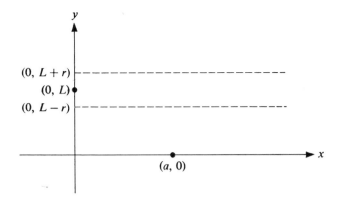

Figure 2.7.2

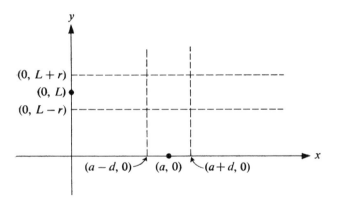

Figure 2.7.3

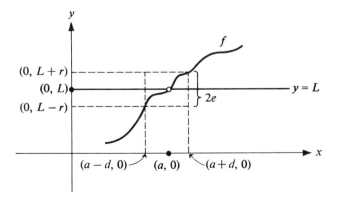

Figure 2.7.4

then there is *some* fixed $r > 0$ such that for *every* $d > 0$ there is *some* x which satisfies $0 < |x - a| < d$ but not $|f(x) - L| < r$.

Let us now look at some examples.

▶ **Example 1.** Verify that $\lim_{x \to 1} (x + 1) = 2$.

Solution. We must show that for every $r > 0$ there is some $d > 0$ such that if $0 < |x - 1| < d$, then $|x + 1 - 2| = |x - 1| < r$. It should be clear that corresponding to any r there is such a d, namely $d = r$, since $0 < |x - 1| < d = r$ implies that $|x - 1| < r$. Hence $\lim_{x \to 1} (x + 1) = 2$.

▶ **Example 2.** Verify that $\lim_{x \to 2} (3x + 1) = 7$.

Solution. We must show that for every $r > 0$, there is some $d > 0$ such that if $0 < |x - 2| < d$, then

$$|3x + 1 - 7| = |3x - 6| = 3|x - 2| < r.$$

Again it should be clear that corresponding to any r, there is such a d—namely $d = \frac{1}{3}r$, since $0 < |x - 2| < \frac{1}{3}r$ implies that $3|x - 2| < r$, which yields $|3x - 6| < r$. Hence, $\lim_{x \to 2} (3x + 1) = 7$.

Now that we have rigorously defined the concept of limit, we can derive all of the Properties 1–8 given in Section 2.4. We illustrate how this is accomplished by deriving Property 4 of that section. The remaining properties could be derived in a similar way, although for Properties 5 and 6 a bit of algebraic manipulation is necessary.

PROPERTY 4. Suppose that f and g are defined for x close to a number a, and assume that

$$\lim_{x \to a} f(x) = L \quad \text{and} \quad \lim_{x \to a} g(x) = M.$$

Then

$$\lim_{x \to a} (f + g)(x) = L + M.$$

Proof: We must show that for every $r > 0$, there is some $d > 0$ such that if $0 < |x - a| < d$, then $|f(x) + g(x) - L - M| < r$. We first observe that

$$|f(x) + g(x) - L - M| = |(f(x) - L) + (g(x) - M)|.$$

Now, by the triangle inequality (Section 1.1, Property 3),

$$|(f(x) - L) + (g(x) - M)| \le |f(x) - L| + |g(x) - M|.$$

Using the fact that

$$\lim_{x \to a} f(x) = L \quad \text{and} \quad \lim_{x \to a} g(x) = M,$$

we know that for $\frac{1}{2}r$, there are numbers d_1 and d_2 such that if $0 < |x - a| < d_1$, then $|f(x) - L| < \frac{1}{2}r$; and if $0 < |x - a| < d_2$, then $|g(x) - M| < \frac{1}{2}r$. Hence, if we choose d to be the smaller of the two numbers d_1, d_2, we see that if $0 < |x - a| < d$, then

$$|(f(x) - L) + (g(x) - M)| \le |f(x) - L| + |g(x) - M| < \frac{1}{2}r + \frac{1}{2}r = r.$$

Thus, the d that "works" is the smaller of the two numbers d_1, d_2, obtained from the hypothesis that $\lim_{x \to a} f(x)$ and $\lim_{x \to a} g(x)$ exist.

The definition of a limit given in this section suggests the following.

GAME. A function f and a number L are furnished.

(1) Player A chooses an $r > 0$.
(2) Player B chooses a $d > 0$.
(3) Player A chooses x, where $0 < |x - a| < d$.

Then $f(x)$ is evaluated, and if $|f(x) - L| < r$, player B wins. If $|f(x) - L| \geq r$, then player A wins. If *every* time the game is played, B wins, then $\lim_{x \to a} f(x) = L$. If B is not able to win every time, then $\lim_{x \to a} f(x) \neq L$. (It is assumed that both players play to win and are infinitely wise.)

We observe that right- and left-hand limits could also be rigorously defined as follows.

DEFINITION 2. $\lim_{x \to a^+} f(x) = L$ *means that for every $r > 0$ there is some number $d > 0$ such that if $a < x < a + d$, then $|f(x) - L| < r$.*

DEFINITION 3. $\lim_{x \to a^-} f(x) = M$ *means that for every $r > 0$ there is some number $d > 0$ such that if $a - d < x < a$, then $|f(x) - M| < r$.*

It is clear from the foregoing definitions that $\lim_{x \to a} f(x)$ exists if and only if $\lim_{x \to a^+} f(x)$ and $\lim_{x \to a^-} f(x)$ exist and they are equal.

Exercises

1. Using the rigorous definition of a limit, derive Property 1 of Section 2.4.

2. Verify that

    (a) $\lim_{x \to 1} (x + 1) = 2$.   (b) $\lim_{x \to 3} (2x - 1) = 5$.

    (c) $\lim_{x \to 1/2} (2x + 1) = 2$.

3. Prove that if c is any constant, and $\lim_{x \to a} f(x)$ exists and is L, then $\lim_{x \to a} [cf(x)] = cL$.

4. Show that, if m and b are constants, then $\lim_{x \to k} (mx + b)$ exists for every k, and a single choice of d to match a given r will apply to all values of k.

*5. Derive Property 5 of Section 2.4. [*Hint:* If $\lim_{x \to a} f(x) = L$ and $\lim_{x \to a} g(x) = M$, write

$$|f(x)g(x) - LM| = |f(x)g(x) - Lg(x) + Lg(x) - LM|.$$

Thus

$$|f(x)g(x) - LM| \leq |g(x)| \, |f(x) - L| + L|g(x) - M|.$$

(Why?) To complete the proof, use the facts that $\lim_{x \to a} f(x) = L$ and $\lim_{x \to a} g(x) = M$.]

117

6. Prove that $\lim_{x \to a} f(x) = \lim_{h \to 0} f(a + h)$.

7. Suppose that $f(x) \le g(x)$ for all x. Prove that $\lim_{x \to a} f(x) \le \lim_{x \to a} g(x)$.

8. (a) If $\lim_{x \to a} f(x)$ exists and $\lim_{x \to a} [f(x) + g(x)]$ exists, must $\lim_{x \to a} g(x)$ exist?

   (b) If $\lim_{x \to a} f(x)$ exists and $\lim_{x \to a} f(x)g(x)$ exists, does it follow that $\lim_{x \to a} g(x)$ exists? (Be careful, this is a bit tricky.)

9. Prove using a rigorous definition of limit that the function

   (a) $f(x) = x^2$ is continuous at $x = 2$.

   (b) $f(x) = x^3$ is continuous at $x = 1$.

Differentiation

3.1 The Derivative

Suppose we wish to measure the steepness or slope of a mountainside at some point, say P_1. (See Figure 3.1.1.) We could obtain a rough value for the slope of the hill at P_1 by sending a friend up the hill to a point P and then calculating the slope of the line joining P_1 to P (a healthy outdoor activity indulged in by surveyors daily).

Figure 3.1.1

The value obtained in this way would be only an approximation, but it is not unreasonable to think that the accuracy would improve as P was brought closer to P_1. Thus, it is tempting to define the slope at P_1 as the limit of the approximations as P approaches P_1 (assuming the limit exists). With the foregoing discussion as motivation, let us consider a similar but more abstract problem.

How can we find the tangent to the curve $y = f(x)$ at a point P? (See Figure 3.1.2.) We have not yet defined the tangent to a curve, so this is the first problem to be overcome.

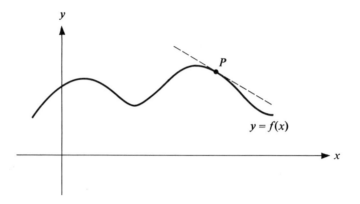

Figure 3.1.2

One might think that it suffices to say that the tangent is simply a line which touches the curve at the point P and nowhere else. (For example, this is a satisfactory definition for the tangent to a circle.) This is clearly unsuitable, as can be seen in Figure 3.1.3. We would hope (relying on our intuition) that we would call L the tangent at P, yet this line cuts the graph of f in three places.

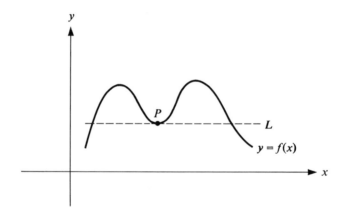

Figure 3.1.3

Now let us try a method for obtaining a tangent and see what happens. Suppose P_0 is a point on the curve $y = f(x)$ at which we wish to find the tangent. (We shall be referring to Figure 3.1.4 in the following discussion.) Let P_1 be another point on the curve and L_1 be the line through P_0 and P_1. This does not seem to be very close to anything we would be willing to call a tangent. Consider the lines $L_2, L_3, L_4, \ldots$ corresponding to the points $P_2, P_3, P_4 \ldots$, where the P's are getting closer to P_0. In this case at least, the lines L_i appear to be approaching a line which is a geometrically appealing candidate for the tangent. We can imagine that the lines $L_1, L_2, L_3, L_4, \ldots$ are successive positions of a line rotating in a counterclockwise position about P_0. Eventually we would hope the line would reach a position where we could call it a tangent line. (See Figure 3.1.4.) Let us look at the situation more closely.

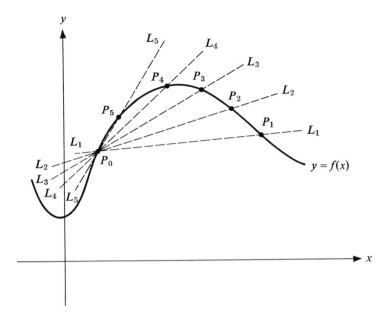

Figure 3.1.4

Think of $P_0 = (a, f(a))$ as a fixed point on the curve at which we want to find the tangent. Choose another point on the curve $(x, f(x))$ and let L be the line through these two points. (See Figure 3.1.5.) Now what we want to know is this: As x approaches a, does the line L_x approach some line L? (It is not very clear just what this last phrase is supposed to mean.) But L_x passes through P_0 in all cases. Certainly

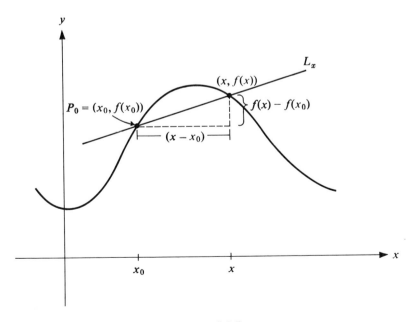

Figure 3.1.5

our tangent line should pass through P_0. Since one point and the slope determine a line, we will ask instead: Does the slope of L_x approach a fixed number as x approaches a? The slope of the line L_x is given by

$$m_{L_x} = \frac{f(x) - f(a)}{x - a}.$$

Hence, we really are asking the following question: Does

$$\lim_{x \to a} \frac{f(x) - f(a)}{x - a}$$

exist? The last expression is so important we shall incorporate it into a definition. Indeed, we have arrived at another of the crucial concepts in calculus.

DEFINITION 1. *If*

$$\lim_{x \to a} \frac{f(x) - f(a)}{x - a}$$

exists, then we say that f is differentiable at a and write

$$f'(a) = \lim_{x \to a} \frac{f(x) - f(a)}{x - a}.$$

The number f'(a) is called the derivative of f at a.

Since we have come this far with the tangent problem we finish it by means of the following definition.

DEFINITION 2. Let $y = f(x)$ be a function which is differentiable at a. Let $f'(a) = M$. We define the tangent line to the curve $y = f(x)$ to be the line whose equation is given by

$$y = f(a) + M(x - a).$$

The equation represents the line with slope M through the point $(a, f(a))$.

The definition of tangent line we have given coincides with the usual geometric definition for circles, ellipses, parabolas, and other standard figures. There are no other serious candidates for the definition.

Let us look at some examples before proceeding further.

▶ Example 1. Consider the curve $y = x^2$. We wish to determine the equation of the tangent line T at $(1, 1)$. (See Figure 3.1.6.)

Solution. First we must determine whether or not the function $f(x) = x^2$ has a derivative at $x = 1$. That is, does

$$\lim_{x \to 1} \frac{f(x) - f(1)}{x - 1}$$

exist? Observe that

$$\frac{f(x) - f(1)}{x - 1} = \frac{x^2 - 1}{x - 1} = \frac{(x - 1)(x + 1)}{x - 1} = (x + 1) \qquad \text{if} \quad x \neq 1.$$

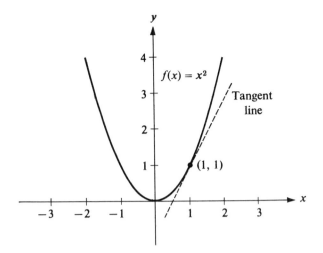

Figure 3.1.6

Hence,

$$\lim_{x \to 1} \frac{f(x) - f(1)}{x - 1} = \lim_{x \to 1} (x + 1) = 2$$

(from our work in Chapter 2). Thus f is differentiable at $x = 1$ and $f'(1) = 2$. The slope of the tangent line is 2; its equation is

$$y = 1 + 2(x - 1) = 2x - 1, \quad \text{since} \quad f(1) = 1.$$

▶ **Example 2.** Find the equation of the line tangent to the curve $y = x^2$ at the point (a, a^2) (where a is any real number).

Solution. First we must determine the limit,

$$\lim_{x \to a} \frac{x^2 - a^2}{x - a} = \lim_{x \to a} \frac{(x + a)(x - a)}{x - a} = 2a.$$

Hence, the slope of the line tangent to $y = x^2$ at (a, a^2) is $2a$. The equation of the line is

$$y = a^2 + 2a(x - a).$$

▶ **Example 3.** Can we determine the tangent to the curve $y = |x|$ at $x = 0$? (Figure 3.1.7.)

Solution. We shall see that the answer is no since f does not have a derivative at $x = 0$. Now,

$$\frac{f(x) - f(0)}{x - 0} = \frac{|x|}{x} = \begin{cases} 1 & \text{for} \quad x > 0, \\ -1 & \text{for} \quad x < 0, \end{cases}$$

since

$$|x| = \begin{cases} x & \text{for} \quad x \geq 0, \\ -x & \text{for} \quad x < 0. \end{cases}$$

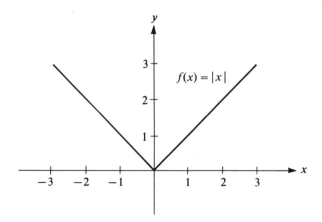

Figure 3.1.7

Thus

$$\lim_{x\to 0^+} \frac{|x|}{x} = 1 \quad \text{and} \quad \lim_{x\to 0^-} \frac{|x|}{x} = -1.$$

Hence $\lim_{x\to 0} |x|/x$ does not exist and $f'(0)$ is not defined.

The limit

$$\lim_{x\to 0^+} \frac{f(x) - f(0)}{x - 0}$$

is called the right-hand derivative of f at 0 and the limit

$$\lim_{x\to 0^-} \frac{f(x) - f(0)}{x - 0}$$

is called the left-hand derivative of f at 0. In this example, the right-hand derivative is 1 and the left-hand derivative is -1.

The derivative is the single most important concept in calculus (with the integral not far behind). In measuring the steepness of a mountain at any given point we were in essence determining the rate at which its shape is changing. In all areas of social and natural science, the study of change is of the utmost importance; change in population, change in the value of the dollar, change in hourly wages, change in interest rate, and so on. Equally important are the *rates* at which these changes occur. For example, if we are told that the consumer price index has gone up by 10%, this does not mean very much until we find out whether the change took place over a week, a month, or a year. The derivative measures the rate of change in the function f when there is a change in x. In economics, the derivative enables us to compute marginal cost (see the following), marginal revenue, and elasticity of demand. It can also be used to solve problems involving the maximization and minimization of quantities and curve tracing.

▶ Example 4. The total cost of producing and marketing x units of a commodity is assumed to be a function of x alone, independent of time and overhead. Designate this function by $C(x)$. We have assumed that if no items are produced, there is no cost: $C(0) = 0$. The average cost to produce x units, $A(x)$, is given by

$$A(x) = \frac{C(x) - C(0)}{x - 0} = \frac{C(x)}{x}.$$

Now, the rate of change of the total cost with respect to the number of units produced is called the *marginal cost*, and is given simply by

$$\lim_{x \to a} \frac{C(x) - C(a)}{x - a},$$

which we recognize as $C'(a)$, the derivative of $C(x)$ at a.

The marginal cost for $x = 12$ is roughly the cost of producing the twelfth item; the marginal cost for $x = 47$ is roughly the cost of producing the forty-seventh item in the production run and so on. In real life situations, one usually expects marginal cost to decrease at first as x increases (various cost savings associated with increased production). Eventually, though, as x continues to increase, marginal cost will begin to increase (increase in expenses as full capacity is reached).

Suppose the total cost is $C(x) = x^2$. The average cost of producing 100 units is

$$A(100) = \frac{(100)^2}{100} = \$100$$

and the marginal cost is

$$\lim_{x \to 100} \frac{x^2 - (100)^2}{x - 100} = \lim_{x \to 100} \frac{(x + 100)(x - 100)}{x - 100} = \$200.$$

Thus, the cost of producing the one hundredth item is approximately \$200.

Exercises

1. Let $f(x) = 2x + 1$.

   (a) What is $\dfrac{f(x) - f(3)}{x - 3}$ equal to?

   (b) Does $\lim\limits_{x \to 3} \dfrac{f(x) - f(3)}{x - 3}$ exist? If so, what is it?

   (c) Does the curve $y = 2x + 1$ have a tangent at the point $(3, 7)$? Write the equation of the tangent line, if the tangent exists.

2. (a) Let $f(x) = ax + b$. Convince yourself that the expression of

   $$\frac{f(x) - f(c)}{x - c}$$

   is independent of the choice of x and c. Give a geometric interpretation of this fact.

   (b) Find the equation of the tangent line to the above curve.

3. Find the equation of the tangent line to the curve $y = 1/x$ at the point $(1, 1)$. Construct the graph of this function first.

4. Find the equation of the tangent line to the curve $y = x^2 - 1$ at the point $(-1, 0)$ and graph both the function and the tangent.

5. Find the derivatives of the following functions at the indicated points, using Definition 1.

   (a) $f(x) = 3$ at $x = 2$.

   (b) $f(x) = 2x$ at $x = 3$.

   (c) $f(x) = 2x - 5$ at $x = 1$.

   (d) $f(t) = t + 3$ at $t = \frac{1}{2}$.

   (e) $f(x) = x^2 - 2$ at $x = 2$.

   (f) $f(x) = 1/x^2$ at $x = 1$.

   (g) $f(x) = \sqrt{x}$ at $x = 1$.

6. (a) Graph the function defined by

   $$f(x) = \begin{cases} x^2 & \text{for} \quad x \geq 0, \\ -x & \text{for} \quad x < 0. \end{cases}$$

   (b) Is f continuous at $x = 0$? Why?

   (c) Does f have a derivative at $x = 0$? Why?

7. Suppose the demand for Double Bubble Bath Soap is expressed by means of the function

   $$d(p) = 100 - 2p,$$

   where p is the price per box (in cents) and d is the weekly demand (in thousands of boxes). What is the instantaneous (or *marginal*) rate of change of the demand for the detergent when the price per box is 25¢?

8. Suppose that the total cost of producing widgets is given by the function $C(x) = x^2 + 1$, where x denotes the number of widgets produced. What is the average cost of producing 50 widgets?

*9. Consider the graph of the function $f(x) = x^2 + cx + d$, where c and d are constants. Find values of c and d such that the line $y = 3x$ is tangent to this graph at the point $(1, 3)$.

3.2 The Relation Between Differentiability and Continuity

Let us now introduce some jargon that is commonly used by mathematicians. If a function f has a derivative at c, we say that it is differentiable at c. If f does not have a derivative at c, we say it is nondifferentiable at c. If f is differentiable at every point in the interval (a, b), we say f is differentiable on (a, b).

It would be convenient at this point to relate the concepts of continuity and differentiability. To say a function is continuous at a means roughly the function knows what its value will be at a and has that value. To say a function is differentiable at a means it also knows what its direction is at a. Note that these are different. Thus it is one thing to predict the free market price of gold on October 10, 1988 and quite another to also predict whether the market will be rising or falling on that day (assume, for the sake of argument, that it cannot do both by taking closing prices).

To illuminate the last remarks we quote the following result. (We shall call any results that are logically deduced from definitions and previously defined properties, theorems.)

THEOREM 1. *If f is differentiable at a, then f is continuous at a.*

Note that the converse is *not* true (that is, f may be continuous, but not differentiable). Consider the function $f(x) = |x|$. We have seen that f is continuous at $x = 0$, but that f does not have a derivative at 0; that is, f is nondifferentiable at 0.

It is a simple matter to establish the validity of the above theorem. For the interested reader, we include a proof.

Proof: If f is differentiable at $x = a$, then we know that

$$\lim_{x \to a} \frac{f(x) - f(a)}{x - a}$$

exists and is $f'(a)$. We must prove that $\lim_{x \to a} f(x) = f(a)$ in order to establish continuity. This is equivalent to showing that

$$\lim_{x \to a} [f(x) - f(a)] = 0.$$

We may write

$$f(x) - f(a) = \frac{f(x) - f(a)}{x - a}(x - a), \qquad \text{for} \quad x \neq a.$$

Hence,

$$\lim_{x \to a} [f(x) - f(a)] = \lim_{x \to a} \left[\frac{f(x) - f(a)}{x - a}(x - a) \right]$$

$$= \lim_{x \to a} \frac{f(x) - f(a)}{x - a} \lim_{x \to a} (x - a) = f'(a) \cdot 0 = 0,$$

by our properties of limits discussed in Chapter 2. Thus, $\lim_{x \to a} f(x) = f(a)$, and f is continuous at $x = a$.

▶ Example. Let

$$f(x) = \begin{cases} x + 2 & \text{if } 0 \le x \le 2, \\ 6 - x & \text{if } 2 < x \le 6. \end{cases}$$

(a) Sketch a graph of f.

(b) Is f continuous at $x = 2$?

(c) Is f differentiable at $x = 2$?

Solution. (a) The graph of f is shown in Figure 3.2.1.

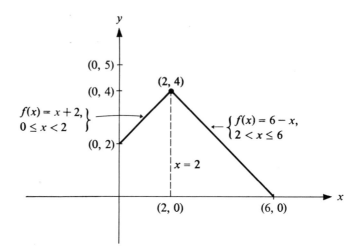

Figure 3.2.1

(b) It is clear that

$$\lim_{x \to 2^+} f(x) = \lim_{x \to 2^-} f(x) = 4.$$

Since $f(2) = 4$, f is continuous at $x = 2$.

(c) It should be clear from the figure that our best strategy here is to calculate the left- and right-hand derivatives for f. Thus

$$\lim_{x \to 2^-} \frac{f(x) - f(2)}{x - 2} = \lim_{x \to 2^-} \frac{(x + 2) - 4}{x - 2} = \lim_{x \to 2^-} \frac{x - 2}{x - 2} = 1.$$

But

$$\lim_{x \to 2^+} \frac{f(x) - f(2)}{x - 2} = \lim_{x \to 2^+} \frac{(6 - x) - 4}{x - 2} = \lim_{x \to 2^+} \frac{-x + 2}{x - 2} = -1.$$

Then the right-hand derivative at 2 is -1 and the left-hand derivative at 2 is $+1$, and since they are not equal, f is not differentiable at 2. We have another example here of a function which is continuous at a point but not differentiable there.

We end this section by briefly discussing the geometric significance of the derivative. As we have already observed, at the point a, the derivative $f'(a)$ can be thought of as the slope of the tangent to the graph of f at the point a. If $f'(a) > 0$, then the value of f at points immediately to the right of a will be greater than the value of f at points immediately to the left of a. Roughly speaking, the curve $y = f(x)$ is "rising" at a. [See Figure 3.2.2(a).] On the other hand, if $f'(a) < 0$, then the value of f at points immediately to the right of a will be less than the value of f at points immediately to the left of a. Roughly speaking, the curve $y = f(x)$ is "sinking" at a, in this case. [See Figure 3.2.2(b).] This discussion will be expanded upon in Chapter 4.

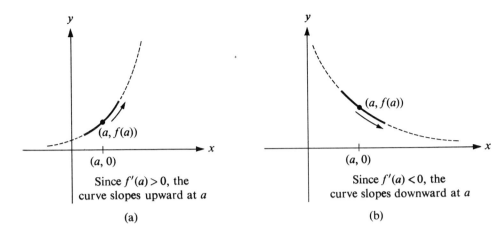

Since $f'(a) > 0$, the
curve slopes upward at a

(a)

Since $f'(a) < 0$, the
curve slopes downward at a

(b)

Figure 3.2.2

Exercises

1. Let

$$f(x) = \begin{cases} 2 & \text{for} \quad x \neq 1, \\ 5 & \text{for} \quad x = 1. \end{cases}$$

(a) Sketch a graph of f.

(b) Is f differentiable at 1? Give two different arguments.

2. Assume

$$f(x) = \begin{cases} x & \text{for} \quad x < 0, \\ 2x & \text{for} \quad x \geq 0. \end{cases}$$

(a) Is f continuous at 0?

(b) Is f differentiable at zero?

$$\left[\text{Hint: Find} \ \lim_{x \to 0^-} \frac{f(x) - f(0)}{x - 0} \ \text{and} \ \lim_{x \to 0^+} \frac{f(x) - f(0)}{x - 0}. \right]$$

3. Let

$$f(x) = \begin{cases} -x & \text{for} \quad x < 0, \\ x^{1/2} & \text{for} \quad x \geq 0. \end{cases} \qquad \text{(See Figure 3.2.3.)}$$

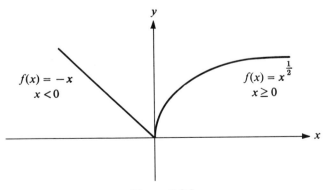

$f(x) = -x$
$x < 0$

$f(x) = x^{\frac{1}{2}}$
$x \geq 0$

Figure 3.2.3

(a) Evaluate (i) $\lim\limits_{x \to 0^-} \dfrac{f(x) - f(0)}{x - 0}$. (ii) $\lim\limits_{x \to 0^+} \dfrac{f(x) - f(0)}{x - 0}$.

(b) Is f continuous at $x = 0$? Is f differentiable at $x = 0$? Why?

4. Consider the function defined by $f(x) = |x - 1|$. Is f differentiable at $x = 1$? Why?

5. Suppose f is defined by

$$f(x) = \begin{cases} 0 & \text{if } x \le 0, \\ x & \text{if } 0 < x \le 1, \\ 1 & \text{if } x > 1. \end{cases}$$

(a) Sketch a graph of f.

(b) Is f continuous at all points in its domain?

(c) At what points is f nondifferentiable? Why?

6. Consider the postage stamp function discussed in Exercise 13 in Section 2.5. At what points is this function nondifferentiable? Why?

*7. Sketch a graph of a function such that

(a) $\lim\limits_{x \to 0^+} \dfrac{f(x) - f(0)}{x}$ exists but $\lim\limits_{x \to 0^-} \dfrac{f(x) - f(0)}{x}$ does *not* exist.

(b) $\lim\limits_{x \to 0^-} \dfrac{f(x) - f(0)}{x}$ exists but $\lim\limits_{x \to 0^+} \dfrac{f(x) - f(0)}{x}$ does *not* exist.

(c) For the functions you give in (a) and (b), can f be differentiable at 0? Why?

[*Hint*: Consider Exercise 3.]

8. Let

$$f(x) = \begin{cases} \dfrac{x^2 - 9}{x - 3} & \text{for } 0 \le x < 3, \\ 7 & \text{for } x = 3, \\ 9 - x & \text{for } 3 < x \le 9. \end{cases}$$

(a) Sketch a graph of f.

(b) Is f continuous at $x = 3$? Why?

(c) Is f differentiable at $x = 3$? Why?

*9. Let f be a function with the following property:

$$|f(b) - f(c)| \le |b - c|^2$$

for *all* real numbers b, c. Find $f'(a)$ for a an arbitrary real number.

3.3 Derivatives of Several Basic Functions

There are several different notations for the derivative of f at a which are useful in applications. Up to now we have used the notation $f'(a)$. This notation, first introduced by J. L. Lagrange (1736–1813), has the advantage that it clearly emphasizes the fact that the derivative f is a new function obtained from f via Definition 1 of Section 3.1. For example, if $a \in$ domain of f and $f(x) = x^2$, we have written $f'(a) = 2a$. (See Example 2 in Section 3.1.) It is just as clear if we write $f'(x) = 2x$. The point is that f' is itself a function. Note that the domain of f' consists of all numbers a for which

$$\lim_{x \to a} \frac{f(x) - f(a)}{x - a}$$

exists. In general, the domain of f' may not be the same as the domain of f.

We introduce several other notations which are widely used in other disciplines. In particular, we may write

$$f'(a) = \left. \frac{df}{dx} \right|_{x=a} = \frac{df}{dx}(a).$$

We read the right-hand side as "dee f dee x evaluated at $x = a$."

Another useful notation [due to G. Leibniz (1646–1716)] is the following:

$$\text{if } y = f(x), \text{ then } f'(x) = \frac{df}{dx}(x) = \frac{dy}{dx}.$$

In order to clear up any future misunderstandings we explicitly point out the meaning of the symbol $\frac{df}{dx}(x)$. Throughout this text we understand that $\frac{df}{dx}(x)$ means the function $\frac{df}{dx}$ evaluated at x (equals the derivative of f at x) and we shall also write this as $\frac{df(x)}{dx}$. Thus, if we write $\frac{df(2)}{dx}$ we mean the derivative of f evaluated at $x = 2$, that is, $\frac{df}{dx}(2)$.

Let us look at some examples illustrating the different notations.

▶ **Example 1.** We have already seen that if $f(x) = x^2$, then $f'(x) = 2x$. If we write $y = x^2$, then

$$\frac{dy}{dx} = 2x.$$

Sometimes we may write

$$\frac{d}{dx}(x^2) = 2x.$$

It is essential to understand that *all* of the notations are equivalent and are defined by Definition 1 of Section 3.1.

We now give the derivatives of several well-known functions. For the sake of convenience, we first list them in tabular form and then briefly discuss how they are obtained.

TABLE 3.3.1	$f(x)$	$f'(x)$
1. k, where k is any constant		0
2. x^n, for *any* n		nx^{n-1}
3. e^x		e^x
4. $\log x$, $x > 0$		$1/x$

The following remarks refer to Table 3.3.1.

Remark 1. It is clear from a geometric point of view that the slope of any line parallel to the x axis is 0. Hence, one would expect $f'(x) = 0$ if $f(x) = k$, a constant. (See Figure 3.3.1.)

Moreover, if a is any number, then

$$\lim_{x \to a} \frac{f(x) - f(a)}{x - a} = \lim_{x \to a} \frac{k - k}{x - a} = \lim_{x \to a} \frac{0}{x - a} = \lim_{x \to a} 0 = 0.$$

Thus, entry 1 can be verified from the definition.

Remark 2. It is not difficult to verify entry 2 if n is a positive integer. We include this as Exercise 12. The formula given in entry 2 is valid whether or not h is a positive integer.

Remark 3. Note that if $f(x) = e^x$, then $f'(x) = e^x$. This simplicity is not just an accident. Indeed, it is fair to say that e was defined with this in mind.

To see more clearly why this is so, consider the function f defined by $f(x) = a^x$. By definition,

$$f'(c) = \lim_{x \to c} \frac{f(x) - f(c)}{x - c}$$

$$= \lim_{x \to c} \frac{a^x - a^c}{x - c}$$

$$= \lim_{x \to c} \frac{a^c[a^{(x-c)} - 1]}{x - c}$$

$$= a^c \lim_{h \to 0} \frac{a^h - 1}{h},$$

where we let $h = x - c$. Note that if $x \to c$, then $x - c \to 0$ and hence $h \to 0$. The expression $\lim_{h \to 0} (a^h - 1)/h$ is not easy to evaluate. However, the limit does exist and

$$\lim_{h \to 0} \frac{a^h - 1}{h} = \log_e a.$$

Thus, if $f(x) = a^x$, then

$$f'(c) = (\log_e a)a^c.$$

When we set $a = e$, the expression becomes considerably simpler, since $\log e = 1$. There is a geometric interpretation to the foregoing discussion. First, note that

$$\lim_{h \to 0} \frac{a^h - 1}{h}$$

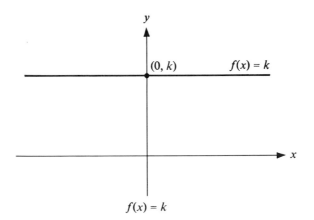

Figure 3.3.1

is just $f'(0)$, where $f(x) = a^x$. Thus, $\log_e a$ is the slope of the curve $y = a^x$ as it crosses the x axis. For a large, the number $\log_e a$ is large; for a small, it is small (in fact negative for $0 < a < 1$). In singling out e^x, we are selecting the curve that crosses the y axis with slope 1. Conversely, we can define e to be that real number a such that the curve $y = a^x$ crosses the y axis with slope 1.

The formula for the derivative of the logarithm can be obtained by using implicit differentiation. This is done explicitly in Chapter 6, Section 6.2, Example 3.

Let us now consider some examples.

▶ **Example 2.** If $f(x) = x^3$, find $f'(5)$.

Solution. Since $f(x) = x^3$, we see from Table 3.3.1 that $f'(x) = 3x^2$. Thus $f'(5) = 3(5^2) = 75$.

▶ **Example 3.** If $f(x) = \log x$, find $f'(\frac{1}{10})$.

Solution. Since $f(x) = \log x$, we see from the table that $f'(x) = 1/x$. Thus

$$f'(\tfrac{1}{10}) = \frac{1}{\frac{1}{10}} = 10.$$

▶ **Example 4.** If $f(x) = e^x$, find $f'(0)$.

Solution. Since $f(x) = e^x, f'(x) = e^x$. Thus $f'(0) = e^0 = 1$.

Note that you can *not* find $f'(c)$ by *first* substituting c. Thus, if $f(x) = x^5$ and we wish to find $f'(2)$, we *must first* write $f'(x) = 5x^4$; *then* substitute $x = 2$ to obtain $f'(2) = 5 \cdot 2^4 = 5 \cdot 16 = 80$. If you first substitute $x = 2$ in the function $f(x) = x^5$ to obtain $f(2) = 32$ (the constant function 32) and then differentiate, you get zero. In other words,

$$\frac{d}{dx}[f(c)] = \frac{d}{dx}(\text{constant}) = 0.$$

▶ **Example 5.** If $f(x) = x^{1/2}$, find $f'(9)$.

Solution. Since $f(x) = x^{1/2}, f'(x) = \frac{1}{2}x^{-1/2}$. Thus,

$$f'(9) = \frac{1}{2}(9^{-1/2}) = \frac{1}{2}(1/\sqrt{9}) = \frac{1}{6}.$$

▶ **Example 6.** If $f(x) = x^{3/2}$, find the tangent to the curve at the point (4, 8).

Solution. Since $f(x) = x^{3/2}, f'(x) = \frac{3}{2}x^{1/2}$. Thus

$$f'(4) = \frac{3}{2}(4^{1/2}) = 3.$$

Hence, $y - 8 = 3(x - 4)$ is the equation of the tangent to the curve at the point (4, 8).

▶ **Example 7.** Does $f(x) = x^{1/3}$ have a derivative at $x = 0$?

Solution. Let us go ahead and simply calculate the derivative from the formula. Since $f(x) = x^{1/3}$, $f'(x) = \frac{1}{3}x^{-2/3} = \frac{1}{3}(1/x^{2/3})$. Thus, if we substitute 0, we obtain

$$f'(0) = \frac{1}{3}(1/0).$$

This tells us something is wrong. Draw the graph of $f(x) = x^{1/3}$. (See Figure 3.3.2.)

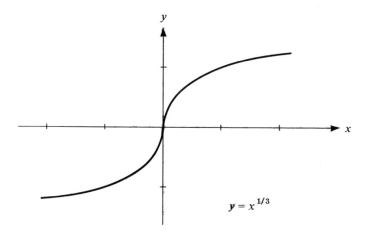

$$y = x^{1/3}$$

Figure 3.3.2

Notice that the curve becomes very steep as we approach 0. It should be clear now that f does not have a derivative at $x = 0$. When a formula yields a bizarre result, we should stop and ask whether we are overlooking something. We remark at this point that the function $f(x) = x^n$ does not have a derivative at $x = 0$ for $-\infty < n < 1$.

Exercises

1. If $f(x)$ is given by
   (a) $f(x) = x^3$, find $f'(1)$ and $f'(-3)$.
   (b) $f(x) = x^2$, find $f'(0)$ and $f'(\frac{1}{2})$.

(c) $f(x) = x^{1/6}$, find $f'(2)$ and $f'(\frac{1}{4})$.

(d) $f(x) = x^{1/2}$, find $f'(8)$ and $f'(4)$.

(e) $f(x) = e^x$, find $f'(2)$ and $f'(0)$.

(f) $f(x) = x^{-1}$, find $f'(-1)$ and $f'(2)$.

(g) $f(x) = \log x$, find $f'(\frac{1}{4})$ and $f'(1)$.

(h) $f(x) = x^4$, find $f'(1)$ and $f'(-1)$.

(i) $f(x) = x^{3/2}$, find $f'(4)$ and $f'(9)$.

(j) $f(x) = x^{4/3}$, find $f'(8)$ and $f'(27)$.

(k) $f(x) = x^{-3}$, find $f'(2)$ and $f'(5)$.

2. Find the equation of the line tangent to the curve $y = 3x^2$ at the point $(2, 12)$. Draw the curve and its tangent at that point.

3. Find the equation of the tangent to the curve $y = x^{1/2}$ at the point $(9, 3)$. Draw the curve and its tangent at that point.

4. Find the equation of the tangent to the curve $y = \log x$ at the point $(e, 1)$. Draw the curve and its tangent at that point.

5. Consider the function defined by $f(x) = x|x|$.

   (a) Construct its graph.

   (b) Is f continuous at 0? Why?

   (c) Is f differentiable at 0? Why?

   (d) Find $f'(-2)$ and $f'(2)$.

   (e) Find the equation of the line tangent to the curve at the point $(1, 1)$.

6. Let
$$f(x) = \begin{cases} x & \text{for} \quad x < 0, \\ x^2 & \text{for} \quad x \geq 0. \end{cases}$$

   (a) Is f differentiable at 0? Why?

   (b) Is f differentiable at 1? -1? 2? Why?

   (c) Calculate f' and carefully indicate the domain of f'.

7. If $f(x)$ is given by the following expressions, find $\dfrac{df}{dx}$ at the points indicated:

   (a) x^{21} at $x = 5$.

   (b) $\log x$ at $x = 4$.

   (c) e^x at $x = -\frac{1}{2}$.

   (d) $x^{-1/3}$ at $x = 8$.

   (e) x^{-2} at $x = -4$.

8. Show that the rate of change of the area of a square field with respect to the length of a side is one-half the perimeter of the field. [*Hint:* Let x denote the length of a side; the area is then given by the function $A(x)$, where $A(x) = x^2$.]

9. Let A be the area of a circle of radius r. Show that the rate of change of A with respect to the radius is given by the circumference of the circle.

10. Suppose that the total costs of producing x boxes of greeting cards is given by $S(x) = 2x^3$. Find the average and marginal costs of producing 100 boxes of greeting cards.

11. The volume of a sphere of radius r inches is given by $V = \frac{4}{3}\pi r^3$. If the radius is changing, find the rate of change of V with respect to r when $r = 2, 4, 6$, and 8 inches, respectively.

12. If $f(x) = x^n$, where n is a positive integer, show that $f'(x) = nx^{n-1}$. [Hint: Observe that if a is any real number, then

$$\frac{f(x) - f(a)}{x - a} = \frac{x^n - a^n}{x - a}.$$

We recall that $x^n - a^n$ may be factored as follows:

$$x^n - a^n = (x - a)(x^{n-1} + x^{n-2}a + x^{n-3}a^2 + \cdots + xa^{n-2} + a^{n-1}).$$

Hence, for $x \neq a$,

$$\frac{x^n - a^n}{x - a} = x^{n-1} + x^{n-2}a + x^{n-3}a^2 + \cdots + xa^{n-2} + a^{n-1}.$$

Thus,

$$f'(a) = \lim_{x \to a} \frac{x^n - a^n}{x - a} = \underbrace{a^{n-1} + a^{n-1} + \cdots + a^{n-1}}_{n \text{ times}} = na^{n-1}.]$$

3.4 Some Basic Differentiation Formulas

So far, while we can differentiate several elementary functions, we do not know how to differentiate such functions as $5x^2$ or $x^2 + x^3$ or $x/(x - 1)$. This difficulty will now be overcome. First, we consider the simplest case, that of differentiating a constant times another function. We state this result in the form of a theorem.

THEOREM 1. *Let f be a differentiable function. Then Cf, where C is a constant, is differentiable and*

$$(Cf)'(x) = Cf'(x).$$

Thus

$$(Cf)' = Cf' \quad or \quad \frac{d}{dx}Cf = C\frac{df}{dx}.$$

We turn immediately to an application of Theorem 1.

▶ Example 1. If $h(x) = 5x^2$, find $h'(x)$.

Solution. Let $h(x) = Cf(x)$, where $C = 5$ and $f(x) = x^2$. Then $f'(x) = 2x$ and so $h'(x) = Cf'(x) = 5(2x) = 10x$; or

$$\frac{dh}{dx} = \frac{d}{dx}(5x^2) = 5\frac{d}{dx}(x^2) = 5(2x) = 10x.$$

As will be seen shortly, the $\dfrac{d}{dx}$ notation affords a very neat, clear, and concise way of simplifying and keeping track of the various steps when differentiating complicated expressions. For this reason it is often preferable.

Next, we consider the derivative of the sum and difference of two functions.

THEOREM 2. *Let f and g be differentiable functions. Then $f \pm g$ is differentiable and*

$$(f \pm g)'(x) = f'(x) + g'(x).$$

Thus

$$(f \pm g)' = f' \pm g' \quad or \quad \frac{d}{dx}(f \pm g) = \frac{df}{dx} \pm \frac{dg}{dx}.$$

Theorem 2 can be summarized in jingle form as "the derivative of the sum is the sum of the derivatives." It is equally valid for 3, 4, or more functions.

We could verify the preceding result directly from the definition of a derivative. We refer those interested to Exercise 13 of this section. We now consider some examples.

▶ **Example 2.** If $h(x) = x^3 + \log x$, find $h'(x)$.

Solution. Write $h(x) = f(x) + g(x)$, where $f(x) = x^3$ and $g(x) = \log x$. Then

$$f'(x) = 3x^2 \quad and \quad g'(x) = 1/x.$$

Hence,

$$h'(x) = f'(x) + g'(x) = 3x^2 + 1/x.$$

Alternatively,

$$\frac{d}{dx}(x^3 + \log x) = \frac{d}{dx}(x^3) + \frac{d}{dx}(\log x) = 3x^2 + \frac{1}{x}.$$

▶ **Example 3.** If $h(x) = 4x^{1/2} + 2e^x$, find $h'(x)$.

Solution.

$$\frac{d}{dx}(4x^{1/2} + 2e^x) = \frac{d}{dx}(4x^{1/2}) + \frac{d}{dx}(2e^x) = 4\frac{d}{dx}(x^{1/2}) + 2\frac{d}{dx}(e^x)$$

$$= 4(\tfrac{1}{2}x^{-1/2}) + 2(e^x) = 2x^{-1/2} + 2e^x.$$

Note that the $\dfrac{d}{dx}$ notation eliminates the need to introduce auxiliary functions while permitting us to do the problem one step at a time.

▶ **Example 4.** (*An Application to Economics*) A manufacturer, after studying the market behavior for his product, comes to the following conclusion: To sell his output of x tons per week he must charge a price of $P = F(x)$ dollars per ton. At a higher price he would sell less; at a lower price he could sell more. His revenue $R(x)$ is given by $R(x) = xF(x)$ (that is, the revenue $R(x)$ is equal to the price per unit multiplied by the number of units sold). In this case $R'(b)$ represents the marginal revenue for b tons. (The marginal revenue gives the rate of increase of revenue per unit increase in output; more bluntly, it tells how fast revenues are increasing.) Obviously the manufacturer's profit is given by the difference between revenue and cost $C(x)$. That is, $T(x) = xF(x) - C(x)$, where T is the profit, C the cost, and F the price he must charge per ton.

(a) Suppose that a manufacturer finds he must charge a price of $P = F(x) = 1 + 1/x + x$ dollars per ton. What is his total revenue for 100 tons? What is the marginal revenue for 100 tons?

(b) In reference to (a), if his total costs are given by $C(x) = x + 1$ (that is, the cost of producing 20 tons is $C(20)$), determine his total profit for selling 100 tons. What is the rate of increase of profit per unit increase of production for 100 tons? This is called the *marginal profit*, and roughly speaking represents the profit on the 100th item produced in this case. To see why this is true observe that

$$\text{profit on the 100th ton} = T(100) - T(99) = \frac{T(100) - T(99)}{100 - 99} \sim T'(100).$$

The accuracy of the last approximation depends on the case in hand.

Solution. (a) The total revenue is

$$R(x) = x\left(1 + \frac{1}{x} + x\right) = (x + 1 + x^2)$$

for x tons. For 100 tons, his total revenue is

$$R(100) = 100 + 1 + (100)^2 = \$10,101.$$

The marginal revenue $R'(x) = 1 + 2x$ for x tons. For 100 tons, $R'(100) = \$201$.

(b) The profit

$$T(x) = R(x) - C(x) = (x + 1 + x^2) - (x + 1) = x^2 \text{ dollars.}$$

His total profit for selling 100 tons is $T(100) = \$10,000$.

The rate of increase of profit per unit increase of production (marginal profit) is $T'(x) = R'(x) - C'(x)$. But $R'(x) - C'(x) = 2x$. The *marginal profit* for selling 100 tons is $\$200$, thus the profit on the 100th ton sold is roughly $\$200$.

Exercises

1. Find the derivative of the following functions:

(a) $2x^3 + \log x$.

(b) $4e^x + x^{-10}$.

(c) $x + x^2 + x^3$.

(d) $4\log x - 30x^2$.

(e) $-e^x + x^{-1} + \log x$.

(f) $e^x - \log x + x^3 - x^4$.

(g) $2x^{1/2} + x^4 + e^x$.

(h) $5x^{-3/2} - \log x$.

(i) $x^4 + 2x^3 - x^2 + 1$.

(j) $1/x + x^{4/3} + 3x^5 + e^x$.

(k) $1/x + e^x - 2x^{10}$.

(l) $x^3 + \log x$.

(m) $e^x + x$.

2. Find $f'(x)$ at a, if

   (a) $f(x) = 14x$ and $a = 2$.

   (b) $f(x) = x - 1$ and $a = -3$.

   (c) $f(x) = 4x^2 + 3x + 2$ and $a = 4$.

   (d) $f(x) = \log x + 10x^{10}$ and $a = 1$.

   (e) $f(x) = 2e^x + x - 3x^4$ and $a = 0$.

   (f) $f(x) = -3 \log x + x^{-3} - 7x^8$ and $a = 1$.

   (g) $f(x) = 2 + 3x + 4x^2 - 5x^3$ and $a = -1$.

3. If $f(x) = -x^2 + x^3$, find the tangent to the curve at $(2, 4)$.

4. If $f(x) = x^{1/2} + 2x^2$, find the tangent to the curve at the point $(1, 3)$.

5. The relation between sales and advertising cost x for a product is given by the formula $S(x) = 200x^2 - 3x$. Find the rate of change in sales for $x = \$500$.

6. Suppose that $f(x) = 3x^4 - 6x^2$. For what values of x is $f'(x) = 0$?

7. A company's total sales revenue is given by the equation $R(x) = \frac{1}{2}x + 2x^2$, where x is the number of years the company has been in business and $R(x)$ is in millions of dollars. At what rate is the company's total sales revenue growing at the end of 5 years?

8. The total cost of producing peanut brittle is given by the function $C(x) = x^2 - 2x$.

   (a) Find the marginal cost function.

   (b) Compute the marginal cost of producing 20 units.

   (c) Find the average cost function.

   (d) Compute the average cost of producing 20 units.

   (e) On one graph plot cost, marginal cost, and average cost.

9. The price of potatoes is estimated to be $220 - 5x$, where x denotes the number of bushels of potatoes.

   (a) Find the revenue and marginal revenue function.

   (b) Compute $R(3)$ and $R'(3)$.

10. Find $f'(x)$ at the indicated point:

    (a) $f(x) = x^3 + 3x^{1/3} + 1$ at $x = 9$.

    (b) $f(x) = x^2 + e^x + x + 1$ at $x = 0$.

    (c) $f(x) = 3 \log x + 2e^x + x^3 + x^2$ at $x = 1$.

    (d) $f(x) = 10e^x - (x^2 + 1)^2 + 5$ at $x = 0$.

    (e) $f(x) = -3x^3 + 9x^2 + \log x - 5e^x$ at $x = 1$.

11. If $f(x) = a_0 + a_1 x + a_2 x + \cdots + a_n x^n$, find $f'(x)$.

12. If $h(x) = Cf(x)$ where f is differentiable, show that $h'(x) = Cf'(x)$.
   [*Hint*:
   $$h'(a) = \lim_{x \to a} \frac{h(x) - h(a)}{x - a} = \lim_{x \to a} \frac{Cf(x) - Cf(a)}{x - a}.]$$

13. Prove Theorem 2.
   [*Hint*: Since $h(x) = f(x) + g(x)$, it follows that
   $$\frac{h(x) - h(a)}{x - a} = \frac{[f(x) + g(x)] - [f(a) + g(a)]}{x - a}$$
   $$= \frac{f(x) - f(a)}{x - a} + \frac{g(x) - g(a)}{x - a}.]$$

3.5 Differentiation of Products and Quotients

We begin with the formula for the derivative of a product of two functions. The result is stated as Theorem 1.

THEOREM 1. *Let f and g be differentiable functions. Then $f \cdot g$ is differentiable and*
$$(f \cdot g)'(x) = f'(x) \cdot g(x) + f(x) \cdot g'(x).$$
Thus
$$(f \cdot g)' = f'g + fg' \quad or \quad \frac{d}{dx} (f \cdot g) = \frac{df}{dx} \cdot g + f \frac{dg}{dx}.$$

The proof of this theorem is a bit tricky and is outlined in Exercise 12 of this section. We immediately proceed to some examples.

▶ Example 1. If $h(x) = x^3 \log x$, find $h'(x)$.

Solution. Let $h(x) = f(x)g(x)$, where $f(x) = x^3$ and $g(x) = \log x$. Then $f'(x) = 3x^2$ and $g'(x) = 1/x$. Hence,
$$f'(x)g(x) + f(x)g'(x) = 3x^2 \log x + x^3(1/x).$$
Alternatively,
$$\frac{d}{dx} (x^3 \log x) = \left[\frac{d}{dx} (x^3)\right] \log x + x^3 \frac{d}{dx} (\log x)$$
$$= 3x^2 \log x + x^3 \left(\frac{1}{x}\right) = 3x^2 \log x + x^2.$$

▶ Example 2. If $h(x) = 4e^x(x^6 - x^{-6})$, find $h'(x)$.

Solution.
$$\frac{d}{dx} [4e^x(x^6 - x^{-6})] = 4 \frac{d}{dx} e^x(x^6 - x^{-6})$$
$$= 4 \left\{\left[\frac{d}{dx} (e^x)\right](x^6 - x^{-6}) + e^x \frac{d}{dx} (x^6 - x^{-6})\right\}$$
$$= 4 \left\{e^x(x^6 - x^{-6}) + e^x \left[\frac{d}{dx} (x^6) - \frac{d}{dx} (x^{-6})\right]\right\}$$
$$= 4\{e^x(x^6 - x^{-6}) + e^x[6x^5 - (-6x^{-7})]\}$$
$$= 4e^x\{x^6 - x^{-6} + 6x^5 + 6x^{-7}\}.$$

(The last step was not really necessary, but there are advantages to algebraic simplification.)

Note that Theorem 1 of Section 3.4 can be obtained from Theorem 1 of this section by applying the product rule to $Cf(x)$. (See Exercise 9 of this section.)

We conclude this section with the formula for the derivative of the quotient of two functions.

THEOREM 2. *Let f and g be differentiable functions where $g(x) \neq 0$. Then f/g is differentiable and*

$$\left(\frac{f}{g}\right)'(x) = \frac{f'(x) \cdot g(x) - f(x)g'(x)}{[g(x)]^2}.$$

Thus

$$\left(\frac{f}{g}\right)' = \frac{f'g - fg'}{g^2} \quad or \quad \frac{d}{dx}\left(\frac{f}{g}\right) = \frac{\dfrac{df}{dx}g - f\dfrac{dg}{dx}}{g^2}.$$

We note that, although $g(x)$ may equal zero for some x's, the formula is still valid at all points where $g(x) \neq 0$.

We observe that if we know the derivative of $1/g(x)$, then the result follows from the product rule. In Exercise 12 of this section, it is shown that

$$\frac{d}{dx}\left[\frac{1}{g(x)}\right] = -\frac{\dfrac{dg(x)}{dx}}{[g(x)]^2}.$$

Hence,

$$\frac{d}{dx}\left[\frac{f(x)}{g(x)}\right] = \frac{df(x)}{dx}\frac{1}{g(x)} + f(x)\frac{d}{dx}\left[\frac{1}{g(x)}\right]$$

$$= \frac{\dfrac{df(x)}{dx}}{g(x)} - \frac{f(x)\dfrac{dg(x)}{dx}}{[g(x)]^2} = \frac{\dfrac{df(x)}{dx}g(x) - f(x)\dfrac{dg(x)}{dx}}{[g(x)]^2}.$$

Thus, we need only know $\dfrac{d}{dx}\left[\dfrac{1}{g(x)}\right]$ and the product rule for differentiation in order to obtain the quotient rule.

We consider some examples.

▶ **Example 3.** If $h(x) = x^3/(1 + x^2)$, find $h'(x)$.

Solution. Note that $1 + x^2 \neq 0$ for any real number x; thus, we can apply the quotient rule. Let

$$h(x) = f(x)/g(x),$$

where $f(x) = x^3$ and $g(x) = 1 + x^2$. Then $f'(x) = 3x^2$ and $g'(x) = 2x$. Hence,

$$h'(x) = \frac{f'(x)g(x) - f(x)g'(x)}{[g(x)]^2}$$

$$= \frac{3x^2(1 + x^2) - x^3(2x)}{(1 + x^2)^2}.$$

It is possible to simplify this expression, but we leave that and the $\dfrac{d}{dx}$ approach to the reader. Alternatively, we could use the remark after the theorem and obtain

$$\frac{d}{dx}\left(\frac{x^3}{1+x^2}\right) = \left[\frac{d}{dx}(x^3)\right]\frac{1}{1+x^2} + x^3\frac{d}{dx}\left(\frac{1}{1+x^2}\right) = \frac{3x^2}{1+x^2} - \frac{x^3 \cdot 2x}{(1+x^2)^2},$$

which is equivalent to the above.

Exercises

1. Find the derivative of the following expressions:

    (a) $x^2 e^x$.

    (b) $\log x - x^{10} \log x$.

    (c) x^3/e^x.

    (d) e^x/x^3.

    (e) $(\log x)/x^5$.

    (f) $5xe^x \log x$.

    (g) $10(\log x) \log x$.

    (h) $e^x e^x$.

    (i) $1 + \dfrac{1}{1 + (1/x)}$.

    (j) $e^x \log x + 5x^{1/2}(x^2 + 2x + 1)$.

    (k) $\dfrac{e^x}{1 + e^x}$.

    (l) $\dfrac{x^2}{1 + x}$.

    (m) $\dfrac{\log x}{1 + e^x}$.

    (n) $\dfrac{e^x}{1 + x^2}$.

    (o) $(x^5 + 1)(e^x + e^{-x})$.

    (p) $\dfrac{e^x}{1 + \log x}$.

2. Find the equation of the tangent line to the following curves at the indicated points:

    (a) $f(x) = xe^x$ at $(1, e)$.

    (b) $f(x) = x^2 e^x$ at $(1, e)$.

    (c) $f(x) = 2x + x \log x$ at $(1, 2)$.

    (d) $f(x) = (x^2 + 2x + 7)(x + 1)$ at $(0, 7)$.

    (e) $f(x) = \dfrac{x}{x^2 + 1}$ at $(0, 0)$ and $(1, \tfrac{1}{2})$.

3. Find the derivative of each of the following functions at the point indicated:

    (a) $f(x) = x^2 e^x$ at $x = 1$.

    (b) $f(x) = x^3 \log x$ at $x = 1$.

    (c) $f(x) = (1 + x^2)e^x$ at $x = 0$.

    (d) $f(x) = \dfrac{\log x}{e^x}$ at $x = 1$.

(e) $f(x) = \dfrac{\log x}{1 + x^2}$ at $x = 1$.

(f) $f(x) = \dfrac{1 + x}{1 - x}$ at $x = -3$.

4. If the total cost of producing x cases of beer is given by

$$C(x) = x^3 - \tfrac{1}{2}x^2 + 7x,$$

find the marginal cost when 10 cases of beer have been produced.

5. Find the points on the graph of the function f, where

$$f(x) = 2x^3 - 3x^2 - 12x + 20,$$

where the tangent line is parallel to the x axis. [*Hint:* It should be clear that at those points x_i, where the tangent line is parallel to the x axis, the slope of the tangent line must be zero. (Why?) Hence, we must find those x_i such that $f'(x_i) = 0$.]

6. Suppose that a New York publishing firm must charge a price of $F(x) = 1/x + \tfrac{1}{2}x + 0.01x^2$ dollars per ton for x tons of paperbacks.

   (a) What is their total revenue for 10 tons?

   (b) What is their marginal revenue for 10 tons?

   (c) Suppose the firm's costs are given by $C(x) = 0.30x + 0.001x^2$ for x tons. Determine their total profit for selling 10 tons. What is their marginal profit?

   (d) For what values of x does the marginal cost equal the marginal revenue? (We shall see in Chapter 4 that for certain economic models total profit is largest when marginal revenue equals marginal cost. In particular, this is true for a competitive market with infinite demand.)

7. Find the values of the constants a, b, and c if the curve $f(x) = ax^2 + bx + c$ passes through the point $(1, 2)$ and is tangent to the line $y = x$ at the origin.

8. Find $f'(x)$ at the indicated point if

   (a) $f(x) = \dfrac{x^3 + 5x^2 - 7x}{x^2 + 1}$ at $x = 1$.

   (b) $f(x) = (x - 3)(2x + 1)(5x - 7)$ at $x = 2$.

   (c) $f(x) = (x^3 + 7x - 1)(x^2 + 2x + 5) - \dfrac{3x + 1}{x + 2}$ at $x = 3$.

   (d) $f(x) = 7e^x \log x + \dfrac{x - 1}{e^x} + x^{5/2}$ at $x = 1$.

   (e) $f(x) = \dfrac{x^3 - 5x + 1}{1 + e^x} + x^2 e^x \log x$ at $x = 1$.

9. Use the product rule (Theorem 1) to show that if $h(x) = Cf(x)$, then $h'(x) = Cf'(x)$. (C is a constant.)

10. If $u(x) = f(x)g(x)h(x)$, find $u'(x)$ in terms of f, f', g, g', h, and h'.

11. If $h(x) = f(x)f(x)$, find $h'(x)$.

12. Let f be differentiable at a when $f(a) \neq 0$, and suppose that $h(x) = 1/f(x)$. Show from first principles that $h'(a) = -f'(a)/[f(a)]^2$.
    [*Hint:* Observe that

$$\frac{h(x) - h(a)}{x - a} = \frac{\dfrac{1}{f(x)} - \dfrac{1}{f(a)}}{x - a}$$

$$= \frac{f(a) - f(x)}{x - a} \, \frac{1}{f(x)f(a)} \, .$$

Now take

$$\lim_{x \to a} \frac{h(x) - h(a)}{x - a}$$

and use the fact that f is continuous at $x = a$.]

13. Prove Theorem 1 of this section.
    [*Hint:* Let a be in the domain of $f \cdot g$. Then,

$$\frac{h(x) - h(a)}{x - a} = \frac{f(x)g(x) - f(a)g(a)}{x - a}$$

$$= \frac{f(x)g(x) - f(a)g(x) + f(a)g(x) - f(a)g(a)}{x - a} \, .$$

(We have simply added and subtracted the quantity $f(a)g(x)$, leaving the numerator unchanged.) Thus,

$$\frac{h(x) - h(a)}{x - a} = \frac{f(x) - f(a)}{x - a} \, g(x) + f(a) \, \frac{g(x) - g(a)}{x - a} \, .$$

Now take

$$\lim_{x \to a} \frac{h(x) - h(a)}{x - a}$$

and use the fact that g is continuous at a.]

3.6 Differentiation of Composite Functions

The preceding sections in this chapter have greatly increased the repertoire of functions that we are able to differentiate. However, we still cannot differentiate such functions as $(x^2 + 1)^5$, e^{x^2}, $\log x^2/(x^2 + 1)$, etc. Recall from Section 1.8 that these may be regarded as functions of functions, or composite functions. In that section we said that h is the composition of two functions f and g, written

$$h = g \circ f, \quad \text{if} \quad h(x) = g[f(x)].$$

Thus, if $f(x) = x^2 + 1$ and $g(x) = x^2$,

$$(x^2 + 1)^2 = g[f(x)] = [g \circ f](x).$$

Although we did not state it explicitly in Chapter 2, it is not difficult to verify the fact that if f and g are continuous then the composition function $g \circ f$ is also continuous. Now we ask a similar question concerning differentiability. Namely, if f and g are differentiable, is $g \circ f$ differentiable? The answer is yes—although the verification of this is a bit subtle. Let us now see how we form the derivative of $g \circ f$.

The composite functions we wish to differentiate will all be of the following form:

$$h(x) = [f(x)]^n, \tag{3.6.1}$$

$$h(x) = e^{f(x)}, \tag{3.6.2}$$

$$h(x) = \log f(x). \tag{3.6.3}$$

The derivatives for these are given in Table 3.6.1.

TABLE 3.6.1	$h(x)$	$h'(x)$
1.	$[f(x)]^n$	$n[f(x)]^{n-1}f'(x)$
2.	$e^{f(x)}$	$e^{f(x)}f'(x)$
3.	$\log f(x)$	$\dfrac{1}{f(x)}f'(x)$

Using the $\dfrac{d}{dx}$ notation we may also write

$$\frac{d}{dx} f^n = nf^{n-1} \frac{df}{dx}, \tag{3.6.1'}$$

$$\frac{d}{dx} e^f = e^f \cdot \frac{df}{dx}, \tag{3.6.2'}$$

$$\frac{d}{dx} \log f = \frac{1}{f} \frac{df}{dx}, \tag{3.6.3'}$$

from Table 3.6.1.

We now illustrate the use of this table by several examples.

▶ Example 1. If $h(x) = (1 - x^2)^{1/2}$, find $h'(x)$.

Solution. The function h is of the form $[f(x)]^n$, where $f(x) = 1 - x^2$ and $n = \frac{1}{2}$. Therefore,

$$h'(x) = \tfrac{1}{2}(1 - x^2)^{(1/2)-1}f'(x)$$
$$= \tfrac{1}{2}(1 - x^2)^{-1/2}(-2x) = -x(1-x^2)^{-1/2}.$$

▶ Example 2. If $h(x) = e^{4x+9x^2}$, find $h'(x)$.

Solution. The function h is of the form $e^{f(x)}$, where $f(x) = 4x + 9x^2$. Thus,

$$\frac{dh}{dx} = \frac{d}{dx} e^{4x+9x^2} = e^{4x+9x^2} \cdot \frac{d}{dx}(4x + 9x^2)$$
$$= e^{4x+9x^2} \cdot (4 + 18x) = 2(2+9x)e^{4x+9x^2}.$$

▶ **Example 3.** If $h(x) = \log(1 + 2x)$, find $h'(x)$.

Solution. The function h is of the form $\log f(x)$, where $f(x) = (1 + 2x)$. Thus,

$$h'(x) = \frac{1}{1 + 2x} \cdot f'(x) = \frac{1}{1 + 2x} \cdot 2.$$

The $\dfrac{d}{dx}$ notation is very convenient for handling more complicated expressions. It allows you to keep track of what you are doing and to proceed one step at a time with a minimum of confusion. Consider the next example.

▶ **Example 4.** If $h(x) = e^{x \log(x^2 + 1)}$, find $\dfrac{dh}{dx}$.

Solution. Let us proceed one step at a time.

$$\frac{dh}{dx} = \frac{d}{dx} \left[e^{x \log(x^2 + 1)} \right]$$

$$= e^{x \log(x^2 + 1)} \cdot \frac{d}{dx} \left[x \log(x^2 + 1) \right]$$

$$= e^{x \log(x^2 + 1)} \left[x \frac{d}{dx} \log(x^2 + 1) + \log(x^2 + 1) \cdot \frac{d}{dx} x \right]$$

$$= e^{x \log(x^2 + 1)} \left[x \cdot \frac{1}{x^2 + 1} \frac{d}{dx} (x^2 + 1) + \log(x^2 + 1) \right]$$

$$= e^{x \log(x^2 + 1)} \left[\frac{2x^2}{x^2 + 1} + \log(x^2 + 1) \right].$$

The last example should convince you that we can differentiate even a very complicated expression by proceeding calmly and carefully, one step at a time.

The formulas in Table 3.6.1 are special cases of a very general theorem.

THEOREM 1. (*The Chain Rule*) *If g and f are differentiable, then so is the function $h = g \circ f$. In addition,*

$$h'(x) = g'[f(x)]f'(x).$$

This holds, of course, for x in the domain of $g \circ f$.

The proof of this theorem appears in Section 3.9. If we use the $\dfrac{d}{dx}$ notation, then the conclusion of Theorem 1 may be written as

$$\frac{dh}{dx} = \frac{dg}{du} \frac{du}{dx},$$

where $u = f(x)$.

Theorem 1 can be used to derive the formulas in Table 3.6.1. Thus, if $h(x) = [f(x)]^n$, then $u = f(x)$ and $g(u) = u^n$;

$$\frac{dh}{dx} = \frac{dg}{du} \frac{du}{dx} = nu^{n-1} \frac{du}{dx} = n[f(x)]^{n-1} \frac{df}{dx}.$$

This checks with entry (1) in Table 3.6.1. Let us also do this for $h(x) = e^{f(x)}$. In this case, $h(x) = g[f(x)]$, where $g(x) = e^x$ and $f(x)$ is just $f(x)$. By Theorem 1,

$$h'(x) = g'[f(x)] \cdot f'(x).$$

Since

$$g(x) = e^x, \qquad g'(x) = e^x.$$

Hence,

$$g'(f(x)) = e^{f(x)}.$$

Substituting, we find

$$h'(x) = e^{f(x)} f'(x),$$

as expected.

We end this section with the following example.

▶ Example 5. Suppose that

$$h(x) = (x^4 - 3)^{1/3}.$$

Find $h'(1)$.

Solution. Let us use Theorem 1 for this example. Let

$$g(x) = x^{1/3} \quad \text{and} \quad f(x) = x^4 - 3.$$

Hence,

$$g'(x) = \tfrac{1}{3}x^{-2/3} \quad \text{and} \quad f'(x) = 4x^3,$$
$$h'(x) = g'[f(x)]f'(x) = \tfrac{1}{3}(x^4 - 3)^{-2/3}4x^3$$
$$= \tfrac{4}{3}(x^4 - 3)^{-2/3}x^3,$$
$$h'(1) = \tfrac{4}{3}(-2)^{-2/3} = \tfrac{4}{3}(1/\sqrt[3]{4}).$$

Exercises

1. Differentiate the following expressions:

- (a) $(1 + x)^2$.
  (b) $\log(2 + 2x)$.

  (c) $(1 + x^2)^{1/2}$.
  (d) e^{2x}.

- (e) $(1 - 2x)^{-1}$.
  -(f) $\log(x^2 + x + 1)$.

- (g) $e^{x^3 + x^2 - x}$.
  -(h) $(x + 1/x)^{3/2}$.

  (i) $(\log x)^2$.
  - (j) $(4 + x)^{10}$.

2. Find the derivatives of the following expressions:

- (a) $\log(2x^2 + x + 1)$.
  (b) $\dfrac{x}{\sqrt{(x^2 - 1)}}$.

-(c) $\dfrac{x^3 + 2x}{(x^2 + 2x + 1)^{3/2}}$.
  (d) $x^2 e^{4x}$.

- (e) $\log(e^x + x)$.
  (f) $(x^2 - 5)^{3/2}$.

  (g) $(x + \log x)^{-4}$.
  (h) $(e^x - 1/x)^7$.

  (i) $\log(xe^x)$.
  (j) $(e^{x^3 - 2x})^{100}$.

  (k) $(\log x)^2 + e^{x^3}$.
  (l) $e^{x^2} + (\log x)^3$.

(m) $\log (x^2 + 2x + 1)^2$.

(n) $\log \left(\dfrac{x^3 - 9}{e^{5x}}\right)^{1/2}$.

(o) $e^{[(\log x)/x]^{10}}$.

-(p) $\log (\log x), \quad x > 1$.

--(q) $\log [\log (\log x)], \quad x > e$.

-(r) $\left(\dfrac{x + \sqrt{1 - x^2}}{x^2 - \sqrt{1 + x^2}}\right)^{1/2}$.

3. Let $f(x) = (x^2 - 4)^2$. Does $f'(b) = 0$ for any b? If so, for what b?

4. Find the derivative of the following functions at the point indicated:

    (a) $f(x) = (1 - 2x)^8 \quad$ at $\quad x = 1$.

    (b) $f(x) = \log (2 + x) \quad$ at $\quad x = 5$.

    (c) $f(x) = e^{x^2} \quad$ at $\quad x = 1$.

    (d) $f(x) = \sqrt{1 - x^2} \quad$ at $\quad x = \frac{1}{2}$.

    (e) $f(x) = (1 + 2x^4)^{-1} \quad$ at $\quad x = 2$.

    (f) $f(x) = (e^x + \log x)^{1/2} \quad$ at $\quad x = 1$.

5. Let $f(x) = \frac{3}{2}x^4 + 4x^3 + 3x^2 + 8$. Find all values of x such that $f'(x) = 0$.

6. Let $h(x) = 1/f(x)$. This can also be written as $h(x) = [f(x)]^{-1}$. Find $\dfrac{dh}{dx}$ by the rule for differentiation of composite functions. Find $\dfrac{dh}{dx}$ by the method for quotients. Compare your answers.

7. Let $f(x) = \log (e^x)$. Find $f'(x)$ by the composite function method. Note that $\log (e^x) = x$. Check your answer.

8. Let $h(x) = f(x)/g(x)$. Rewrite this as $h(x) = f(x)[g(x)]^{-1}$. Find $\dfrac{dh}{dx}$ by using the chain rule. Check your answer against the standard method of differentiating f/g.

9. Find the equation of the tangent to the curve

$$y = 2(x^3 + 1)^{1/2}$$

    at the point $(2, 6)$.

10. If $f'(x) = \sqrt{3x^2 - 1}$ and $y = f(x^2)$, find $\dfrac{dy}{dx}$.

*11. Previously we have verified that if n is *any* positive integer, then the derivative of x^n is nx^{n-1}. Now we can, with the use of the chain rule, extend this result to any rational number n, where $n = p/q$ with p, q integers.

    (a) First, use the quotient rule to show that if $n < 0$,

$$\frac{d}{dx}(x^n) = nx^{n-1}.$$

    [*Hint:* Let $n = -m$, where m is positive. Then,

$$\frac{d}{dx}(x^{-m}) = \frac{d}{dx}\frac{1}{x^m} = -\frac{mx^{m-1}}{x^{2m}} = -mx^{-m-1}.]$$

(b)   Now, we show that

$$\frac{d}{dx}(x^n) = nx^{n-1}$$

if $n = 1/p$, p an integer and nonzero. [*Hint:* Let $g(x) = x^{1/p}$, thus (see Chapter 1) $[g(x)]^p = x$. Since $[g(x)]^p = x$, the derivative of $[g(x)]^p$ must be 1. (Why?) Thus,

$$\frac{d}{dx}[g(x)]^p = 1$$

and thus $p[g(x)]^{p-1}g'(x) = 1$. Hence,

$$g'(x) = \frac{1}{p}[g(x)]^{1-p} = \frac{1}{p}(x^{1/p})^{1-p} = \frac{1}{p}x^{[(1/p)-1]}.]$$

(c)   Next show that

$$\frac{d}{dx}(x^n) = nx^{n-1},$$

where $n = p/q$, p, q integers and $q \neq 0$. (Use the chain rule.) [*Hint:*

$$\frac{d}{dx}(x^{p/q}) = \frac{d}{dx}(x^{1/q})^p = p(x^{1/q})^{p-1}\frac{d}{dx}(x^{1/q})$$

$$= px^{(p/q)-(1/q)}\frac{1}{q}x^{(1/q)-1}.$$

Complete the verification.]

*12.   (a)   Consider the function $f(x) = x^x$ for $x > 0$. One might think that $f'(x) = x \cdot x^{x-1}$. Explain why the chain rule (Theorem 1) does *not* yield this result.

(b)   If $f(x) = x^x$ for $x > 0$, find $f'(x)$. [*Hint:* Recall that $x^x = e^{x \log x}$.]

(c)   If $f(x) = x^{(x^2)}$, find f'.

(d)   If $f(x) = x^{\log x}$, find f'.

3.7   Higher Derivatives

Let f be a differentiable function. Since f' is a function in its own right, one can ask whether $g(x) = f'(x)$ has a derivative at b, say. If g is differentiable at b, then we write $g'(b) = f''(b)$. In this case we say that f has a second derivative at b, which can be denoted by

$$f''(b) = f''(x)|_{x=b} = \frac{d^2f}{dx^2}\bigg|_{x=b} = \frac{d^2f}{dx^2}(b).$$

It is obviously possible (formally) to define derivatives of higher orders in this manner. The third derivative is usually written $f'''(x)$ or $\dfrac{d^3f(x)}{dx^3}$. However, we will need only first and second derivatives in our application. (Second and higher derivatives can also be defined in terms of limits, but there is really no advantage to doing this.) Let us now consider some examples.

▶ Example 1. If $f(x) = -4x^3 + 3x^2 + x - 1$, find $f''(x)$.

Solution. There is no fancy method for finding second derivatives; you simply differentiate and then differentiate again. Thus,

$$f'(x) = -12x^2 + 6x + 1 \quad \text{and} \quad f''(x) = -24x + 6.$$

▶ Example 2. If $f(x) = x^2 e^x$, find $f''(x)$.

Solution.

$$\frac{df}{dx} = 2xe^x + x^2 e^x.$$

$$\frac{d^2 f}{dx^2} = \frac{d}{dx}(2xe^x + x^2 e^x) = \frac{d}{dx}(2xe^x) + \frac{d}{dx}(x^2 e^x)$$

$$= (2e^x + 2xe^x) + (2xe^x + x^2 e^x) = e^x(2 + 4x + x^2).$$

We can also give a geometric interpretation to the second derivative. The first

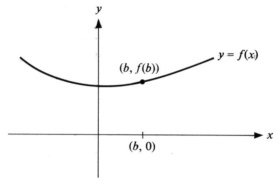

$f''(b) > 0$ and $|f''(b)|$ is small.

Figure 3.7.1

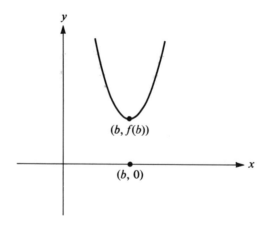

$f''(b) > 0$ and $|f'(b)|$ is large. The curve is turning upward (rather sharply)

Figure 3.7.2

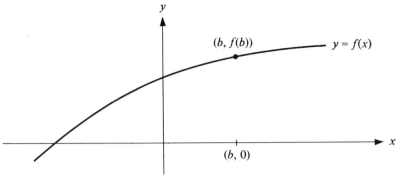

$f''(b) < 0$ and $|f''(b)|$ is small.

Figure 3.7.3

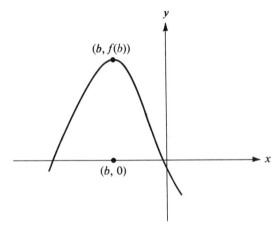

$f''(b) < 0$ and $|f'(b)|$ is large. Curve is bending
downward (rather sharply)

Figure 3.7.4

derivative, roughly speaking, tells you whether the curve is rising or sinking to the right, and how steeply. The second derivative tells you how fast the curve is bending (turning) and whether it is turning up or down. For example, if $f''(x) > 0$ then, as in Section 3.2, the graph of f' is rising to the right. This means that the curve is bending upward (since the slope of the tangent is always increasing). In like manner, if $f''(x) < 0$, then the graph of f' is sinking to the right, which means that the curve is bending downward. These ideas are illustrated in Figures 3.7.1–3.7.4.

The foregoing will be discussed in much greater detail in Chapter 4, where other applications of the second derivative are given.

Exercises

1. Find the second derivative of the following expressions:

    (a) $x^3 - 9x + 1$.   (b) $(x^2 - 1)^{1/2}$.

(c) $x \log x$.

(d) $(\log x) e^x$.

(e) $(x + 1)e^{4x^2}$.

(f) $\dfrac{x + 1}{x - 1}$.

(g) $\log (\log x)$.

(h) $e^{(1-x)/(1+x)}$.

(i) $\dfrac{x}{(x^2 - 1)^{1/3}}$.

(j) $\log (x^3 + x^2 e^x + 1)$.

(k) $e^{\log (x^2+1)}$.

(l) $(x^3 - 9x^2 + 1)^3$.

(m) $(x^2 + 1)^{1/2} \log (x^3 - 2)$.

(n) $\dfrac{(\log x)^{1/2}}{x^2 - 1}$.

2. If $f(x) = \frac{1}{12}x^4 - x^3 + \frac{9}{2}x^2 + 2x + 7$, find all points where $f''(x)$ equals 0.

3. If $y = \dfrac{1 + x}{1 - x}$, find the third derivative of y, $y'''(x)$, at the point $x = \frac{1}{2}$.

4. (a) If $y = a_0 + a_1 x + a_2 x^2 + a_3 x^3$, find $y'''(x)$ and $y^{(iv)}(x)$ (that is, the fourth derivative of y.)

(b) Suppose that $y(x) = a_0 + a_1 x + a_2 x^2 + \cdots + a_n x^n$, where n is a positive integer. Find $y'(x)$, $y''(x)$, $y'''(x)$, $y^{(iv)}(x)$. Can you guess at what $y^{(n)}(x)$, where $y^{(n)}(x)$ denotes the nth derivative of x, might be? What do you think the value of $y^{(n+1)}(x)$ is?

(c) Referring to (b), what is $y(0)$? $y'(0)$? $y''(0)$? ... $y^{(n)}(0)$? [in terms of the a_i's $(i = 0, 1, 2, \ldots, n)$]. Rewrite $y(x)$ by substituting the various derivatives of y, evaluated at $x = 0$.

5. If $h(x) = \dfrac{1}{f(x)}$, find $h''(x)$, in terms of f, f', f''.

6. If $h(x) = f(x)g(x)$, find $h''(x)$.

7. If $f(x) = Ax^2 + Bx + C$, where A, B, and C are constants, show that $f''(x) = 2A$ no matter what A, B, and C are.

8. If $y = Ax^4 + Bx^2$, where A and B are constants, find the first four derivatives of y with respect to x, and show that, whatever the values of A and B,

$$x^2 \frac{d^2 y}{dx^2} - 5x \frac{dy}{dx} + 8y = 0.$$

9. Show that if $y = C_1 e^{-x} + C_2 e^{-2x}$, where C_1 and C_2 are constants, then $y'' + 3y' + 2y = 0$, independent of the values of C_1 and C_2.

3.8 Chapter 3 Summary

1. (a) The derivative of $f'(b)$ of f at a point b in the domain of f is defined by

$$\lim_{x \to b} \frac{f(x) - f(b)}{x - b} = f'(b)$$

provided that this limit exists. Other notations for the derivative are:

$$\frac{df}{dx}(b), \quad \frac{df(x)}{dx}\bigg|_{x=b}, \quad \text{and} \quad \frac{df(b)}{dx}.$$

(b) If f has a derivative at a point b, we say that f is differentiable at b.

(c) If f is differentiable at b, then f is continuous at b. The converse does *not* hold. However, if f is *not* continuous at b, then f is not differentiable at b.

2. The derivative of a function f measures the rate of change in $f(x)$ when there is a change in x. Thus, if f represents cost, f' denotes marginal cost. If f represents position, f' denotes velocity.

3. We list here all the important formulas discussed in this chapter. We assume all functions are differentiable.

(a1) $\dfrac{d}{dx}(C) = 0$ when C is a constant.

(a2) $\dfrac{d}{dx}(x^n) = nx^{n-1}$ for any n.

(a3) $\dfrac{d}{dx}(e^x) = e^x$.

(a4) $\dfrac{d}{dx}(\log x) = 1/x$ for $x > 0$.

(b1) $\dfrac{d}{dx}(Cf) = C\dfrac{df}{dx}$ when C is a constant.

(b2) $\dfrac{d}{dx}(f \pm g) = \dfrac{df}{dx} \pm \dfrac{dg}{dx}$.

(b3) $\dfrac{d}{dx}(f \cdot g) = g\dfrac{df}{dx} + f\dfrac{dg}{dx}$.

(b4) $\dfrac{d}{dx}\left(\dfrac{f}{g}\right) = \dfrac{g\dfrac{df}{dx} - f\dfrac{dg}{dx}}{(g)^2}$.

(b5) $\dfrac{d}{dx}[(g \circ f)(x)] = \dfrac{d}{dx}[g(f(x))] = g'(f(x))f'(x)$.

(c1) $\dfrac{d}{dx}[f(x)]^n = n[f(x)]^{n-1}\dfrac{df(x)}{dx}$.

(c2) $\dfrac{d}{dx}(e^{g(x)}) = e^{g(x)}\dfrac{dg(x)}{dx}$.

(c3) $\dfrac{d}{dx}[\log g(x)] = \dfrac{1}{g(x)}\dfrac{dg(x)}{dx}$.

Review Exercises

1. (a) If $f(x) = x^5$, then evaluate

$$\lim_{x \to 2} \frac{f(x) - f(2)}{x - 2}.$$

   [*Hint:* There is an easy way to do this using Section 3.1.]

   (b) Suppose that $f(x) = \dfrac{1}{x + 2}$. Find $f'(2)$, using the *definition* of a derivative.

   (You must use the definition given in Section 3.1, not the quotient rule.)

2. If $f(x)$ is given as follows, determine $f'(x)$:

   (a) $f(x) = \dfrac{x^2}{\sqrt{x^2 - 4}}.$

   (b) $f(x) = \dfrac{x^3 - 1}{x - 1}.$

   (c) $f(x) = 2x^2 + x^3 e^x.$

   (d) $f(x) = e^{(x^3 + x^2 + 1)^2}.$

   (e) $f(x) = \log (x^2 + 2x + 1)^5.$

   (f) $f(x) = (\log x)^8.$

   (g) $f(x) = \log (x^8).$

   (h) $f(x) = \left(\dfrac{x + 1}{x - 1}\right)^3.$

   (i) $f(x) = x^2 e^{-x^2} + e^{-x} \log x.$

   (j) $f(x) = \sqrt{x + \sqrt{x}}.$

3. Find the equation of the line tangent to the curve $y = (x + 1)^{-1}\log x$ at the point $(1, 0)$.

4. Determine the constant c such that the straight line joining the points $(0, 3)$ and $(5, -2)$ is tangent to the curve $y = c/(x + 1)$.

5. Find the points on the curve

$$y = \tfrac{1}{3}x^3 - 3x^2 + 9x + 5,$$

   where the tangent is parallel to the x axis.

6. Consider the function defined by

$$f(x) = \begin{cases} \dfrac{x^2 - 9}{x - 3} & \text{if } 0 \le x < 3, \\ 2 & \text{if } x = 3, \\ 9 - x & \text{if } 3 < x \le 9. \end{cases}$$

   (a) Sketch the graph of f and indicate the domain and range of the function.

   (b) Does $\lim_{x \to 3} f(x)$ exist? Why?

   (c) Is f continuous at $x = 3$? Why? If not, how can f be defined at $x = 3$ in order to make it continuous at that point?

   (d) Is f differentiable at $x = 2$? Why?

7. A Budd car will hold 100 people. If the number x of persons per trip who use the Budd car is related to the fare charged (p dollars) by $p = (5 - \tfrac{1}{20}x)^2$,

write the function expressing the total revenue per trip received by the Budd car company. What is the number x of people per trip that will make the marginal revenue equal to zero? What is the corresponding fare? (See Example 4, Section 3.4.)

8. If s represents the distance a body moves in time t seconds, determine (a) the velocity, $v = \dfrac{ds}{dt}$, (b) the acceleration, $a = \dfrac{d^2s}{dt^2}$, if

$$s(t) = 250 + 40t - 16t^2.$$

9. Find $f''(b)$ if

(a) $f(x) = 5x + 9x^2$ and $b = 4$.

(b) $f(x) = xe^x$ and $b = 0$.

(c) $f(x) = (1 + x^2) \log x$ and $b = 1$.

(d) $f(x) = x\sqrt{(1 - x)}$ and $b = \frac{1}{2}$.

*10. Suppose we are given a function f satisfying the following conditions for all x and y:

(i) $f(x + y) = f(x)f(y)$.
(ii) $f(x) = 1 + xh(x)$, where $\lim_{x \to 0} h(x) = 1$.

Prove that

(a) The derivative $f'(x)$ exists.

(b) $f'(x) = f(x)$.

[*Hint:* Let $x - a = h$ in the definition of a derivative, so that as $x \to a$, $h \to 0$. Thus,

$$\lim_{x \to a} \frac{f(x) - f(a)}{x - a} = \lim_{h \to 0} \frac{f(a + h) - f(a)}{h}.$$

Now, use properties (1) and (2).]

11. The demand for Superlux, the wonder detergent, is expressed by means of the equation

$$d(p) = 520 - 45p + p^2,$$

where p is the price per box (in cents) and d is the weekly demand (in thousands of boxes). What is the marginal demand (that is, the rate of change of demand) for Superlux when the price per box is 25¢?

12. Let

$$f(x) = \begin{cases} x^2 & \text{for} \quad x \geq 0, \\ x & \text{for} \quad x < 0. \end{cases}$$

Does $f'(0)$ exist? Why? Sketch a graph of f.

13. Find $f'(x)$ if

$$f(x) = \left[\frac{(x - 5)^{2/5}(1/x + 2)^{1/3}}{(x - 5)^2} \right]^{1/2}.$$

14. Consider the function

$$f(x) = \begin{cases} \frac{1}{2}x^2 & \text{for } x \leq 0, \\ \frac{1}{4}x^2 & \text{for } x > 0. \end{cases}$$

Determine the functions $f'(x)$ and $f''(x)$ and state the domain for each. Sketch the graphs of $f(x)$, $f'(x)$, and $f''(x)$.

15. Determine where the first and second derivatives of $y = 2x/(1 + x^2)$ vanish.

16. If $f(x) = e^{x^2 - 2x}$, what is $f'(0)$? $f''(0)$? $f'''(0)$? $f^{(iv)}(0)$?

17. Given that $F'(x) = G(x)$, show that

$$\frac{d}{dx} F(ax + b) = aG(ax + b)$$

provided that F is differentiable at $ax + b$. (a, b are arbitrary constants.) [*Hint:* Use the chain rule.]

A P P E N D I X

3.9 Proof of the Chain Rule

We now verify Theorem 1 of Section 3.6, the chain rule. It is necessary for us to show that if $a \in$ domain of $f \circ g$, then

$$\frac{d}{dx} [f \circ g(a)] = f'[g(a)]g'(a).$$

If we let $h(x) = f[g(x)]$, then

$$\frac{h(x) - h(a)}{x - a} = \frac{f[g(x)] - f[g(a)]}{x - a}.$$

Now we multiply the right-hand side of the preceding by

$$\frac{g(x) - g(a)}{g(x) - g(a)},$$

obtaining

$$\frac{h(x) - h(a)}{x - a} = \frac{f[g(x)] - f[g(a)]}{g(x) - g(a)} \frac{g(x) - g(a)}{x - a}.$$

Taking limits,

$$\lim_{x \to a} \frac{h(x) - h(a)}{x - a} = \lim_{x \to a} \frac{f[g(x)] - f[g(a)]}{g(x) - g(a)} \lim_{x \to a} \frac{g(x) - g(a)}{x - a}.$$

It looks like we might be finished since

$$\lim_{x \to a} \frac{f[g(x)] - f[g(a)]}{g(x) - g(a)} = f'[g(a)]$$

and thus

$$h'(x) = f'[g(a)]g'(a).$$

However, there is a flaw in this argument. The foregoing method is valid *only if* $g(x) - g(a) \neq 0$. If $g(x) = C$, a constant, then clearly the method is in error since $g(x) - g(a) = 0$. Thus, we must somehow overcome this difficulty.

We must define an auxiliary function $F(t)$ as follows:

$$F(t) = \begin{cases} \dfrac{f(t) - f[g(a)]}{t - g(a)} & \text{if } t \neq g(a), \\ f'[g(a)] & \text{if } t = g(a). \end{cases}$$

First observe that F is continuous at $g(a)$ since

$$\lim_{t \to g(a)} F(t) = \lim_{t \to g(a)} \frac{f(t) - f[g(a)]}{t - g(a)} = f'[g(a)],$$

the value of F at $g(a)$.

For $y \neq a$, we have

$$\frac{f[g(y)] - f[g(a)]}{y - a} = F[g(y)] \frac{g(y) - g(a)}{y - a},$$

since if $g(y) = g(a)$, then both sides are 0, and if $g(y) \neq g(a)$,

$$F[g(y)] = \frac{f[g(y)] - f[g(a)]}{g(y) - g(a)}.$$

Now,

$$\lim_{y \to a} \frac{f[g(y)] - f[g(a)]}{y - a} = F[g(a)]g'(a)$$

since F is continuous at $g(a)$ *and* g is differentiable at that point.

Hence, substituting for $F[g(a)]$ and for the left-hand side of the foregoing equation, we have

$$h'(a) = f'[g(a)]g'(a).$$

Applications
of Derivatives

4.1 Maxima and Minima. Definitions and Basic Results

In this chapter we shall consider the problem of determining maximum and minimum values of a function. Such problems frequently arise in applications. For example, consider the following classic problem. A farmer has 6,000 ft of fence. He wishes to fence in a rectangular portion of land along a river in such a way as to enclose the largest possible area (Figure 4.1.1). How long should the sides of the rectangle

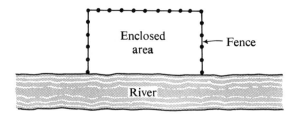

Figure 4.1.1

be to accomplish this? The first thing to do is rephrase the problem in a more mathematical but equivalent form. Let x and y be the length of the sides of the rectangle. (See Figure 4.1.2.) Since the perimeter of the fence is 6,000 ft, we see that $2x + y = 6,000$, or $y = 6,000 - 2x$. Let $A(x)$ represent the area of the fenced-in plot when the length of the land is x. Then,

$$A(x) = xy = x(6,000 - 2x) = 6,000x - 2x^2.$$

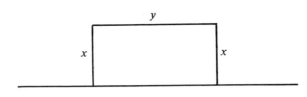

Figure 4.1.2

By the nature of the problem, $x \geq 0$ and $x \leq 3{,}000$. (There are only 6,000 ft of fence.)

A problem equivalent to the original one is the following: Find the maximum value of the function $A(x) = 6{,}000x - 2x^2$ for $0 \leq x \leq 3{,}000$. This problem probably doesn't look any easier than the first. Let us suggest one way of obtaining an approximate solution. Simply graph the function $A(x) = 6{,}000x - 2x^2$ at the integers between 0 and 3,000. (This would be time consuming but not impossible.) What you would get would look like Figure 4.1.3. However, it is not clear that the

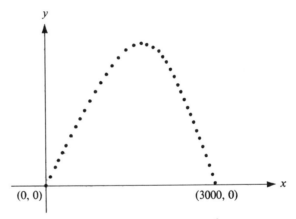

Function graphed for integer values

Figure 4.1.3

graph might not look like that in Figure 4.1.4 if more points were plotted.

We now present a method for attacking a wide range of problems of a similar nature. First, it is necessary to carefully define the terms "maximum" and "minimum."

Generally we say that f has a local maximum at c if $f(c)$ is greater than or equal to f evaluated at *any* nearby point (see Figure 4.1.5). More precisely, we have

DEFINITION 1. *A function f, defined on an open interval (a, b) or a closed interval $[a, b]$, has a* local maximum *at $c \in (a, b)$ if there exists a number $d > 0$ such that $f(c) \geq f(x)$ for all $x \in (c - d, c + d)$.*

DEFINITION 2. *A function f, defined on an open interval (a, b) or a closed interval $[a, b]$, has a* local minimum *at $c \in (a, b)$ if there exists a $d > 0$ such that $f(c) \leq f(x)$ for $x \in (c - d, c + d)$. (See Figure 4.1.6.)*

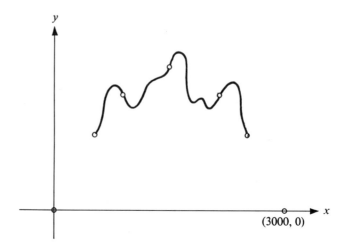

Figure 4.1.4

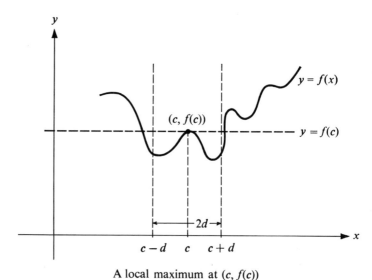

A local maximum at $(c, f(c))$

Figure 4.1.5

Sometimes a local maximum or minimum is called a relative maximum or minimum.

A word of caution is in order at this point. Note that we have defined local maximum and local minimum *only* at interior points of an interval. We have *not* defined them for the *endpoints* of an interval. We could do this, but it is not useful for our purposes.

Observe that the function $f(x) = 2$ for all real x has a local maximum and a local minimum at every point.

The next theorem tells us how to go about finding local maxima and minima. Before stating the theorem, we clarify what is meant by saying a function is

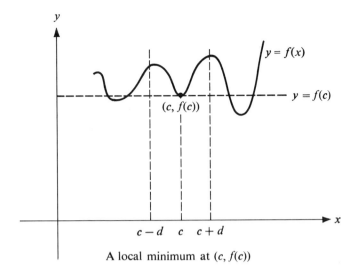

A local minimum at $(c, f(c))$

Figure 4.1.6

differentiable at an endpoint of a closed interval. If f is defined on $[a, b]$, then to say that f is differentiable at a means that

$$\lim_{x \to a^+} \frac{f(x) - f(a)}{x - a}$$

exists; f is differentiable at b means that

$$\lim_{x \to b^-} \frac{f(x) - f(b)}{x - b}$$

exists. In this case, we write $f'(a)$ and $f'(b)$ to denote these limits.

THEOREM 1. *Let f be a function differentiable on the open interval (a, b) or on the closed interval $[a, b]$. Let f have either a local maximum or a local minimum at $c \in (a, b)$. Then $f'(c) = 0$.*

The verification is not difficult and is included as Exercise 5 in this section. Note that the theorem does *not* say that if $f'(c) = 0$, then c is a local maximum or local minimum, as can be seen by Example 2. It does say that, if f has a local maximum (or minimum), the point where it has the local maximum (or minimum) will be included in the set of points where $f'(x) = 0$.

▶ **Example 1.** Find the local maxima and local minima for the function $f(x) = x^2$ defined on the interval $(-1, 1)$.

Solution. We first note that $f'(x) = 2x$ and thus $f'(x) = 0$ precisely when $x = 0$. Since $f(0) = 0$ and $f(x) = x^2 > 0$ for all $x \neq 0$, it is clear that 0 is a local minimum. Since 0 is the only point for which the derivative vanished we see that f has one local minimum (at 0) and no local maxima on the interval $(-1, 1)$. (See Figure 4.1.7.)

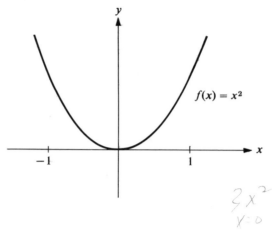

$f(x) = x^2$

$3x^2$
$x = 0$

Figure 4.1.7

▶ Example 2.   Find the local maximum and local minimum of the function $f(x) = x^3$ on $[-1, 2]$.

Solution. $f'(x) = 3x^2$ so that 0 is the only place where $f'(x) = 0$. One might think that 0 must be a local maximum or local minimum, but it is not. It is impossible to find any $d > 0$ such that $f(0) \geq f(x)$ for $x \in (-d, d)$. In like manner we cannot find a $d_1 > 0$ such that $f(0) \leq f(x)$ for any $x \in (-d_1, d_1)$. For example, if $d = \frac{1}{2}$, then we see that for $x \in (-\frac{1}{2}, 0)$ we have $f(0) = 0 > x^3$, while for $x \in (0, \frac{1}{2})$, $0 < x^3$.

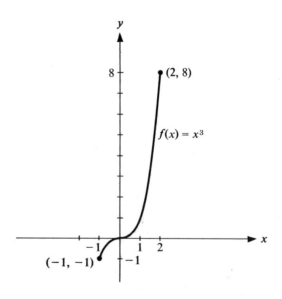

$(2, 8)$

$f(x) = x^3$

$(-1, -1)$

Figure 4.1.8

Thus, even though $f'(0) = 0$, 0 is still not a local maximum or a local minimum. Thus we conclude that the function in question does not have a local maximum or a local minimum at a point in $(-1, 2)$. Consider the graph of $f(x)$ as shown in Figure 4.1.8. The fact that $f'(0) = 0$ means only that the function is "flat" at zero.

We now define the maximum and minimum of a function over a closed interval $[a, b]$. Sometimes this is called the absolute maximum and absolute minimum.

DEFINITION 3. *The function* f *has a maximum* [*minimum*] *on* $[a, b]$ *at* c *if* $f(c) \geq f(x)$ [$f(c) \leq f(x)$] *for all* $x \in [a, b]$.

Thus, in Example 2, $x = -1$ and $x = 2$ are the minimum and maximum points, respectively, since $f(-1) = -1 \leq f(x)$ for $x \in [-1, 2]$ and $f(2) = 8 \geq f(x)$ for $x \in [-1, 2]$.

We quote the following results, which will be useful for further applications.

THEOREM 2. *Let* f *be a continuous function on* $[a, b]$. *Then*

(1) f *has a maximum and a minimum on* $[a, b]$;
(2) *this maximum* [*minimum*] *occurs at*
    (i) *either* a *or* b,
    (ii) *one of the numbers* $x_i, i = 1, 2, \ldots, n$, *where* $f'(x_i) = 0$,
or (iii) *at a point where* f *is not differentiable.*

Since the verification of this theorem is fairly difficult, we omit it. One can see that this theorem is at least plausible by observing that if f is continuous on $[a, b]$, it must have a value for every $x \in [a, b]$. Thus, there must be a smallest and largest value; that is, it has a maximum and minimum on $[a, b]$. However, the proof of this statement is quite subtle. Stop and ask yourself if the function in Figure 4.1.9

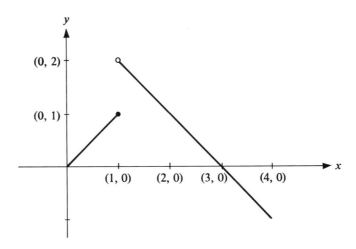

Figure 4.1.9

has a maximum on the interval $[0, 4]$. A certain amount of reflection should convince you that it does not. [Note first that $f(x) \leq 2$ for $x \in [0, 4]$, and second that although the functional values get arbitrarily close to 2, there is no c for which $f(c) = 2$.] Clearly, f is *not* differentiable at 1. Indeed, f is not even continuous at 1. This example shows that just because a function is defined on $[a, b]$ does not mean it has a maximum (or minimum) on $[a, b]$.

Let us now complete the problem we began in this chapter. We were faced with the task of maximizing the function $A(x) = 6{,}000x - 2x^2$ for $0 \leq x \leq 3{,}000$. Since the function is differentiable, it *must* have a maximum on the closed interval in question. Now we will locate the zeros of the derivative $f'(x) = 6{,}000 - 4x$. Thus, setting the derivative equal to zero, $6{,}000 - 4x = 0$ or $x = 1{,}500$. Hence the maximum must occur at either 0, 1,500, or 3,000. Since $f(0) = 0$, $f(3{,}000) = 0$, and $f(1{,}500) = 4{,}500{,}000$, it is clear that the maximum occurs at $x = 1{,}500$. Thus the area to be fenced in should have the dimensions shown in Figure 4.1.10.

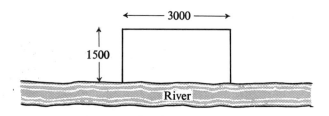

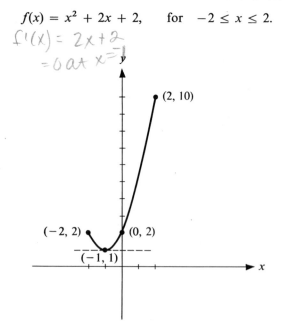

Figure 4.1.10

▶ **Example 3.** Find the maximum and minimum for the function

$$f(x) = x^2 + 2x + 2, \qquad \text{for} \quad -2 \leq x \leq 2.$$

Figure 4.1.11

Solution. Clearly, f is differentiable on $[-2, 2]$. Hence, the maximum and minimum is attained at either $x = -2$, $x = 2$, or those points x for which $f'(x) = 0$. Now, $f'(x) = 2x + 2$. Thus $f'(x) = 0$ precisely when $x = -1$. We see that $f(-2) = 2$; $f(-1) = 1$ and $f(2) = 10$. Hence the maximum occurs for $x = 2$ and the minimum for $x = -1$. The graph of f is shown in Figure 4.1.11.

▶ **Example 4.** Find the maximum and minimum of the function $f(x) = 2x^3 - 15x^2 + 36x + 1$ on the interval $1 \leq x \leq 5$. $6 \times 2 - 30x - 36$

Solution. Note that since f is differentiable on $[1, 5]$, it must have both a maximum and a minimum. Since $x^2 - 5x - 6$

$$f'(x) = 6x^2 - 30x + 36 = 6(x - 2)(x - 3), \quad (x - 2)(x - 3)$$

the zeros of f are $x = 2$ and $x = 3$. By simply evaluating f we find that $f(1) = 24$, $f(2) = 29$, $f(3) = 18$, $f(5) = 156$. Hence, the minimum occurs at $x = 3$ and the maximum at $x = 5$.

Exercises

1. Find all the local and absolute maxima and minima for the following functions by graphing and inspection:

(a) $f(x) = \begin{cases} |x| & \text{for} \quad x \neq 0, \\ 2 & \text{for} \quad x = 0, \end{cases} \quad -1 \leq x \leq 1.$

(b) $f(x) = x^2 + 1, \quad 0 \leq x \leq 2.$

(c) $f(x) = \begin{cases} x & \text{for} \quad -1 \leq x < 0, \\ 2 & \text{for} \quad x = 0, \\ x - 2 & \text{for} \quad 0 < x \leq 1. \end{cases}$

(d) $f(x) = 5x^2, \quad -3 \leq x \leq 3.$

(e) $f(x) = \begin{cases} x + 2 & \text{for} \quad -5 \leq x \leq -1, \\ x^2 & \text{for} \quad -1 \leq x \leq 1, \\ -2x + 3 & \text{for} \quad 1 \leq x \leq 4. \end{cases}$

(f) $f(x) = \begin{cases} 0 & \text{for} \quad x \neq \pm 1, \pm 2, \pm 3, \pm 4, \ldots, \\ x & \text{for} \quad x = \pm 1, \pm 2, \pm 3, \pm 4, \ldots. \end{cases}$

Exercises (g) and (h) are represented in Figure 4.1.12 and 4.1.13.

(g)

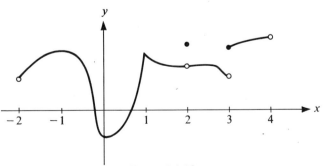

Figure 4.1.12

(h)

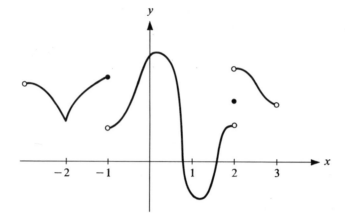

Figure 4.1.13

2. Find the maximum and minimum for the following functions:

(a) $f(x) = x^2 + 1$, $-1 \leq x \leq 10$.

(b) $f(x) = x^3 - 3x$, $0 \leq x \leq 2$.

(c) $f(x) = (x + 1)(x + 2)$, $-1 \leq x \leq 2$.

(d) $f(x) = \frac{1}{3}x^3 + x^2 + x + 11$, $|x| \leq 4$.

(e) $f(x) = \frac{1}{3}x^3 - x^2 - 15x + 9$, $|x| \leq 6$.

(f) $f(x) = \log x$, $1 \leq x \leq 10$.

(g) $f(x) = \log x + (1 - x)^2$, $1 \leq x \leq 4$.

(h) $f(x) = xe^{2x}$, $0 \leq x \leq 3$.

(i) $f(x) = \dfrac{x + 1}{1 + x^2}$, $|x| \leq 4$.

(j) $f(x) = \log (3x)$, $\frac{1}{3} \leq x \leq 3$.

(k) $f(x) = \dfrac{x^2 + 2x + 1}{1 + x^2}$, $|x| \leq 4$.

(l) $f(x) = x + 1$, $|x| \leq 8$.

(m) $f(x) = x \log x$, $1 \leq x \leq e$.

(n) $f(x) = \dfrac{\sqrt{x + 1} - \sqrt{x}}{x}$, $1 \leq x \leq 4$.

(o) $f(x) = e^{-x^2}$, $|x| \leq 1$.

(p) $f(x) = x^2 e^x$, $-10 \leq x \leq 1$.

(q) $f(x) = x - \log x$, $\frac{1}{2} \leq x \leq 4$.

3. A photographer has a thin piece of wood 32 inches long. How should he cut the wood to make a rectangular picture frame that encloses the maximum area? [*Hint:* Call the length of the required frame x and the width y. We wish to maximize the area enclosed, $A = xy$. Now, $2x + 2y = 32$. Thus, $x + y = 16$, $y = 16 - x$, and $A = x(16 - x)$. Remember that we want the area as a function of x alone. Clearly, $0 \leq x \leq 16$.]

4. Find two positive numbers whose sum is 20 and such that their product is as large as possible. [*Hint:* Let x denote one number and y the other. Then, $x + y = 20$ and we wish to maximize xy.]

5. Show that if f has a local maximum at $c \in (a, b)$, then $f'(c) = 0$. [*Hint:* First consider the expression

$$\frac{f(x) - f(c)}{x - c}$$

for $x < c$. Note that $f(x) \leq f(c)$. (Why?) Thus,

$$\frac{f(x) - f(c)}{x - c} \geq 0.$$

(Why?) Hence,

$$\lim_{x \to c^-} \frac{f(x) - f(c)}{x - c} \geq 0.$$

(Why?) Now consider

$$\frac{f(x) - f(c)}{x - c}$$

for $c > x$. Note that $f(x) \leq f(c)$. (Why?) Thus,

$$\frac{f(x) - f(c)}{x - c} \leq 0.$$

(Why?) This yields

$$\lim_{x \to c^+} \frac{f(x) - f(c)}{x - c} \leq 0.$$

(Why?) We may then conclude that $f'(c) = 0$. (Why?)]

4.2 Some Applications of Maxima and Minima

We illustrate how one uses the material developed in Section 4.1 for solving problems involving the maximization or minimization of functions.

▶ Example 1. A man wishes to travel from point A to point B. (See Figure 4.2.1.) To do this he must rent a car and drive to the railroad and then take the train the rest of the way. It costs $2 a mile to travel by car and $1 to travel by rail. Find the least expensive route from A to B, assuming the man drives in a straight line and meets the railroad some place between B and C. We also assume that the distance between A and C is 3 miles and between B and C is 5 miles.

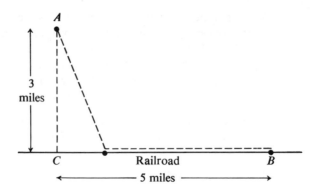

Figure 4.2.1

Solution. Since the problem asks us to minimize the cost of the trip, we should set up a function for the cost in terms of some variable. In this problem, a convenient variable is the distance from C to the point where he meets the railroad. Thus, if $f(x) = $ cost of traveling from A to B,

$$f(x) = 2\sqrt{9 + x^2} + 1(5 - x) \qquad \text{for} \quad 0 \le x \le 5.$$

(See Figure 4.2.2.) Now we determine $f'(x)$. Thus

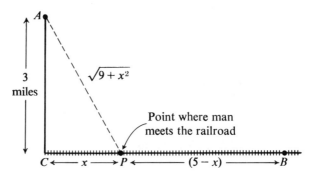

Figure 4.2.2

$$f'(x) = \frac{2x}{\sqrt{9 + x^2}} - 1.$$

Next, we find the zeros of $f'(x)$; that is, we find all solutions to the equation

$$0 = \frac{2x}{\sqrt{9 + x^2}} - 1.$$

Multiplying both sides by $\sqrt{9 + x^2}$, we obtain

$$\sqrt{9 + x^2} = 2x,$$

and thus $9 + x^2 = 4x^2$, yielding $x^2 = 3$ and $x = \sqrt{3} \simeq 1.732$. Note that we reject the solution $x = -\sqrt{3}$ since the distance from C to P must be positive. Thus, the minimum *must* occur at either $x = 0$, $x = \sqrt{3}$, or $x = 5$. A simple calculation reveals that $x = \sqrt{3}$ yields the minimum cost which is $f(\sqrt{3}) = 5 + 3\sqrt{3}$.

If we look at the first examples in Sections 4.1 and 4.2, we should be able to extract a general scheme for attacking problems of this kind.

Given a max-min problem, we proceed as follows:

1. Draw a clear picture describing the situation and label it carefully.
2. Make up a function $f(x)$ for the quantity you are trying to maximize (minimize) in terms of a convenient variable x. (It may be necessary to express other quantities in terms of x to do this.)
3. Find $f'(x)$.
4. Find all points x_i such that $f'(x_i) = 0$, $i = 1, \ldots, n$. That is, find all the zeros of f'.
5. Since we are assuming the variable x ranges over some closed interval $[a, b]$, the maximum (minimum) must occur at a, b, or one of the x_i's.
6. Evaluate f at a, b, and x_i, $i = 1, \ldots, n$. Pick out the x values that yield the largest and smallest values for $f(x)$.

Let us now look at some additional examples.

▶ **Example 2.** The Hookem and Fleecem Realty Company handles an apartment house with 100 units. When the rent of each of the units is $80 per month, all of the units are filled. Experience shows that, for each $5 per month increase in rent, five units become vacant. The cost of servicing a rented apartment is $20 per month. What rent should be charged to maximize profit?

Solution. Let us choose for our unknown, x, the number of $5 per month increases in rent. Thus, the amount of rent charged per month is $80 + 5x$ dollars. We see that the number of units rented for each value of x is $100 - 5x$ units. Hence, in order to determine the profit, $f(x)$, we must subtract the operating costs from the amount taken in from the rented apartments, which is $(80 + 5x)(100 - 5x)$. Thus, we wish to maximize the function,

$$f(x) = (80 + 5x)(100 - 5x) - 20(100 - 5x),$$

where $0 \leq x \leq 20$. (We can leave the rent unchanged, or at the other extreme, after 20 such increases all apartments are empty.) Now

$$f(x) = (60 + 5x)(100 - 5x)$$

and hence,

$$f'(x) = 5(100 - 5x) - 5(60 + 5x) = 5(40 - 10x).$$

Letting $f'(x) = 0$, we obtain $x = 4$. We claim that this value of x maximizes the profit. By our theorem, since f is differentiable, it must have a maximum at either $x = 0$, $x = 20$, or $x = 4$. But, $f(0) = \$6,000$, $f(20) = 0$, and $f(4) = \$6,400$. Thus the rent which should be charged is $\$80 + \$20 = \$100$ per month. At that rent, 80 units will be rented, producing a profit of $6,400 per month.

▶ **Example 3.** Economists define a *perfect competitor* as a firm that has no control over price. This situation usually occurs when a large number of small companies sell a product for which there is a large demand. In such a situation no company produces enough to influence the price. There are few good examples of "perfect competitor," but wheat farmers and the grain market come fairly close, because one farmer cannot affect the market by raising or lowering his price. Assume, then, that we have a market with a perfect competitor, producing a single product. Assume further that the selling price is fixed and that the company can sell all it produces.

How much should the firm produce to maximize its profit?

Solution. Let x be the quantity of the product produced and suppose that the revenues generated on selling x units amount to $R(x)$. Let $C(x)$ be the total cost of producing the x units. Then, it is clear that the profit obtained as a result of producing x units is

$$P(x) = R(x) - C(x).$$

Suppose that M is the maximum quantity that can be produced. Then we wish to find an x satisfying $0 \leq x \leq M$ which maximizes $P(x)$. If we assume that R and C are differentiable on $[0, M]$, then the maximum occurs at either 0, M, or those points x such that $P'(x) = 0$. Now,

$$P'(x) = R'(x) - C'(x)$$

and $P'(x) = 0$ implies that

$$R'(x) = C'(x).$$

Thus, if x_1, the quantity of the product produced that yields maximum profit, is not 0 or M, it satisfies the equation $R'(x_1) = C'(x_1)$; that is, the quantity x_1 such that marginal revenue is equal to marginal cost. (See Example 4, Section 3.4.) In textbooks on economics, the statement is often found that production should be adjusted to the point where marginal revenue is equal to marginal cost.

▶ Example 4. There is a relation between the price of a product and the amount people will buy at that price called the demand function. Thus, if g is the demand and x is the price, then $g(x)$ is the amount demanded at price x (see Figure 4.2.3).

Usually, as the price increases the demand for the product decreases; it has become too expensive for some. Similarly, as the price drops the demand increases; more people can now afford the product (or the same people will each purchase more).

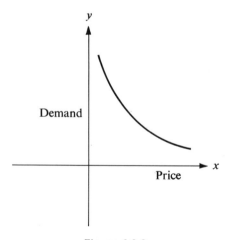

Figure 4.2.3

Economists usually put demand on the x axis and price on the y axis. We have reversed the order on this occasion to make the idea easier to grasp.

Let us consider a situation where we have a single company, the A.B.C. Company, which manufactures widgets. Assume the product is simple to produce and thus the cost is $4 per widget, independent of the number manufactured. (This assumption is reasonable for a product such as lead pencils or bobby pins.) Assume the demand function for widgets is $g(x) = 10^6(10 - x)$. Thus, if the price is $8, then $10^6(10 - 8) = 2 \times 10^6$ widgets are demanded. How many widgets should the A.B.C. Company produce and how much should they charge for each widget in order to maximize their profits?

Solution. Let x be the selling price. Then $(x - 4)$ is the profit on each widget sold. Thus the profit $P(x)$ as a function of price is equal to the profit per widget times the number of widgets sold. In other words,

$$P(x) = (x - 4)[10^6(10 - x)].$$

We can assume $x \geq 4$ (Why?) and $x \leq 10$ (since no one will pay $10 for a widget).

To maximize P we apply the techniques of this section. Thus,

$$P'(x) = 10^6[(10 - x) - (x - 4)]$$
$$= 10^6[14 - 2x].$$

Hence, $P'(x) = 0$ precisely for $x = 7$. It is not hard to see that $x = 7$ yields maximum P. Thus the company should charge $7 per widget and manufacture 3×10^6 widgets.

Clearly the situation we have just considered, namely a monopoly in a certain industry, differs from the competitive model mentioned in Example 3. In the monopolistic setup it was advantageous to the A.B.C. Company to produce only as many widgets as could be sold at the optimum price. (Stated in a more dramatic manner, A.B.C. keeps widgets scarce in order to support the price.)

▶ **Example 5.** Now let us suppose the government places a $1 tax on each widget manufactured by the A.B.C. Company (in Example 4). The new cost to A.B.C. is $5 per widget. Should the company now charge $8 per widget and thus pass the entire cost of the tax on to the consumer?

Solution. The demand function is still $g(x) = 10^6(10 - x)$. The consumer either does not know or does not care about the tax on widgets. Thus the profit function now becomes

$$P(x) = 10^6(10 - x)(x - 5).$$

Hence,

$$P'(x) = 10^6[(10 - x) - (x - 5)]$$
$$= 10^6(15 - 2x).$$

Thus, $P'(x) = 0$ precisely when $x = 7.50$. Therefore the A.B.C. Company can now maximize its profits by producing only 2.5×10^6 widgets and charging $7.50 per widget. In the previous example the total profit was 21×10^6 for the A.B.C. Company. In the present case it is 18.75×10^6. There is no way A.B.C. can maintain its original profits in the face of the new tax. Moreover, even though it is a monopoly, the A.B.C. Company must absorb $.50 of the tax itself and can pass only $.50 of the tax on to the consumer by following the optimal strategy. It is a harsh world in which monopolistic corporations are subjected to such cruel indignities.

Exercises

1. The sum of two positive numbers is 16. How small can the sum of their squares be?

2. Find the maximum possible area of a rectangle whose perimeter is 32 in.

3. The Grass Roots Political Party determines that it can sell 1,000 memberships per year if the cost is $5 per membership. It also observes that, for each 1¢ reduction in the membership fee, 10 more members can be found. Under these conditions, what is the maximum possible income the party can obtain through its member fee, and what fee yields this income? [*Hint:* Let x be the number of memberships over 1,000. Hence, the total number of members is $1,000 + x$. The price for each x is $5 less 1¢ (0.01 dollars) for each block of 10 members over 1,000. Thus, the price paid by each member x is $5 - 0.01(x/10) = 5 - 0.001x$ dollars. The income is then given by $I = (1,000 + x)(5 - 0.001x)$.]

4. The International Glass Company sells glass plates for $10. The monthly cost of production is given by $C = 1 + 2x + 0.1x^2$, where x is the number of glass plates produced. Determine how many plates should be produced in a month in order to maximize the profit. (*Note:* profit per plate = sale price − cost of production.)

5. A photographer has a thin piece of wood 16 in. long. How should he cut the wood to make a rectangular picture frame that encloses the maximum area?

6. A rancher has a total of 12 miles of fencing with which to enclose a rectangular pasture. He plans to fence the entire area and then to subdivide it by running a fence across the width. What dimensions should the pasture have in order to enclose the maximum area with the available fencing?

7. Find the area of the largest rectangle with lower base on the x axis and upper vertices in the curve

$$y = 6 - x^2.$$

8. A triangular area is enclosed on two sides by a fence and on the third side by the straight edge of a river. The two sides of the fence have equal length, 50 ft. Find the maximum area enclosed. [*Hint:* The area enclosed is given by $\frac{1}{2}hx$. Thus, you must find h in terms of x. (See Figure 4.2.4.)]

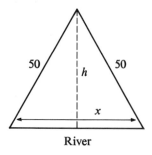

River

Figure 4.2.4

9. A retailer knows that if he charges x dollars for an alarm clock, he will be able to sell $480 - 40x$ clocks. The alarm clock costs him \$4 per clock. How much should he charge per clock in order to maximize his total profit? (Note that $480 - 40x$ is the demand function.) [*Hint:* His profit per clock is obviously $x - 4$ dollars. Hence, his total profit is $(480 - 40x)(x - 4)$. Thus, it is this last quantity which should be maximized over the interval $4 \leq x \leq 12$.]

10. Referring to Exercise 9, the government now adds a tax of \$1 to the price of each clock. How much should the retailer now charge in order to maximize profit?

11. A rectangular trough is formed from a piece of metal 12 inches wide and of indeterminate length. (See Figure 4.2.5.) What should the dimensions be so as to maximize the amount of water carried? Thus, what dimensions maximize the cross-sectional area of the trough?

Figure 4.2.5

12. A trough is to be formed as in Exercise 11, but this time in the shape of an isosceles triangle. What should the dimensions be so as to maximize the amount of water carried? Again the metal is 12 inches wide. A convenient variable is the distance labeled x in Figure 4.2.6. Find the dimensions that make the cross-sectional area of the trough largest.

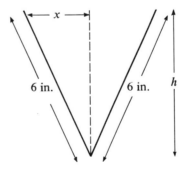

Figure 4.2.6

13. A box is to be formed from a piece of paper 16 inches square by cutting out squares in the corners and folding as shown in Figure 4.2.7. What should the dimensions of the box be so as to maximize its volume?

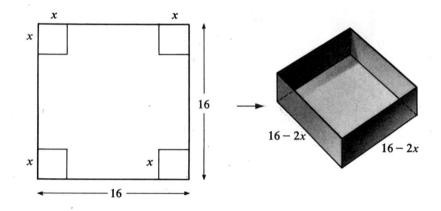

Figure 4.2.7

14. The sum of two positive integers is 75. How large can the square of the first times the second be?

15. According to U.S. Postal Service regulations, the girth plus length of a package cannot exceed 72 in. Assuming the package has square ends, what is the maximum volume permissible under this regulation? (See Figure 4.2.8.)

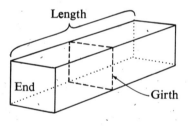

Figure 4.2.8

16. You wish to construct a tomato can out of 100π sq in. of metal. What is the volume of the largest can that may be so constructed? (See Figure 4.2.9.) [*Hint:*

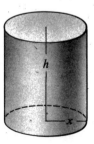

Figure 4.2.9

Let x = radius of can,

h = height of can.

Then   Area of top = πx^2,

Area of side = $2\pi xh$,

Volume of can = $\pi x^2 h$.

(You may assume that $x \le \sqrt{50}$ or else all the metal would be used in the construction of the top and bottom.)]

17. Analogous to Exercise 16, you now wish to construct a beer can. This means the top and bottom must be double thickness. What is the largest can that can be constructed out of the 100π square inches of metal?

18. A man wishes to walk from his present point to the river and then to his horse (Figure 4.2.10). Assuming he walks in a straight line and meets the river somewhere between A and B, find the shortest distance from the man to the river to the horse.

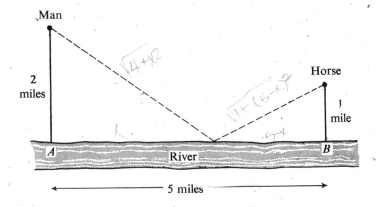

Figure 4.2.10

19. Find the area of the largest rectangle that can be inscribed in a semicircle as shown in Figure 4.2.11.

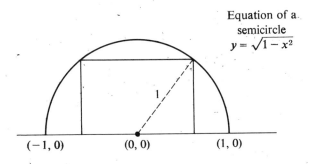

Equation of a
semicircle
$y = \sqrt{1 - x^2}$

Figure 4.2.11

20. A firm's revenue and cost functions are given by

$$R(x) = -\tfrac{1}{4}x^2 + 2x - 1, \qquad 2 \leq x \leq 5,$$

and

$$C(x) = \tfrac{3}{4}x + 1, \qquad 2 \leq x \leq 5,$$

respectively; x is the output in thousands of units. Find the quantity of output that minimizes the cost, the quantity that maximizes the revenue, and the quantity that maximizes the profit. Is there any relationship between these values?

4.3 Increasing and Decreasing Functions: Geometrical Significance of the Derivative

So far we have only used information on the zeros of f' to locate maxima and minima. However, it is possible to learn a great deal more about a function by studying the behavior of its derivative. We begin by defining what is meant by an increasing or decreasing function.

DEFINITION 1. Let f be defined on (a, b). If $f(c) < f(d)$ whenever $c < d$ (and $c, d \in (a, b)$), then f is strictly increasing. If $c < d$, where $c, d \in (a, b)$, implies that $f(c) > f(d)$, we say that f is strictly decreasing on (a, b). (See Figure 4.3.1.)

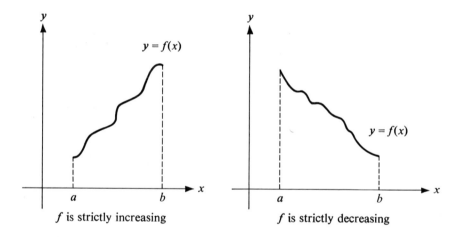

f is strictly increasing f is strictly decreasing

Figure 4.3.1

It is easy to see that if f is a strictly increasing function on (a, b) and if f exists at $c \in (a, b)$, then $f'(c) > 0$. This seems reasonable from a geometric standpoint and is not hard to see analytically. Consider the quantity

$$\frac{f(x) - f(c)}{x - c};$$

if $x > c$, then $f(x) - f(c) > 0$ and

$$\frac{f(x) - f(c)}{x - c} > 0.$$

Hence,

$$\lim_{x \to c^+} \frac{f(x) - f(c)}{x - c} \geq 0,$$

and a similar argument shows that

$$\lim_{x \to c^-} \frac{f(x) - f(c)}{x - c} \geq 0.$$

Hence, $f'(c) \geq 0$.

What is important for us is that a converse to the preceding statement is true if we slightly strengthen the hypotheses.

THEOREM 1.   *Let f be differentiable on (a, b).*

(1)   *If f'(x) > 0 for x ∈ (a, b), then f is strictly increasing on (a, b).*
(2)   *If f'(x) < 0 for x ∈ (a, b), then f is strictly decreasing on (a, b).*

Before discussing the verification of this result, let us look at some examples.

▶ Example 1.   If

$$f(x) = 2x^3 - 21x^2 + 60x - 9,$$

find all intervals on which f is strictly increasing or strictly decreasing.

Solution.   First we differentiate f to obtain

$$\begin{aligned} f'(x) &= 6x^2 - 42x + 60 \\ &= 6(x^2 - 7x + 10) \\ &= 6(x - 2)(x - 5). \end{aligned}$$

Now a simple analysis reveals that

$$\begin{aligned} f'(x) &> 0 \quad \text{for} \quad x < 2, \\ f'(x) &< 0 \quad \text{for} \quad 2 < x < 5, \\ f'(x) &> 0 \quad \text{for} \quad 5 < x. \end{aligned}$$

Clearly $f'(x) = 0$ for $x = 2, 5$. The results are displayed in Figure 4.3.2. Thus f is strictly increasing on $(-\infty, 2)$ and $(5, \infty)$ and strictly decreasing on $(2, 5)$. [Recall that $(-\infty, a)$ is the set $\{x | x < a\}$, while (b, ∞) is the set $\{y | y > b\}$, where a and b are real numbers.]

We can use the information in Example 1 to make a very rough sketch of the graph of f (Figure 4.3.3). First, we find that $f(2) = 43$ and $f(5) = 16$. We sketch

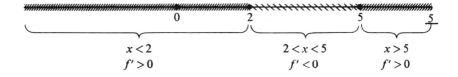

Figure 4.3.2

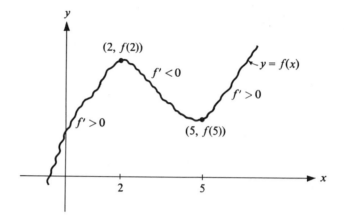

Figure 4.3.3

the graph in a wiggly manner to indicate that we don't know what it's doing more precisely. However, just from the information we have, it is clear that f has a local maximum at 2 and a local minimum at 5.

Theorem 1 is a consequence of the mean value theorem. The latter is one of the most important and useful theorems of all calculus and we now state it.

THEOREM 2. (*Mean Value Theorem*) *Let f be a function continuous on $[a, b]$ and differentiable on (a, b). Then, there exists a number $c \in (a, b)$ such that*

$$f'(c) = \frac{f(b) - f(a)}{b - a}.$$

At first glance the theorem may appear technical and uninteresting, but it really says a very simple and intuitively appealing thing. Suppose you join the points $P_1 = (a, f(a))$ and $P_2 = (b, f(b))$ in Figure 4.3.4 with a straight line L. The slope of L is

$$m = \frac{f(b) - f(a)}{b - a}.$$

Then, some place between a and b the tangent to f has the same slope m. (There may, in fact, be several points where the tangent to f has slope m, but the important thing is there is *always* at least one such point.)

If we take a few liberties with the notion of functions, the theorem could be paraphrased as follows: You can't get from P_1 to P_2 without going in the right direction at least once.

Someone tells you he left Chicago at 1:00 P.M. and arrived at Indianapolis 170 miles away at 3:00 P.M. When you tell him that he must have been going 85 mph at some point, you are using a disguised version of the mean value theorem. (This will become clearer in a later chapter.)

We omit the proof of this theorem since it requires techniques out of the mainstream

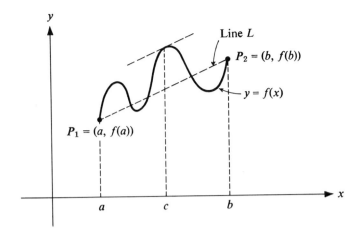

Figure 4.3.4

of the course. The interested reader can find a proof in any of a number of calculus books.*

Let us briefly outline how Theorem 1 follows from the mean value theorem. Suppose x_1 and x_2 are *any* two numbers in (a, b) with $x_1 < x_2$. If f is differentiable on (a, b), then f is certainly differentiable on $[x_1, x_2]$. Hence, there exists a number $c \in (x_1, x_2)$ such that

$$\frac{f(x_2) - f(x_1)}{x_2 - x_1} = f'(c).$$

But by (1) of Theorem 1, $f'(c) > 0$. Hence

$$\frac{f(x_2) - f(x_1)}{x_2 - x_1} > 0,$$

and since $x_2 > x_1$, we see that $f(x_2) > f(x_1)$. Thus f is strictly increasing on (a, b). The second part follows in a similar manner.

As another consequence of the mean value theorem, we have the following result, which is extremely useful in applications, particularly in Chapter 5.

THEOREM 3. *Let f be differentiable on (a, b) and let $f'(x) = 0$ for all $x \in (a, b)$. Then $f(x) = k$ (k constant) for $x \in (a, b)$.*

This result is easily verified. Let x_1, x_2 be any elements of (a, b). Then the mean value theorem states that there exists a point $c \in (x_1, x_2)$ such that

$$\frac{f(x_2) - f(x_1)}{x_2 - x_1} = f'(c) = 0.$$

Hence, for any $x_1, x_2 \in (a, b)$, $f(x_1) = f(x_2)$, implying $f(x) = k$ on (a, b).

---

* For example, T. Apostol, *Calculus* (Xerox, Lexington, Mass., 1967), 2nd ed., Vol. I, p. 83.

Let us summarize what we have so far.

1. If $f'(x) > 0$ for $x \in (a, b)$, then $f(x)$ is strictly increasing.
2. If $f'(x) < 0$ for $x \in (a, b)$, then $f(x)$ is strictly decreasing.
3. If $f'(x) = 0$ on (a, b), then f is constant.

We end this section with an application to economics.

▶ Example 2. (*Elasticity of Demand*) From experience it is known that changes in price and demand are generally in opposite directions and moreover, for different commodities, changes in price affect demand in different degrees. For example, a rise or fall in the price of a staple item such as wheat only slightly changes the demand while a change in the price of mink coats or snowmobiles may cause a sharp change in their demand.

Alfred Marshall (1842–1924) first defined *elasticity of demand* as a measure of the influence of variation in price on demand. According to his definition

$$\text{elasticity of demand} = \frac{\text{percentage change in demand}}{\text{percentage change in price}}.$$

For example, suppose that when wheat is selling for \$10 a bushel, the demand is 100 bushels. When the price is increased to \$12 a bushel, the demand falls to 90 bushels. Thus

$$\text{percentage change in demand} = \frac{90 - 100}{100} = -\frac{10}{100} = -10\%$$

while

$$\text{percentage change in price} = \frac{12 - 10}{10} = \frac{2}{10} = 20\%.$$

In this case

$$\text{elasticity of demand} = \frac{-10}{20} = -\frac{1}{2}.$$

If $Q(p)$ is the quantity demanded at price p then we see that the formula for elasticity of demand at price a can be written as

$$E = - \frac{\dfrac{Q(p) - Q(a)}{Q(a)}}{\dfrac{p - a}{a}}. \tag{1}$$

A few comments are needed at this point. First, following Marshall, we have introduced a minus sign in the formula to ensure that the elasticity of demand will always be positive. Second, one might wonder why we do not define E to be

$$\frac{\text{change in demand}}{\text{change in price}}.$$

If this definition were used, then E would depend on whether wheat were measured in bushels or pecks and whether the price were in dollars, marks, or pounds, whereas (1) avoids this difficulty. Formula (1) can be rewritten as

$$E = - \frac{a}{Q(a)} \frac{Q(p) - Q(a)}{p - a}.$$

To compute E when the price is a, the question arises as to how p should be chosen. In our preceding example we took $a = 10$ and $p = 12$. Economists in Marshall's time did not make much use of calculus. However, the modern-day economist lets p approach a, and hence we obtain

$$E = -\frac{a}{Q(a)} \lim_{p \to a} \frac{Q(p) - Q(a)}{p - a} = -\frac{a}{Q(a)} Q'(a).$$

Thus we take

$$E(a) = -\frac{a}{Q(a)} Q'(a)$$

to be our final definition of the elasticity of demand at $p = a$.

The elasticity of demand is important in pricing policies. Assume that $Q(p)$ is the number of widgets demanded at price p. The Universal Widget Company would like to know what will happen to its total revenue $p \cdot Q(p)$ if it raises or lowers the price of widgets. To answer this question we investigate the first derivative of the quantity $pQ(p)$ with respect to the price p. Thus

$$\frac{d}{dp}(pQ(p)) = p \cdot Q'(p) + Q(p)$$

$$= Q(p)\left[1 - \frac{P}{Q(p)} Q'(p)\right]$$

$$= Q(p)[1 - E(p)].$$

(1) If $E(p) = 1$, $(d/dp)[PQ(p)] = 0$ *and* by Theorem 3 $PQ(p) = c$, a constant. That is, the revenue $pQ(p)$ is a constant, independent of the changes in price. In this case, we say that the demand is unitary.

(2) If $E > 1$, then $(d/dp)[pQ(p)] < 0$, which means that $pQ(p)$ is decreasing and a rise in price leads to a fall in revenue. In this case, we say that the demand is *elastic*.

(3) If $E < 1$, then $(d/dp)[pQ(p)] > 0$, which means that $pQ(p)$ is increasing and a rise in price leads to an increase in revenue. In this case, we say that the demand is inelastic.

The case $E = 1$ is of key importance in the determination of expected revenue. If $E = 1$, then the demand Q is inversely proportional to the price p.

If $E > 1$, the demand falls more than proportionally with respect to increase in price and the total revenue $[pQ(p)]$ decreases.

If $E < 1$, the demand falls less than proportionally with respect to increase in price and the total revenue $[pQ(p)]$ increases.

In summary, E, the elasticity of demand, tells us whether a change in the price will result in a large, small, or comparable change in demand. This is of particular significance for monopolists who fix market prices. It is also important to note that E *does not* depend on the slope of the demand curve alone.

Exercises

1. Determine by inspection when f' and f'' are positive and negative in Figures 4.3.5–4.3.8.

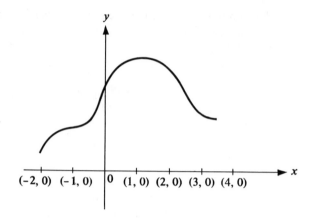

Figure 4.3.5

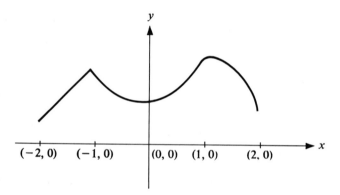

Figure 4.3.6

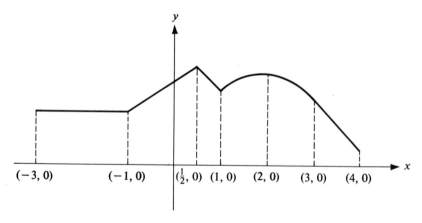

Figure 4.3.7

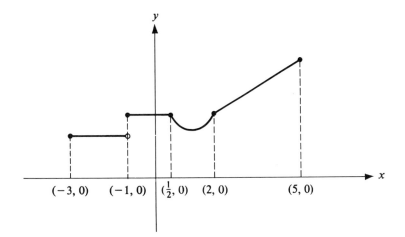

Figure 4.3.8

2. Determine where the following functions are increasing and where they are decreasing. Also decide where all relative maxima and minima occur. Sketch *as much* of the graph as you can. (Similar to Figure 4.3.3.)

(a) $f(x) = x^2 - 5x + 6$.

(b) $f(x) = x^2 - x + 1$.

(c) $f(x) = x^3 + x^2 - 8x + 1$.

(d) $f(x) = 2x^3 - 3x^2 + 3$.

(e) $f(x) = x - x^2$.

(f) $f(x) = xe^{-x}$.

(g) $f(x) = x + \dfrac{1}{x}$.

(h) $f(x) = (4 - x^2)^{1/2}$.

(i) $f(x) = x^4 + 32x + 32$.

(j) $f(x) = \sqrt{1 - x^2}$.

(k) $f(x) = (x + 2)(x - 3)^2$.

(l) $f(x) = x - \log x$.

(m) $f(x) = xe^{-x^2}$.

(n) $f(x) = \dfrac{x}{1 + x^2}$.

(o) $f(x) = \sqrt{x^2 - 1}$.

3. Sketch as much of the graph as you can of a curve having the following characteristics: $f(0) = 2$; $f'(x) < 0$ for $x < 0$; $f'(x) > 0$ for $x > 0$.

4. For which values of b will the graph of

$$y = 2x^3 - 3x^2 + bx + 1$$

be always increasing?

5. In each of the following problems, a, b, c refer to the equation

$$\frac{f(b) - f(a)}{b - a} = f'(c),$$

which expresses the mean value theorem. Given $f(x)$, a, and b, find c.

(a) $f(x) = x^2 + 2x - 1, \quad a = 0, b = 1.$

(b) $f(x) = x^3, \quad a = 0, b = 3.$

(c) $f(x) = x^{2/3}, \quad a = 0, b = 1.$

(d) $f(x) = \sqrt{x - 1}, \quad a = 1, b = 3.$

[*Hint to* (a):

$$\frac{f(b) - f(a)}{b - a} = \frac{2 + 1}{1} = 3,$$

and

$$f'(x) = 2x + 2.$$

Thus we must find a c such that

$$3 = f'(c) = 2c + 2.$$

Clearly $c = \frac{1}{2}$ in the desired solution.]

6. Can the mean value theorem be applied to the function $f(x) = |x|$ on the interval $[-1, 1]$? Why?

7. Let the demand for cartons of cigarettes with respect to a price p be given by

$$Q(p) = Kp^{-\alpha},$$

where K and α are constants. Find the elasticity of demand for Q at any price p.

8. Suppose that the demand curve is given by

$$Q(p) = 30 - 4p - p^2.$$

Find the elasticity of demand for $p = 3$. Determine E for *any p*.

4.4 Concavity and Graphing: Significance of the Second Derivative

By looking at the second derivative, we can obtain more information on the behavior of the function. First we introduce the concept of concavity, which is of great use in graphing functions.

DEFINITION 1. *A curve is concave upward on* (a, b) *if*

(1) the chord joining any two points on the curve lies above the curve, or, equivalently,
(2) at every point on the curve there exists a tangent line which lies below the curve.
(See Figure 4.4.1.)

DEFINITION 2. *A curve is concave downward on* (a, b) *if*

(1) the chord joining any two points on the curve lies below the curve, or, equivalently,
(2) at every point on the curve there exists a tangent line which lies above the curve.
(See Figure 4.4.2.)

▶ Example 1. Consider the graph of the function $f(x) = x^2$ on $[-1, 1]$. (See Figure 4.4.3.) It is intuitively clear that this curve is concave upward on $(-1, 1)$, since at every point the tangent line lies below the curve.

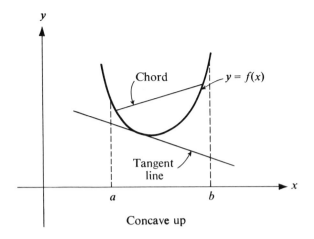

Concave up

Figure 4.4.1

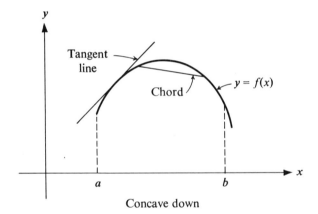

Concave down

Figure 4.4.2

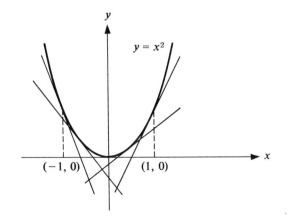

Figure 4.4.3

▶ Example 2. Consider the graph of the function $f(x) = x^3$ on $[-1, 1]$. (See Figure 4.4.4.) Observe that in this case the curve is concave upward in $(0, 1]$ and concave downward in $[-1, 0)$. Thus, a change in concavity occurs at $x = 0$.

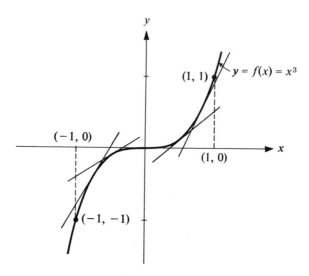

Figure 4.4.4

Fortunately, the second derivative stands willing and able to tell us whether the curve is concave upward or downward. More precisely, we have the following result:

THEOREM 1. If $f''(x) > 0$ on (a, b), then $f(x)$ is concave upward on (a, b).

Although we shall not give a rigorous proof of this theorem, we remark that geometrically it is easy to observe its verification. If $f''(x) > 0$ on (a, b), then by Theorem 1 of Section 4.3, $f'(x)$ is strictly increasing on (a, b) and the curve is bending upward (since the slope of the tangent is always increasing.) In like manner, if $f''(x) < 0$, then $f'(x)$ is strictly decreasing on (a, b) and the curve is bending downward. For example, if $f(x) = x^2$, then $f'(x) = 2x$ and $f''(x) = 2$. In this case, the curve must be concave up since $2 > 0$. We show the graph of f, f', and f'' in Figure 4.4.5. You can see that the derivative is a strictly increasing function of x. However, the function f itself is strictly increasing for $x \geq 0$ and is strictly decreasing for $x < 0$.

Let us now consider the function in Example 1 of Section 4.3 in more detail. The function under discussion is

$$f(x) = 2x^3 - 21x^2 + 60x - 9.$$

So far we know the graph appears as shown in Figure 4.4.6. Now, $f'(x) = 6x^2 - 42x + 60$ and $f''(x) = 12x - 42 = 12(x - \frac{7}{2})$. Thus $f''(x) < 0$ for $x < \frac{7}{2} = 3\frac{1}{2}$ and $f''(x) > 0$ for $x > \frac{7}{2} = 3\frac{1}{2}$. Therefore, the curve is concave down for $x < \frac{7}{2}$ and concave up for $x > \frac{7}{2}$. At $x = 3\frac{1}{2}$, $f''(x) = 0$, which means that at $x = 3\frac{1}{2}$ the

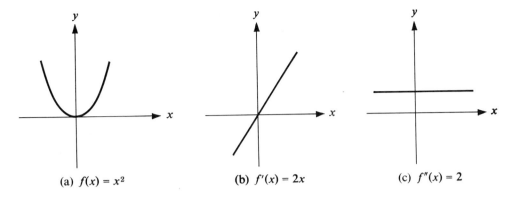

(a) $f(x) = x^2$ (b) $f'(x) = 2x$ (c) $f''(x) = 2$

Figure 4.4.5

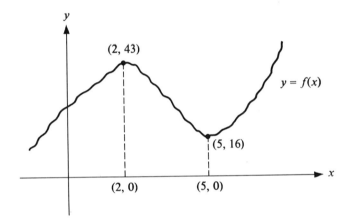

Figure 4.4.6

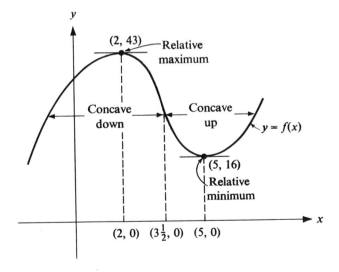

Figure 4.4.7

curve changes from being concave down to concave up. This point is called a *point of inflection*. Using the information we have just obtained, we are able to draw the more detailed graph shown in Figure 4.4.7.

▶ Example 3.   Draw the graph of

$$f(x) = x^4 - 8x^3 + 18x^2 + 1.$$

Solution.   First we find $f'(x)$ and $f''(x)$ and write them in a convenient form.

$$f'(x) = 4x^3 - 24x^2 + 36x = 4x(x^2 - 6x + 9)$$
$$= 4x(x - 3)^2.$$
$$f''(x) = 12x^2 - 48x + 36 = 12(x^2 - 4x + 3)$$
$$= 12(x - 3)(x - 1).$$

It is easy to see that $f'(0) = f'(3) = 0$ and 0, 3 are the only points where $f'(x)$ equals zero.   Also

$$f'(x) < 0 \quad \text{for} \quad x < 0,$$
$$f'(x) > 0 \quad \text{for} \quad x > 0.$$

Thus, the curve is decreasing for $x < 0$ and increasing for $x > 0$. Let us now investigate the second derivative, f''. It is easy to see that $f''(x) > 0$ for $x < 1$, $f''(x) < 0$ for $1 < x < 3$, and $f''(x) > 0$ for $x > 3$. Hence, for $x < 1$ and $x > 3$ the curve is concave up, and for $x \in (1, 3)$, the curve is concave down. We summarize the results on the real line in Figure 4.4.8. In Figure 4.4.9, we draw the graph of

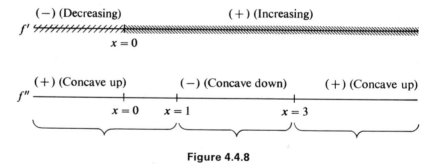

Figure 4.4.8

the function. We remark that even though $f'(3) = 0$, $x = 3$ does not yield a local maximum or minimum. Actually, the curve is said to have a "point of inflection" at 3; that is, the curve changes its concavity at this point. It is important to remember that a point c for which $f'(c) = 0$ *and* where the curve changes its concavity is neither a minimum or a maximum.

We summarize the work in this section by giving the following rules for graphing $y = f(x)$.

RULE 1.   Find $f'(x)$ and $f''(x)$.

RULE 2.   Find the zeros of f' and f''.

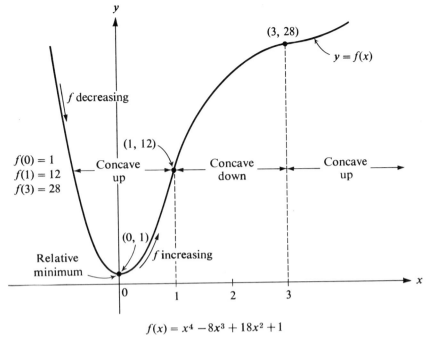

$$f(x) = x^4 - 8x^3 + 18x^2 + 1$$

(Different scales are used on x and y axes)

Figure 4.4.9

RULE 3. Find the intervals in which (1) $f' < 0$ and (2) $f' > 0$. (Two algebraic methods for doing this are described in an appendix to this chapter.)

RULE 4. Find the intervals on which (1) $f'' < 0$ and (2) $f'' > 0$.

RULE 5. Use Rules 1–4 with Table 4.4.1 to graph the function.

TABLE 4.4.1

1. $f'(x) > 0$ on (a, b)	f is strictly increasing
2. $f'(x) < 0$ on (a, b)	f is strictly decreasing
3. $f''(x) > 0$ on (a, b)	f is concave up
4. $f''(x) < 0$ on (a, b)	f is concave down

We end this section by considering one more example.

▶ Example 4. One of the most important functions used in the application of statistics to many fields is the so-called *normal density function*, defined by

$$f(x) = e^{-x^2}.$$

Find all relative maximum and minimum points; all intervals where the function is increasing and decreasing; all intervals where it is concave up and concave down. Also, sketch a graph of the function for $-2 \leq x \leq 2$.

Solution. Now,

$$f'(x) = -2xe^{-x^2},$$

and

$$f''(x) = -2e^{-x^2} + 4x^2e^{-x^2} = 2(2x^2 - 1)e^{-x^2}.$$

Since e^{-x^2} is always positive, $f'(x) < 0$ for $x > 0$ and $f'(x) > 0$ for $x < 0$. Thus, $f'(x) = 0$ precisely for $x = 0$. Observe that at 0, the function changes from an increasing function to a decreasing function. Thus, since $f(0) = 1$, $f(2) = e^{-4} \simeq 0.02$, and $f(-2) = e^{-4} \simeq 0.02$, we see that on $[-2, 2]$, the point $(0, 1)$ is a *relative maximum*. It is not hard to see that $(0, 1)$ is also an absolute maximum point of the curve. Now $f''(x) > 0$ for $2x^2 - 1 > 0$, which means that for $x^2 > \frac{1}{2}$ or $|x| > 1/\sqrt{2}$ the curve is concave upward, while for $2x^2 - 1 < 0$ or $|x| < 1/\sqrt{2}$, the curve is concave down. On $[-2, 2]$, the absolute minima occur at $x = -2$ and $x = +2$. We summarize as follows.

1. $x < 0$ f is strictly increasing

2. $x > 0$ f is strictly decreasing

3. $x = 0$ a maximum point

4. $|x| < \dfrac{1}{\sqrt{2}}$ curve is *concave down*

5. $|x| > \dfrac{1}{\sqrt{2}}$ curve is *concave up*

6. $x = \pm \dfrac{1}{\sqrt{2}}$ inflection points

7. $x = -2, x = 2$ absolute minima on $[-2, 2]$.

This information is used to sketch the graph of the function in Figure 4.4.10. (Note that the scales used on the x and y axes are different.)

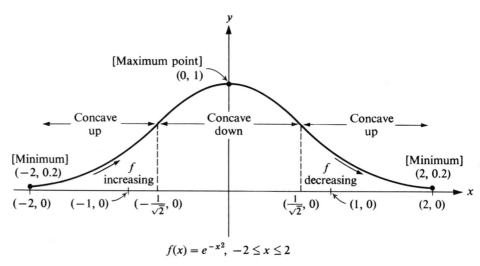

$$f(x) = e^{-x^2}, \quad -2 \leq x \leq 2$$

Figure 4.4.10

Exercises

1. Find the maxima and minima and the intervals where the graph is increasing, decreasing, concave up, and concave down, of the following functions. Use the preceding information to graph the functions.

   (a) $f(x) = x^2 - 5x + 6$.

   (b) $f(x) = x^2 + 4$.

   (c) $f(x) = x^3 + x^2 - 8x + 1$.

   (d) $f(x) = x^3 - 3x$.

   (e) $f(x) = 2x^3 - 15x^2 - 36x - 1$.

   (f) $f(x) = 4x^3 + 7x^2 - 10x + 2$.

   (g) $f(x) = 4 + 3x - x^3$.

   (h) $f(x) = (x - 2)^2(x + 3)$.

   (i) $f(x) = x^3 - 3x^2 - 9x + 1$.

   (j) $f(x) = xe^{-x}$.

   (k) $f(x) = x^2 - x^3$.

   (l) $f(x) = \dfrac{x}{x^2 + 1}$.

   (m) $f(x) = (x + 1)(x - 2)^2$.

   (n) $f(x) = (x - 1)^2(x^2 + 1)$.

   (o) $f(x) = x^2 + 1/x$.

   *(p) $f(x) = x - \log x$ for $0 < x < \infty$.

   *(q) $f(x) = (x - 1)^3(x + 4)^2$.

   *(r) $f(x) = \dfrac{x - 1}{x - 2}$.

   *(s) $f(x) = xe^{-x^2}$.

2. Sketch a curve $y = f(x)$ having the following properties: $f(2) = 0; f''(x) < 0$ for $x < 2; f''(x) > 0$ for $x > 2$.

3. Sketch a curve $y = f(x)$ having the following characteristics: $f(1) = 2;$ $f'(x) < 0$ for $x < 1; f'(x) > 0$ for $x > 1; f''(x) > 0$ for all x.

4. It is conjectured from experience that the ability A to memorize in the early years obeys a law of the form

$$A(x) = x \log x + 1,$$

when $x \in (0, 4)$, $A(0) = 1$, where x is measured in years. Where does A attain its local minimum? Sketch a graph of A.

5. Sketch a graph of a function f, for $x > 0$, if $f(1) = 0$ and $f'(x) = 1/x$ for all $x > 0$. Is such a curve necessarily concave upward or concave downward?

6. Sketch a smooth curve $y = f(x)$, illustrating $f(1) = 0; f''(x) < 0$ for $x < 1;$ $f''(x) > 0$ for $x > 1$.

7. Make a diagram showing how the graph of $y = f(x)$ might appear if f has a second derivative for each x, given that $f(-3) = 4; f(-1) = 1; f(0) = 2;$ $f''(x) > 0$ when $x < 0$; and $f''(x) < 0$ when $x > 0$; supposing in addition, (a) that $f(2) = 0$ and (b) that $f'(x) > 0$ and $f(x) < 4$ when $x > 0$. Why is $f'(0) = 0$ impossible in both cases?

4.5 Tests for Maxima and Minima

The second derivative can be used in conjunction with the first derivative to determine local maxima and minima. The following theorem yields a very useful test.

THEOREM 1.   *Let* $f'(c) = 0$, *where* f *is defined on* (a, b) *and* $c \in (a, b)$.

(1)   *If* $f''(c) > 0$, *then* f *has a local minimum at* c.
(2)   *If* $f''(c) < 0$, *then* f *has a local maximum at* c.
(3)   *If* $f''(c) = 0$, *then it is unclear whether* c *is a local maximum or minimum* (*it may be neither*) *and other tests must be made.*

We omit the proof of this theorem but note from some of our previous examples that when a maximum occurred, the curve was always concave down in an interval about the maximum, and always concave up near a minimum. Thus, the theorem agrees with our intuitive expectations.

▶ Example 1.   If $f(x) = x^3 - x^2 - 8x + 1$, find all local maxima and minima of f on $(-\infty, \infty)$ and graph the function. The symbol $(-\infty, \infty)$ means $\{x | x$ is a real number$\}$.

Solution.   First, we find those points where $f' = 0$. Now,

$$f'(x) = 3x^2 - 2x - 8 = (3x + 4)(x - 2),$$

and thus $f'(x) = 0$ at $x = 2$ and $x = -\frac{4}{3}$. It is clear that $f''(x) = 6x - 2 = 2(3x - 1)$, $f''(2) = 10$, and $f''(-\frac{4}{3}) = -10$. Hence, by Theorem 1, f has a local minimum at $x = 2$ and a local maximum at $x = -\frac{4}{3}$. To sketch the graph, we observe the following facts:

1. $f'(x) > 0$ for $x > 2$ and $x < -\frac{4}{3}$. Therefore, f is increasing for these values of x.
2. $f'(x) < 0$ for $-\frac{4}{3} < x < 2$. Hence for $x \in (-\frac{4}{3}, 2)$, f is decreasing.
3. $f''(x) > 0$ for $x > \frac{1}{3}$. Thus, f is concave up for $x > \frac{1}{3}$.
4. $f''(x) < 0$ for $x < \frac{1}{3}$ implying that f is concave down for $x < \frac{1}{3}$.
5. f has three zeros x_1, x_2, x_3 (which we need not find exactly) and the curve must cross the y axis at the point $(0, 1)$.

Combining 1–5 and the facts that 2 is a local minimum and $-\frac{4}{3}$ is a local maximum, we obtain the graph shown in Figure 4.5.1.

So far, we have been considering maximum-minimum (as opposed to local max and min) problems on closed intervals only, where the situation is relatively simple. Recall that if f is continuous on a closed interval, f has both a maximum and minimum there. On open intervals there are more possibilities. In particular, f may have a max and not a min, or conversely. Let us look at an example.

▶ Example 2.   Suppose we consider the function in Example 4 of Section 4.4, $f(x) = e^{-x^2}$. This time, we consider the entire domain of the function—the open interval $(-\infty, \infty)$. Does this function have a maximum and minimum on $(-\infty, \infty)$?

Solution.   From our work in Example 4, Section 4.4, we see that a maximum occurs at $(0, 1)$. The second-derivative test also immediately implies that $(0, 1)$ is a local maximum. Since $f''(x) = 2e^{-x^2}(2x^2 - 1)$, $f''(0) = -2$, and thus by Theorem 1, $x = 0$ yields a maximum. We claim that f does *not* have a minimum any place. Consider any point x_1. If $x_1 > 0$, then for $x_1 < x_2$ we see that $f(x_1) > f(x_2)$. (Why?) Thus x_1 is not a minimum. On the other hand, if $x_1 < 0$, then for $x_3 < x_1$

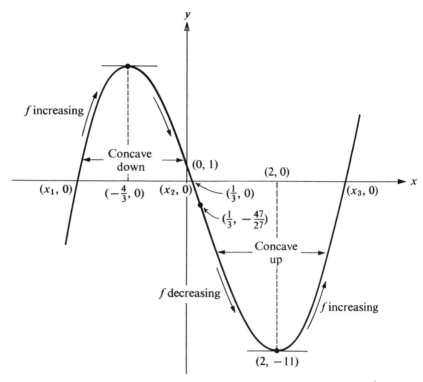

(Different scales are used on x and y axes)

Figure 4.5.1

we recognize that $f(x_3) < f(x_1)$. (Why?) So again x_1 does not yield a minimum. By collecting the information we have and adding a little more, it is not hard to graph $f(x) = e^{-x^2}$ on $(-\infty, \infty)$. Figure 4.4.10 shows $f(x) = e^{-x^2}$ on $[-2, 2]$. Now,

1. $f > 0$ for all x. (Why?)
2. f is symmetric about the y axis.
3. f is increasing on $(-\infty, 0)$ and it is decreasing on $(0, \infty)$.
4. $f''(x) = -2e^{-x^2} + 4x^2 e^{-x^2} = 2e^{-x^2}(2x^2 - 1)$.

Note that $2e^{-x^2} > 0$ for all x and $2x^2 - 1 > 0$ for $|x| > 1/\sqrt{2}$ and $2x^2 - 1 < 0$ for $|x| < 1/\sqrt{2}$. Hence $f''(x) > 0$ for $|x| > 1/\sqrt{2}$, and $f''(x) < 0$ for $|x| < 1/\sqrt{2}$. Therefore, f is concave up on $|x| > 1/\sqrt{2}$ and f is concave down on $|x| < 1/\sqrt{2}$. The graph is shown in Figure 4.5.2.

The clearest way to decide whether a function on an open interval has a maximum or minimum (and where) is to carefully graph the function as we have done here. Then these questions can be answered by inspection.

We will, however, state one rule which is simple and very useful for open-interval problems. This is sometimes referred to as the first-derivative test for determining extrema.

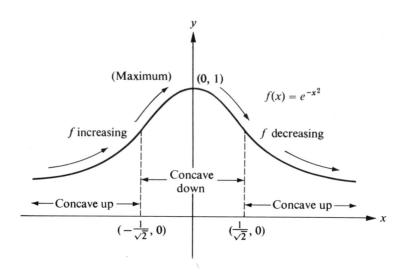

Figure 4.5.2

THEOREM 2.   *Let f be a differentiable function on (a, b) and f'(p) = 0, where a < p < b.*

(1) *If f'(x) < 0 for x < p and f'(x) > 0 for x > p, then f has a minimum at p (and no maximum).*

(2) *If f'(x) > 0 for x < p and f'(x) < 0 for x > p, then f has a maximum at p (and no minimum).*

This theorem is easily verified. For example, if $f'(x) < 0$ for $x < p$, then f is *decreasing* there, and $f(p) < f(x)$ for every $x < p$. Since $f'(x) > 0$ for $x > p$, f is increasing there, and $f(x) > f(p)$ for $x > p$. Hence, in either case, $f(p) < f(x)$ for every x. Thus, $f(p)$ represents a true minimum. Part (2) is verified in exactly the same way. We now give an example which illustrates the use of this rule.

▶ Example 3.   Find the shortest distance from the point $(5, 1)$ to the parabola $y = 2x^2$.

Solution.   In keeping with our recipe for finding maxima, we first draw a picture (Figure 4.5.3). If $p = (x, 2x^2)$ is a general point on the parabola, then the distance from $(5, 1)$ to the point p is given by

$$d[(5, 1), p] = \sqrt{(5 - x)^2 + (1 - 2x^2)^2}.$$

Let

$$f(x) = \sqrt{(5 - x)^2 + (1 - 2x^2)^2}.$$

We wish to minimize f. Now,

$$f'(x) = \tfrac{1}{2}[-2(5 - x) - 8x(1 - 2x^2)][(5 - x)^2 + (1 - 2x^2)^2]^{-1/2}.$$

Observe that

$$[(5 - x)^2 + (1 - 2x^2)^2]^{-1/2} \neq 0,$$

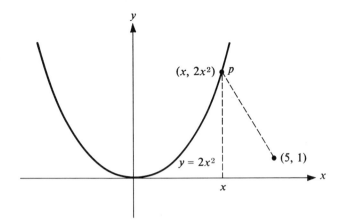

Figure 4.5.3

and moreover this quantity is always positive by definition of the square root. Thus $f' = 0$ precisely when

$$-2(5 - x) - 8x(1 - 2x^2) = 0$$

or equivalently

$$16x^3 - 6x - 10 = 0.$$

It is easy to see that 1 is a root of this equation; when factored, it becomes

$$(x - 1)(16x^2 + 16x + 10) = 0.$$

The expression $16x^2 + 16x + 10$ has no real roots. (The roots are, in fact,

$$\frac{-16 \pm \sqrt{16^2 - 4 \cdot 10 \cdot 16}}{32} .)$$

Thus this expression must always be positive. (Since there is no real x such that $16x^2 + 16x + 10 = 0$ and if $x > 0$, $16x^2 + 16x + 10 > 0$, it follows that $16x^2 + 16x + 10$ can never be negative; otherwise its graph must cross the axis and we would have a contradiction.) Since

$$f'(x) = \tfrac{1}{2}[(x - 1)(16x^2 + 16x + 10)][(5 - x)^2 + (1 - 2x^2)^2]^{-1/2},$$

positive

it is clear that $f'(x) < 0$ for $x < 1$ and $f'(x) > 0$ for $x > 1$. Hence, f has a minimum at $x = 1$ and no maximum. This should be clear since there are points on $y = 2x^2$ which are arbitrarily far from $(5, 1)$. Thus, the minimum distance is equal to

$$f(1) = \sqrt{(5 - 1)^2 + (1 - 2)^2} = \sqrt{17}.$$

In this example, it would be far more complicated to use the second-derivative test.

In our discussion of maxima and minima, we have mainly considered differentiable functions, although Theorem 2 of Section 4.1 admits other functions. Recall from that theorem that if f is a continuous function on $[a, b]$ which is differentiable except at $r_1, \ldots, r_n$ and $f'(x) = 0$ precisely for $x = s_1, \ldots, s_n$, then f has an absolute maximum and minimum on $[a, b]$. The maximum or minimum must occur at one of the following points:

1. $a, b,$
2. $r_1, \ldots, r_n,$
3. $s_1, \ldots, s_n.$

We illustrate this by an example.

▶ Example 4.   Find the maximum and minimum of $f(x) = |x|$ on $[-1, 2]$.

Solution.   The function f is differentiable except at $x = 0$. For $-1 \leq x < 0$, $f'(x) = -1$. For $0 < x \leq 2$, $f'(x) = 1$. Thus, the maximum and minimum must occur at -1, 0, or 2. Since $f(-1) = 1$, $f'(0) = 0$, and $f(2) = 2$, it is clear that 0 is where the minimum occurs and 2 is where the maximum occurs. The graph of f is shown in Figure 4.5.4.

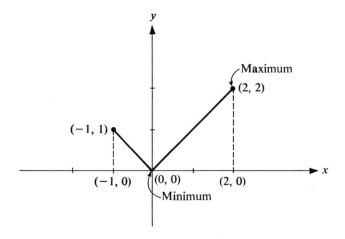

Figure 4.5.4

We shall not discuss this situation for open intervals but the same method is valid. The interested student may wish to write it out in more detail on his own.

Exercises

1. Use the second-derivative test to find the local maxima and minima of the following functions and then sketch a graph of the functions:

    (a) $f(x) = x^2 - 6x + 9$.

    (b) $f(x) = \frac{1}{3}x^3 + \frac{1}{2}x^2 - 12x + 9$.

    (c) $f(x) = x^3 - 3x^2 + 3x - 1$.

    (d) $f(x) = x^4 + 4x^3 + 6x^2 + 4x + 1$.

    (e) $f(x) = e^{-(x/2)^2}$.

    (f) $f(x) = 2x^3 + 3x^2 - 60x + 9$.

    (g) $f(x) = x^4 - 8x$.

    (h) $f(x) = x^2(x - 1)$.

    (i) $f(x) = 2x(x^2 - 1)$.

    (j) $f(x) = 3xe^{-x}$.

    (k) $f(x) = \log(1/x)$.

2. Find the maxima and minima for the following functions. Also sketch a graph of the functions.

    (a) $f(x) = 2|x - 1|$ on $[-1, 2]$.

    (b) $f(x) = (x - 2)|x - 2|$ on $[-2, 2]$.

    (c) $f(x) = |16 - x^2|$ on $[-3, 3]$.

3. If the hypotenuse of a right triangle is 10, what must the sides be in order that the area is a maximum?

4. The slope of a curve at any point (x, y) is given by the equation

    $$\frac{dy}{dx} = 6(x - 1)(x - 2)^2(x - 3)^3(x - 4)^4.$$

    (a) For what value (or values) of x is y a local maximum? Why?

    (b) For what value (or values) of x is y a local minimum? Why?

5. Find the shortest distance from the point $(0, 1)$ to the parabola $y = 4x^2$.

6. A rectangular field to contain a given area A is to be fenced off along a straight river. If no fencing is needed along the river, show that the least amount of fencing will be required when the length of the field is twice its width.

7. Determine a, b, c, d so that the curve whose equation is $y = ax^3 + bx^2 + cx + d$ has a local maximum at $(-1, -3)$ and a local minimum at $(0, -5)$.

8. Find two *positive* numbers whose sum is 50 and such that their product is as large as possible. Can the problem be solved if the product is to be as small as possible? Explain.

*9. Given $f(x) = ax^2 + bx + c$ with $a > 0$. By considering the minimum, show that $f(x) \geq 0$, for all real x, if and only if $b^2 - 4ac \leq 0$. [*Hint:* Suppose that $b^2 - 4ac \leq 0$. Then, $f'(x) = 2ax + b$ and $f'(x) = 0$ yields $x = -b/2a$. Now, $f''(x) = 2a > 0$ for all x. Thus, by the second-derivative test, the point $(-b/2a, f(-b/2a))$ is a minimum point. But,

    $$f\left(-\frac{b}{2a}\right) = -\frac{b^2 - 4ac}{4a} \geq 0 \quad \text{and} \quad f(x) \geq f\left(-\frac{b}{2a}\right) \qquad \text{for all } x.$$

    Hence, $f(x) \geq 0$ for all x. The converse is not difficult to show.]

10. Let $f(x) = (\log x)/x$, if $x > 0$. Describe intervals in which f is increasing, decreasing, concave up, and concave down. Sketch a graph of f.

11. Find the maximum and minimum points for the following functions. Also sketch a graph of the functions.

    (a) $f(x) = 5|x|$ on $[-1, 3]$.

    (b) $f(x) = |x - x^2|$ on $[-2, 2]$.

    (c) $f(x) = \dfrac{x}{1 + |x|}$ on $(-\infty, \infty)$.

(d) $f(x) = \dfrac{|x|}{1 + |x|}$ on $(-\infty, \infty)$.

(e) $f(x) = |x^3|$ on $[-2, 3]$.

(f) $f(x) = |x^3|$ on $(-2, 3)$.

(g) $f(x) = \dfrac{1}{x - x^2}$ on $(0, 1)$.

12. Find the point on the curve $y = \sqrt{x}$ nearest the point $(a, 0)$

  (a) if $a \geq \frac{1}{2}$. (b) if $a < \frac{1}{2}$.

4.6 More Applications of Maxima and Minima

In this section, we further illustrate the theorems dealing with maxima and minima by considering several examples which appear in various applications.

▶ Example 1. Suppose that firm A has a contract to supply 1,000 items a month at a uniform daily rate, and that each time a production run is started, it costs $50. In order to avoid high production costs, the firm decides to produce a large quantity at a time and to store it until the contract calls for delivery. Unfortunately, even storage can be expensive; hence firm A does not want to store so many items that the storage costs exceed the savings due to large production runs. If the cost of storage is 1¢ per item per month, how many items should be made per run so as to minimize total average costs?

  Solution. Let x denote the number of items made in each run and assume that a graph of inventory against time appears as in Figure 4.6.1. The time period required

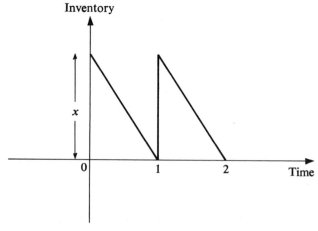

Figure 4.6.1

to sell x items is called an *inventory cycle*. If we assume that demand—and sales—are constant throughout the cycle, then the average inventory during any cycle is $\frac{1}{2}x$ and the average cost of holding inventory is $(\frac{1}{2}x)(0.01)$ dollars per month. Now, the batches of x items will last $\frac{1}{1000}x$ months (since 1,000 items must be supplied per month). The average setup cost will be $50 \div \frac{1}{1000}x$ (since each time a production run is started it costs \$50 and every $\frac{1}{1000}x$ months we have a production run). Thus, the total average cost $C(x)$ is given by

$$C(x) = \frac{x}{2}(0.01) + \frac{50}{x/1,000}$$

$$= \frac{x}{200} + \frac{50,000}{x}.$$

Thus,

$$C'(x) = \frac{1}{200} - \frac{50,000}{x^2},$$

and hence $C'(x) = 0$ when $x^2 = 10,000,000$. Therefore $C'(x) = 0$ when $x \simeq 3,160$ (where $\simeq$ means approximately). Also observe that $C''(x) = 100,000x^{-3}$ and $C''(3,160) > 0$. Thus, by Theorem 1 of Section 4.5, the cost is minimized when 3,160 items are made. This means that firm A should make a little over 3 months' supply at a time.

▶ Example 2. A sheet of paper for a poster contains 18 sq ft. The margins at the top and bottom are 9 in. and at the sides 6 in. What are the dimensions if the printed area is a maximum?

Solution. First, we draw a picture and let x be the length of the poster and y the width of the poster. (See Figure 4.6.2.) Now the area of the total poster is xy;

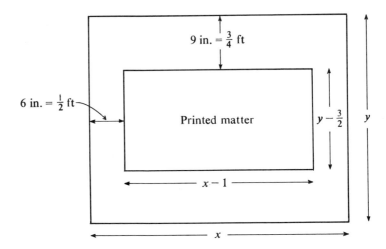

Figure 4.6.2

however, we are told that this is 18 sq ft and thus, $xy = 18$ or $y = 18/x$. We wish to maximize the area A of the printed material. From Figure 4.6.2, we see that

$$A = (x - 1)\left(y - \frac{3}{2}\right) \quad \text{and thus} \quad A(x) = (x - 1)\left(\frac{18}{x} - \frac{3}{2}\right).$$

Thus, our problem has been reduced to maximizing the function $A(x)$ where $1 \le x \le 18$. Now,

$$A'(x) = \frac{18}{x} - \frac{3}{2} + (x - 1)\left(-\frac{18}{x^2}\right)$$

$$= \frac{18}{x} - \frac{3}{2} - \frac{18}{x} + \frac{18}{x^2}$$

$$= -\frac{3}{2} + \frac{18}{x^2},$$

and $A'(x) = 0$ implies that

$$3x^2 = 36,$$

$$x^2 = 12,$$

or $\qquad x = \pm 2\sqrt{3}.$

It is also clear that $A''(x) = -36x^{-3}$ and $A''(2\sqrt{3}) < 0$ and $A''(-2\sqrt{3}) > 0$. Hence, the point $x = 2\sqrt{3}$ maximizes $A(x)$ by Theorem 1 of Section 4.5. (One should check the endpoints $x = 1$ and $x = 18$ to be sure.) The dimensions should be $x = 2\sqrt{3}$ ft and $y = 9/\sqrt{3}$ ft $= 3\sqrt{3}$ ft. Since the numbers $2\sqrt{3}$ and $3\sqrt{3}$ are irrational, in practice one would approximate them in making the poster.

▶ **Example 3.** The cost of fuel in running a locomotive is proportional to the square of the speed and is $25 per hour for a speed of 25 mph. Other costs amount to $100 per hour, regardless of the speed. Find the speed that will make the cost per mile a minimum.

Solution. Let v be the speed and let C denote the total cost per mile. The fuel cost per hour is kv^2, where k is a constant to be determined. Since the cost is $25 per hour for a speed of 25 mph, we see that $25 = k(25)^2$ or $k = \frac{1}{25}$. Thus, the fuel cost per hour $C_{\text{fuel}} = \frac{1}{25}v^2$.

The total cost C in dollars per mile is given by

$$C(v) = \frac{\text{cost in dollars per hour}}{\text{speed in miles per hour}}$$

$$= \frac{\frac{1}{25}v^2 + 100}{v}.$$

Thus, we want to minimize the quantity $C(v) = \frac{1}{25}v + 100/v$, where $v > 0$. Now, $C'(v) = \frac{1}{25} - 100/v^2$, and thus $C'(v) = 0$ implies that

$$v^2 = 2,500 \quad \text{or} \quad v = \pm 50.$$

[We reject -50 since speed is positive.] Since $C''(v) = +200v^{-3}$ and $C''(50) > 0$, it follows that $v = 50$ produces a minimum value for C by Theorem 1 of Section 4.5. Hence the most economical speed at which to run the locomotive is 50 mph.

▶ **Example 4.** In experiments dealing with radioactive substances, it is often necessary to consider the *exponential distribution function f* defined by

$$f(t) = 1 - e^{-kt},$$

where the time $t > 0$ and where $k > 0$. Find the local maxima and minima of f, if any, and draw a graph of the function when $k = 1$. Also construct a graph of the derivative of f.

Solution. If $f(t) = 1 - e^{-kt}$, then $f'(t) = ke^{-kt}$. Now $f'(t) \neq 0$ for any value of t. We do note that as t becomes very large, $f'(t)$ tends toward 0. Since f' exists for all t, the local maxima and minima must occur at a point where $f'(t) = 0$. Thus, *no* local maxima and minima exist. We observe that $f(0) = 0$ and if $k = 1$, $f'(t) = e^{-t}$. Since $e^{-t} > 0$ for all t, the function $f(t) = 1 - e^{-t}$ is increasing for all values of t. Also, as $t \to \infty$, $f(t) \to 1$. The graph of $f(t)$ for $k = 1$ is shown in Figure 4.6.3.

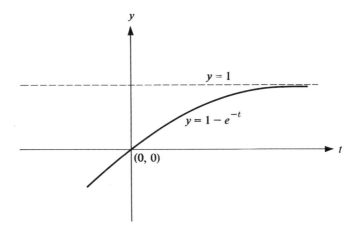

Figure 4.6.3

For $k = 1$, $f'(t) = e^{-t}$ and $f''(t) = -e^{-t}$. Notice that $f''(t) < 0$ for *all* values of t; hence the curve f' is strictly decreasing and does not have any local maximum or minimum points. The graph of f' is shown in Figure 4.6.4.

▶ Example 5. (*An Application to Biology**) By use of delicate machines it can be demonstrated that, during coughing, the diameter of the trachea and main bronchi decreases. B. F. Visser devised a theoretical scheme to further verify this phenomenon. The first assumption made is that after a deep inspiration the glottis is closed. He assumes that the radius of the airways (i.e., trachea and bronchi) is a linear function of the pressure. That is,

$$r = r_0 - \alpha P, \qquad (4.6.1)$$

where r is the radius at a pressure P above atmospheric pressure, r_0 is the radius at atmospheric pressure, and α is a constant. A law due to Poiseuille states that the resistance R offered by the air passages is inversely proportional to the fourth power of the radius r, that is

$$R = k/r^4,$$

---

* An example taken from J. G. Defares and I. N. Sneddon, *The Mathematics of Medicine and Biology* (North-Holland, Amsterdam, 1973), 2nd edition.

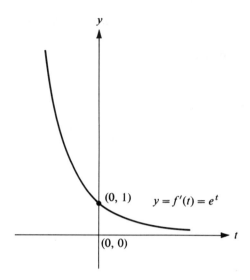

Figure 4.6.4

where k is the constant of proportionality. After the glottis is opened, the amount of air passing a certain point per unit time (known as the rate of flow and denoted by G) is given by

$$G = \frac{P}{R} = \frac{Pr^4}{k} = \frac{r^4(r_0 - r)}{\alpha k} \qquad \text{(from (4.6.1)).} \qquad (4.6.2)$$

The rate of flow is *not* the velocity, which is related to G by

$$v = G/\pi r^2. \qquad (4.6.3)$$

Substituting for G, we obtain

$$v = \frac{r^2(r_0 - r)}{\pi \alpha k}. \qquad (4.6.4)$$

Observe that

$$\frac{dv}{dr} = \frac{2rr_0 - 3r^2}{\pi \alpha k} = -\frac{3r}{\pi \alpha k}(r - \tfrac{2}{3}r_0) \qquad (4.6.5)$$

and

$$\frac{d^2v}{dr^2} = \frac{2r_0 - 6r}{\pi \alpha k}. \qquad (4.6.6)$$

From (4.6.5) we see that when $r = \tfrac{2}{3}r_0$,

$$\frac{dv}{dr} = 0 \quad \text{and} \quad \frac{d^2v}{dr^2} = \frac{-2r_0}{\pi \alpha k} < 0.$$

(α, k, and r_0 must be positive.) Hence, v has its maximum value when $r = \tfrac{2}{3}r_0$. We may interpret this result as follows: with decreasing radius the velocity increases during coughing until a maximum velocity is reached at a radius having a value two-thirds of the resting value. When there is a further decrease of the radius the velocity decreases (because of the increased resistance offered by the smaller radius). We might point out that maximum velocity in coughing is clearly desirable from a functional standpoint.

This example illustrates a common phenomenon in the application of calculus to biology, namely the use of mathematics to theoretically verify a result already conjectured on the basis of experimental data.

Exercises

1. What is the *maximum slope* of the curve $y = 6x^2 - x^3$?

2. A poster is to contain 136 sq ft of printed matter with margins of 1 ft at the top and bottom and 2 ft on each side. Find the dimensions of the poster if the total area is to be a minimum.

3. Find the number that exceeds its square by the greatest amount. [*Hint:* Let x be the number. Set up a function which expresses the difference between x and its square, x^2.]

4. Three repetitions of an experiment yield the numbers a_1, a_2, a_3 for a certain physical quantity x. What value should we take for x in order to minimize the sum of the squares of the deviations,
$$(x - a_1)^2 + (x - a_2)^2 + (x - a_3)^2?$$

5. The product of two positive numbers is 36. Find the numbers if their sum is to be a minimum.

6. What is the area of the largest rectangle that can be inscribed in the first quadrant under the parabola $y = 4 - x^2$?

7. A study showed that the reaction time of individuals to a particular stimulus varied with the age of the individual according to the function
$$t = 0.02(x^2 - 40x + 500)^{1/2},$$
where t is the reaction time in seconds and x is the age of the individual in years $(5 < x < 60)$. Determine the minimum reaction time and the age at which it occurs.

8. A snowmobile company finds that there is a net profit of $10 for each of the first 1,000 snowmobiles produced each week. For each snowmobile over 1,000 produced, there is 0.02¢ less profit per snowmobile. How many snowmobiles should be produced each week to net the greatest profit? (The company can sell all the machines it produces.) [*Hint:* See the hint for Exercise 3, Section 4.2.]

9. The total cost of producing x radio sets per day is $\$(\frac{1}{4}x^2 + 35x + 25)$, and the price per set at which they may be sold is $\$(50 - \frac{1}{2}x)$.

   (a) What should be the daily output to obtain a maximum total profit?

   (b) Show that the cost of producing a set has a local minimum. (Thus $p = 50 - \frac{1}{2}x$ is the demand function with price as a function of the amount.)

10. A piece of wire 100 in. long is to be cut in two and the two pieces bent to form a square and a circle. (See Figure 4.6.5.) What is the smallest area that

Figure 4.6.5

can be formed in this manner? What is the largest area? (Be careful.)

11. What is the largest area that can be achieved by a rectangle fitted under the curve $y = e^{-x^2}$ as shown in Figure 4.6.6?

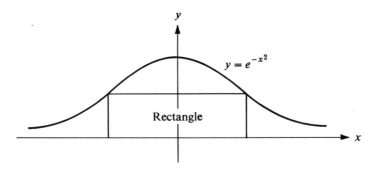

Figure 4.6.6

12. A flower bed is to be formed in the shape of a semicircle on top of a rectangle (Figure 4.6.7). The flower bed is then to be fenced in with 200 ft of available

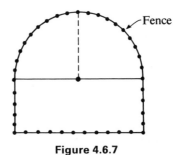

Figure 4.6.7

fence. What should the dimensions of the flower bed be to yield the largest area?

13. A printing company plans to have 40 sq in. of printed matter per page in a particular book. Each page is to have margins of $1\frac{1}{4}$ in. on the sides and $1\frac{1}{2}$ in. at the top and bottom. What are the most economical dimensions for the pages in terms of the cost of the paper?

14. A school group makes plans for an excursion from their home town in New Jersey to New York City. The bus company agrees to take the contract if they are guaranteed that at least 30 students will go. There is a possibility that as

many as 200 could go. The fare is to be \$50 a person if 30 go, and will decrease by 40¢ per person for every person above the minimum of 30 who goes. What number of people will give the bus company maximum revenue? Will they be able to get maximum revenue? What is the revenue if everyone eligible goes?

15. We wish to consider a simple model of inventory. Let D equal the total number of items to be ordered in a year (known). Let x be the number of items in each order. Let A, B, C be positive constants. Then

$$\text{purchase cost} = AD,$$
$$\text{order cost} = B(D/X),$$
$$\text{holding cost} = Cx.$$

The total cost of keeping inventory for a year, $T(x)$, is equal to purchase cost + order cost + holding cost when $1 \leq x \leq D$. Find the order size (in terms of A, B, C, D) that makes the total cost smallest.

16. A manufacturer of widgets does a cost control analysis on his product and discovers the following. Since the government is furnishing the widget press and depreciation allowances are generous, there is no overhead. If x is the number of widgets produced, then the cost per widget, $f(x)$, due to labor costs, raw materials, maintenance, and miscellaneous production costs is $f(x) = 400 \log (1 + x) + (29 - x)^2$. If production capacity is 25 widgets, at what point is the cost per widget lowest? (Use the logarithm table in the Appendix, Table A2.)

17. An architect wishes to incorporate 100 ft of existing stone wall, laid in a straight line, into a boundary around a rectangular garden. He has an additional 200 ft of fence. How should he lay out the remaining fence in order to maximize the area of the garden? (Be careful.)

18. The "information content" or "entropy" of a binary source (such as a telegraph that submits dots and dashes) whose two values occur with probability p and $1 - p$ is defined as

$$f(p) = -p \log p - (1 - p) \log (1 - p),$$

where $0 < p < 1$. Show that f has a maximum at $p = \frac{1}{2}$. The practical significance of this result is that for maximum flow of information per unit time, dots and dashes should, in the long run, appear in equal proportions.

19. The demand q for bellbottom trousers is given by $q = 1/x^3$, where x is the selling price. If the article costs 20¢ apiece to manufacture, find the selling price that yields a maximum profit.

20. Show that the sum of the square of any number and the square of its reciprocal is always greater than or equal to 2.

21. Find the area of the largest rectangle with lower base on the x axis and the upper vertices on the curve $y = 1 - x^2$.

*22. Show that for fixed a, the function

$$y = \left(a - \frac{1}{a} - x \right) (4 - 3x^2)$$

has just one maximum and one minimum, and that the difference between them is

$$\frac{4}{9}\left(a + \frac{1}{a}\right)^3.$$

If a is allowed to vary, find the least value of this difference.

4.7 Chapter 4 Summary

In this section we summarize the most important results of this chapter. Let f be defined on the interval specified below.

1. If f is continuous on $[a, b]$, it has both a maximum and a minimum on $[a, b]$. Moreover, the maximum and minimum occur either at

    (a) Points where f' does not exist.

    (b) Points where $f'(x) = 0$.

    (c) The endpoints a, b.

2. (a) If $f'(x) > 0$, then f is strictly increasing.

    (b) If $f'(x) < 0$, then f is strictly decreasing.

3. (a) If $f''(x) > 0$, then f is concave up.

    (b) If $f''(x) < 0$, then f is concave down.

4. *Second-derivative test.* Let $f'(c) = 0$.

    (a) If $f''(c) < 0$, then f has a local maximum at c.

    (b) If $f''(c) > 0$, then f has a local minimum at c.

    (c) If $f''(c) = 0$, no information. (See 5.)

5. *First-derivative test.* Let f be differentiable on (a, b) where $f'(c) = 0$ and $a < c < b$.

    (a) If $f'(x) < 0$ for $x < c$ and $f'(x) > 0$ for $x > c$, then f has a minimum at c.

    (b) If $f'(x) > 0$ for $x < c$ and $f'(x) < 0$ for $x > c$, then f has a maximum at c.

Review Exercises

In all exercises, you must use some test to justify whether or not you have a maximum or minimum.

1. For each of the following functions: (i) find the intervals where the function is increasing and decreasing; (ii) find the intervals where the function is concave up and concave down; (iii) find all maxima and minima; (iv) sketch a graph of the function.

(a) $f(x) = x^3 + 6x^2 - 15x + 5.$

(b) $f(x) = 3 + x^{2/3}.$

(c) $f(x) = 1 - x^{2/3}, \quad -1 \le x \le 1.$

(d) $f(x) = \frac{1}{4}x^4 - \frac{1}{2}x^2 + 1, \quad -2 \le x \le 2.$

(e) $f(x) = e^{-(x/3)^2}.$

(f) $f(x) = (\log x)/x, \quad x > 0.$

2. Suppose $P(x)$, the price for the quantity x, is given by $P(x) = 6 - x^2$. Find the maximum total revenue if demand varies so that $0 \le x \le \frac{5}{2}$. [*Hint:* Recall from Example 4, Section 3.4 that the total revenue is given by $xP(x)$.]

3. Determine the maximum total revenue if $P(x) = 8/(4 + x^2)$, $x \ge 0$. Sketch the graph of $P(x)$.

4. An apple grower observes that if 25 apple trees are planted per acre, the yield is 450 apples per tree, and that the yield per tree decreases by 10 for each additional tree per acre. How many trees should be planted per acre to obtain the maximum crop?

5. A telephone company has a profit of \$2 per telephone when the number of telephones in the exchange is not over 10,000. The profit per telephone decreases by 0.01¢ for each telephone over 10,000. What is the largest possible profit?

6. If all inputs (i.e., labor, material, machinery, plant) to the production process are held fixed in amount except one, called the *variable input* x, and if x is increased over the interval $a \le x \le b$, the *law of variable proportions* states that the total output $Q(x)$ will increase first with positive acceleration (i.e., $\frac{d^2Q}{dx^2} > 0$) and then with negative acceleration (i.e., $\frac{d^2Q}{dx} < 0$). The graph of Q will be concave upward as x increases from 0 to an *inflection point*, from which point it will be concave downward. This inflection point is called, in economics, *the point of diminishing returns.* Suppose that

$$Q(x) = 0.1x + 0.35x^2 - 0.01x^3, \qquad 0 \le x \le 23.$$

How many units of x can be employed before diminishing returns set in?

7. At a certain point c where y' and y'' are continuous, we have $y' = 0$ and $y'' = 2$. Must $y = f(x)$ have a local maximum or local minimum at c? What if $y'' = 0$? What if $y'' = -2$? Explain.

8. Determine the coefficients a, b, c, d so that the curve whose equation is

$$y = ax^3 + bx^2 + cx + d$$

has a maximum at $(-1, 10)$ and an inflection point at $(1, -6)$.

9. A company wishes to manufacture at minimum cost a closed wooden box with square base and a volume of 12 cu ft. Find the dimensions of the box if the lumber is 5¢ per square foot. [*Hint:* Volume of box = length × width × height.]

10. A farmer wishes to fence off a piece of land in the shape of a triangle, using the river as one side. Assuming that the two sides of the fence have equal length and that the farmer has only 100 ft of fence, find the maximum area that he can enclose.

11. The Profoundly Pessimistic Political Party has found that for each demonstration staged, they receive $3,000 in contributions. They also notice that the cost for a demonstration rises cubically (more precisely, x demonstrations cost $250x^3$). What is the most profitable number of demonstrations to hold, given that they can hold at most 5?

12. Suppose that it costs a manufacturer $y = 2 + 3x$ dollars to produce x units per week. Assume that the price, P dollars per item, at which he can sell x items per week is $P = 15 - x$.

   (a) What level of production maximizes his profit?

   (b) What is the corresponding price?

   (c) What is his profit (per week) at this level of production?

   (d) If a tax of $1 per item sold is imposed on this product, and the manufacturer still wishes to maximize his profit, at what price should he sell each item? Comment on the difference between this price and the price before tax.

13. (a) Let $f(x) = x^{1/2}$. Determine the equation of the secant line passing through the points $(0, 0)$ and $(4, 2)$. Find the point c, $0 < c < 4$, such that the slope of curve at c is the slope of this secant line. Illustrate geometrically.

   (b) Can the mean value theorem be applied to the function $f(x) = |2x - 1|$ over the interval $[-3, 3]$? Why?

A P P E N D I X

*4.8 Note on Determining the Sign of a Function

In the graphing problems one is faced with the task of deciding where f' and f'' are positive and negative. We present two methods for doing this.

(1) A METHOD FOR POLYNOMIALS

Let

$$g(x) = (x - r_1)^{n_1}(x - r_2)^{n_2} \cdots (x - r_k)^{n_k}.$$

We are assuming that the r's are all distinct (different) and that they have been placed in increasing order, i.e., $r_1 < r_2 < \cdots < r_k$. Then the sign of g changes at r_1 if n_1 is odd and does not change if n_1 is even. The same is true at a general r_i. The sign of g changes as we cross r_i if n_i is odd and does not change if n_i is even. Let us consider an example. Let

$$g(x) = [x - (-1)]^5(x - 2)^6(x - 7)(x - 12)^3(x - 14)^4.$$

It is easy to see that $g(-10) < 0$ and that $g(x) < 0$ for $x < -1$. [If $x < -1$, the terms $(x - 2)^6$ and $(x - 12)^4$ are positive and the three remaining terms negative, hence the result.] Since $n_1 = 5$ is odd, the sign of g changes at (-1), and hence $g(x) > 0$ for $-1 < x < 2$. The sign of $n_2 = 6$ is even, and so the sign of g does not change at 2. Thus, $g(x) > 0$ for $2 < x < 7$. The reader can check the rest of the points against Figure 4.8.1.

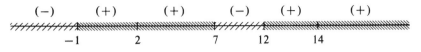

Figure 4.8.1

If this means of determining the sign seems complicated, we may use a second method, which involves a little more work, but is simpler to apply.

(2) A METHOD FOR GENERAL FUNCTIONS

Let $g(x) = 0$ for $x = x_1, \ldots, x_n$ and at no other points. We shall assume g is continuous. Moreover, we assume the zeros of g are in increasing order, i.e., $x_1 < x_2 < \cdots < x_n$. See Figure 4.8.2. Now, simply evaluate g at points a_i, where

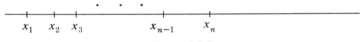

Figure 4.8.2

$x_i < a_i < x_{i+1}$; i.e., evaluate g somewhere between each adjacent pair of x_i's. The sign of g does *not necessarily* change at each x_i, but it cannot change anyplace else. So to decide whether g is positive or negative between x_i and x_{i+1}, we need to check it at a single point. [You also have to check the interval $(-\infty, x_i)$ and the interval (x_n, ∞).]

▶ Example. If
$$g(x) = (e^x - 1) \log\left(\frac{1 + x^2}{10}\right),$$
determine the sign of $g(x)$ on $(-\infty, \infty)$.

Solution. First we note that g is defined and continuous on $(-\infty, \infty)$. It should be clear by inspection and a simple calculation that $g(x) = 0$ precisely for $x = -3, 0, 3$. Since

$$g(-4) = \left(\frac{1}{e^4} - 1\right) \log\left(\tfrac{17}{10}\right) < 0; \qquad g(-1) = \left(\frac{1}{e} - 1\right) \log\left(\tfrac{1}{5}\right) > 0;$$

$$g(1) = (e - 1) \log\left(\tfrac{1}{5}\right) < 0; \qquad g(5) = (e^5 - 1) \log\left(\tfrac{13}{5}\right) > 0;$$

g must behave as in Figure 4.8.3.

Figure 4.8.3

Integration

5.1 Antiderivatives

Until now, we have been dealing with the problem of finding the derivative f' of a given function f. It is also natural to ask the following question: Given any function f, can we find a function g whose derivative is f? For example, if $f(x) = x^2$, does there exist a function g such that $g'(x) = x^2$? Clearly, if $g(x) = x^3/3$, then $g'(x) = x^2 = f(x)$. Now, is this the only function g satisfying $g'(x) = x^2$? A little thought yields a negative answer. Note that if $g(x) = x^3/3 + 8$, then $g'(x) = x^2$ (since the derivative of the constant function is zero). In fact, we observe that there are infinitely many functions satisfying $g'(x) = x^2$, namely, functions of the form $x^3/3 + c$, where c is an arbitrary constant. This problem brings us into the realm of *integral calculus*. As we shall see in later sections, the techniques we are now developing have applications to a wide range of important problems in physics, biology, economics, psychology, and business. For example, in a sense that we shall make clear, the antiderivative of acceleration is velocity; of velocity is distance; of marginal revenue is total sales revenue; of marginal cost is total cost; of area is volume; of learning rate is a knowledge function. In particular, antiderivatives will provide us with an easy method for calculating the area under a curve $y = f(x)$. See Figure 5.1.1.

Let us make the following definition.

DEFINITION 1. The function F is called an antiderivative *of the function f on an interval I if and only if F is differentiable on I and $F'(x) = f(x)$ for $x \in I$. (I may be a closed, open, or half-open interval; see Section 1.1.)*

▶ **Example 1.** Let $f(x) = 4x^2$. We assert that $F(x) = \frac{4}{3}x^3$ is an antiderivative of $f(x)$, since $F'(x) = 4x^2$. Note that if c is any arbitrary constant, then $F(x) = \frac{4}{3}x^3 + c$ is also an antiderivative of $4x^2$, since $F'(x) = 4x^2$.

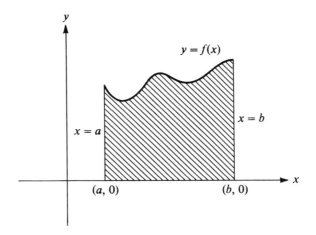

Figure 5.1.1

▶ Example 2. Let $f(x) = ax^n$ for $n \neq -1$, where a is a constant. It is clear that $F(x) = ax^{n+1}/(n + 1)$ is an antiderivative of $f(x)$ since

$$F'(x) = (n + 1)ax^n/(n + 1) = ax^n.$$

In like manner, if c is any arbitrary constant, then

$$F(x) = ax^{n+1}/(n + 1) + c$$

is also an antiderivative. Thus there are infinitely many antiderivatives of $f(x)$.

▶ Example 3. Let $f(x) = 1/x$. If $x > 0$, then $F(x) = \log x$ is well defined and $F(x)$ is an antiderivative for $1/x$ when $x > 0$. The function $f(x) = 1/x$ is not defined at zero. However, it is defined for $x < 0$, and we can ask whether there is a function G, defined for $x < 0$, with the property that $G'(x) = 1/x$ for $x < 0$. Let us make a guess and try the function $G(x) = \log(-x)$ for $x < 0$. Since the log is defined for positive numbers, G makes sense. If we differentiate, using the methods of Section 3.6, we find that $G'(x) = -1/(-x) = 1/x$. Thus G is indeed an antiderivative of $1/x$ on $(-\infty, 0)$. At this point we have found an antiderivative $F(x) = \log x$ for $1/x$ on $(0, \infty)$ and an antiderivative $G(x) = \log(-x)$ for $1/x$ on $(-\infty, 0)$.

In the interests of economy, simplicity, and organization, we may combine F and G as follows. Set $H(x) = \log |x|$ for $x \neq 0$. Then $H'(x) = 1/x$ for $x \neq 0$. In other words, $H(x) = \log |x|$ is an antiderivative of $1/x$ for $x \neq 0$.

▶ Example 4. Let $f(x) = e^x$. Then $F(x) = e^x$ is an antiderivative of $f(x)$ since $F'(x) = e^x = f(x)$. Clearly $e^x + c$, where c is an arbitrary constant, is also an antiderivative of e^x.

▶ Example 5. Let $f(x) = x^2 + x^3 + 1/x + e^x$. Find an antiderivative of $f(x)$. Note that

$$F(x) = x^3/3 + x^4/4 + \log |x| + e^x$$

is an antiderivative of $f(x)$ since

$$F'(x) = x^2 + x^3 + 1/x + e^x.$$

If we add any arbitrary constant to $F(x)$, we still have an antiderivative of f. Observe that we could find F by first finding an antiderivative of x^2, x^3, $1/x$, and e^x, and then adding them together. In the next section, we show that this is a general property for antiderivatives; namely, if F, G, H are antiderivatives of f, g, h, then $F + G + H$ is an antiderivative of $f + g + h$.

It is now natural to ask if there is a relation between the antiderivatives of a function. For example, $x^3/3 + 7$ and $x^3/3 + 2$ are both antiderivatives of x^2. Notice that these two antiderivatives differ by the constant, 5. We shall see now that this is true in general; any two antiderivatives of a function on a given interval can differ only by a constant. Thus, apart from constant terms, a function can have only one antiderivative on a given interval. We state this result in the form of a theorem.

THEOREM 1. If F and G are two antiderivatives of a function f on an interval I, then there is a constant C such that

$$F(x) = G(x) + C, \qquad x \in I.$$

This theorem simply states that any two antiderivatives of a function differ by a constant. Observe that it also tells us that if F is any antiderivative of f, then *all* antiderivatives of f must be of the form $F + C$.

The verification of the above theorem is quite straightforward. Let

$$H(x) = F(x) - G(x),$$

where $x \in I$. Since $F'(x) = G'(x) = f(x)$, then $H'(x) = 0$. By the corollary to the mean value theorem (see Section 4.3), $H(x) = C$, where C is a constant. Hence, $H(x) = F(x) - G(x) = C$, the desired result.

At this stage, we point out why there is good reason for us to be concerned with intervals when discussing antiderivatives. Consider the following function:

$$F(x) = \begin{cases} \dfrac{1}{x} + 5, & x > 0, \\ \dfrac{1}{x}, & x < 0. \end{cases}$$

It is easy to check that F is an antiderivative of $-1/x^2$ for $x \neq 0$. However, $G(x) = 1/x$ for $x \neq 0$ is also an antiderivative of $-1/x^2$. Moreover, the functions F and G do not differ by a constant. [Of course, the functions F and G do differ by one constant on $(-\infty, 0)$ and they differ by another constant on $(0, \infty)$. However $F(x) \neq G(x) + c$ for all $x \neq 0$ where c is a constant.] Hence $-1/x^2$ has two anti-derivatives which do not differ by a constant. Thus intervals do play a crucial role in the last theorem.

You should also be aware of the fact that there are functions which have *no* antiderivatives. An example is the function

$$f(x) = \begin{cases} -1, & -1 \leq x < 0, \\ 0, & x = 0, \\ +1, & 0 < x \leq 1. \end{cases}$$

We discuss this topic more fully in an appendix to this chapter.

At this point, it would be convenient to have a notation for antiderivatives.

DEFINITION 2. *If F is an antiderivative of f, we write*

$$\int f(x)\, dx = F(x) + C.$$

For example,

$$\int x^2\, dx = \frac{x^3}{3} + C,$$

$$\int t\, dt = \frac{t^2}{2} + C,$$

$$\int e^x\, dx = e^x + C.$$

Sometimes the symbol $\int f(x)\, dx$ is called an *indefinite integral* of f. The symbols dx and dt contribute little, if anything, in these examples. However, they become useful when the expression for a function involves unspecified constants. We refer to $f(x)$ as the *integrand*.

Hence,

$$\int x^2 t\, dx = \frac{x^3 t}{3} + C,$$

since dx singles out x as the variable, while

$$\int x^2 t\, dt = \frac{x^2 t^2}{2} + C,$$

since dt singles out t as the variable. Thus, dx tells us what the variable is in each case. In the above example, we assume that t is *not* a function of x and x is *not* a function of t.

We now construct a table similar to Table 3.3.1 giving us antiderivatives of several well-known functions. (Table 5.1.1.) Note that each entry can be checked by verifying that the derivative of the right-hand side yields the function on the left.

TABLE 5.1.1	$f(x)$	$\int f(x)\, dx$
1.	K, where K is a constant	$Kx + C$
2.	$x^n, n \neq -1$	$\dfrac{x^{n+1}}{n+1} + C$
3.	$1/x$	$\log |x| + C$
4.	e^x	$e^x + C$

Exercises

1. Determine whether the following functions have antiderivatives on the intervals indicated. If so, find them.

   (a) $f(x) = x^3$ on the interval $[-1, 1]$.

   (b) $f(x) = 1/x^2$ on the interval $(0, 1)$.

   (c) $f(x) = x^{-1/2}$ on the interval $(0, 5]$.

   (d) $f(x) = e^x$ on the interval $[-5, 5]$.

   (e) $f(x) = 5x$ on the interval $(-\infty, \infty)$.

   (f) $f(x) = 1/x^4$ for $x \neq 0$.

2. Calculate each of the indicated antiderivatives.

   (a) $\displaystyle\int x^{3/2} \, dx.$

   (b) $\displaystyle\int x^{-3/2} \, dx.$

   (c) $\displaystyle\int (t^2 + t) \, dt.$

   (d) $\displaystyle\int \left(x^2 + \frac{1}{x} + e^x \right) dx.$

   (e) $\displaystyle\int (1 - x)x^{1/2} \, dx.$

   (f) $\displaystyle\int e^{\log x} \, dx.$

   (g) $\displaystyle\int (x^2 + 1)^2 \, dx.$

   (h) $\displaystyle\int \log (e^x) \, dx.$

   (i) $\displaystyle\int \left(12x^5 - \frac{5}{x^2} + 3 \right) dx.$

   (j) $\displaystyle\int (3x - 5e^x) \, dx.$

   (k) $\displaystyle\int (x^4 - 2e^x) \, dx.$

   (l) $\displaystyle\int (x^2 - 1)x^{5/2} \, dx.$

3. Let $h(x) = \sqrt{1 - x^2}$.

   (a) Find $h'(x)$.

   (b) What is $\int [-x(1 - x^2)^{-1/2}] \, dx$?

4. Let $g(x) = e^{4x + 9x^2}$.

   (a) Find $g'(x)$.

   (b) Evaluate $\int (4 + 18x)e^{4x + 9x^2} \, dx$.

5. Evaluate the following:

   (a) $\displaystyle\int \frac{x^3 + 1}{x^2} \, dx.$

   (b) $\displaystyle\int \frac{dt}{t\sqrt{t}}.$

   (c) $\displaystyle\int (x^2 - \sqrt{x}) \, dx.$

   (d) $\displaystyle\int \left(\sqrt{t} + \frac{1}{\sqrt{t}} \right) dt.$

(e) $\displaystyle\int \left(t^{5/2} - \frac{1}{t^{2/5}} \right) dt.$

(f) $\displaystyle\int at^{9000}\, dt$, where a is a constant.

6. Let $g(x) = \frac{1}{3}(2x + 1)^{3/2}$.

   (a) Find $g'(x)$.

   (b) Evaluate $\displaystyle\int (2x + 1)^{1/2}\, dx$.

*7. Let

$$f(x) = \frac{1 + |x|}{x} \quad \text{and} \quad g(x) = \frac{1}{x}.$$

   (a) Show that f and g are antiderivatives of $-1/x^2$, $x \neq 0$.

   (b) Show that there does not exist a constant k such that $f(x) = g(x) + k$ whenever $x \neq 0$.

   (c) Does this contradict the theorem about antiderivatives stated in this section? Why?

5.2 Properties of Antiderivatives

In general, the problem of finding the antiderivative of a given function is more difficult than the problem of differentiating a given function. It is convenient to introduce, at this time, some techniques for finding antiderivatives.

First, we discuss properties of antidifferentiation analogous to those given in Sections 3.4–3.6 for derivatives.

PROPERTY 1.   If F is an antiderivative of f, then kF is an antiderivative of kf. Equivalently,

$$\int kf(x)\, dx = k \int f(x)\, dx.$$

Proof: Simply observe that

$$\frac{d}{dx}(kF) = k\frac{dF}{dx} = kf,$$

since $F'(x) = f(x)$. Hence kF is an antiderivative of kf.

▶ Example 1.   Find $\int 3x^2\, dx$.

   Solution.

$$\int 3x^2\, dx = 3 \int x^2\, dx = 3 \cdot \frac{x^3}{3} + C = x^3 + C,$$

where C is an arbitrary constant.

PROPERTY 2. If F and G are antiderivatives of f and g, respectively, then $F \pm G$ is an antiderivative of $f \pm g$. Equivalently,

$$\int (f \pm g)(x)\, dx = \int f(x)\, dx \pm \int g(x)\, dx.$$

[*Note:* This rule is valid for 3, 4, or more functions as illustrated in Example 2.]

Proof: We must show that

$$\frac{d}{dx}[F(x) \pm G(x)] = f(x) \pm g(x).$$

However,

$$\frac{d}{dx}[F(x) \pm G(x)] = \frac{d}{dx}F(x) \pm \frac{d}{dx}G(x) = f(x) \pm g(x),$$

since $F' = f$ and $G' = g$.

▶ Example 2. Find $\int (3x^2 + 2x^3 + x)\, dx$.

Solution. By Property 2,

$$\int (3x^2 + 2x^3 + x)\, dx = \int 3x^2\, dx + \int 2x^3\, dx + \int x\, dx$$

$$= 3 \int x^2\, dx + 2 \int x^3\, dx + \int x\, dx.$$

By Property 1,

$$= 3 \cdot \frac{x^3}{3} + 2 \cdot \frac{x^4}{4} + \frac{x^2}{2} + C$$

$$= x^3 + \frac{x^4}{2} + \frac{x^2}{2} + C,$$

where C is an arbitrary constant.

▶ Example 3. Find a function f satisfying the conditions that $f'(x) = x^2 + x + 1$ and $f(0) = 1$.

Solution. In order to find f, we must first find an antiderivative of $x^2 + x + 1$. Now,

$$\int (x^2 + x + 1)\, dx = \int x^2\, dx + \int x\, dx + \int dx = \frac{x^3}{3} + \frac{x^2}{2} + x + C.$$

Thus, $f(x) = x^3/3 + x^2/2 + x + C$. In order to find the function satisfying $f(0) = 1$, we observe, after substituting 0 for x, that $f(0) = C$, and hence $C = 1$. The function of f satisfying the two given conditions is then given by

$$f(x) = \frac{x^3}{3} + \frac{x^2}{2} + x + 1.$$

There is really no direct analog to the theorems concerning differentiation of products and quotients for antiderivatives. Note, however, that if F is an antiderivative of f and G is an antiderivative of g, then $F \cdot G$ is not necessarily an antiderivative of $f \cdot g$. The following example demonstrates this.

▶ Example 4.   Let $f(x) = x^2$ and $g(x) = x$. Then $F(x) = x^3/3$ and $G(x) = x^2/2$. Hence, $F(x)G(x) = x^5/6$. Now, $f(x) \cdot g(x) = x^3$ and an antiderivative of x^3 is $x^4/4$. Clearly, $x^5/6$ and $x^4/4$ do not differ by only a constant. Thus, $F \cdot G$ is certainly not an antiderivative of $f \cdot g$.

We end this section with an example typical of the use of antiderivatives for applications.

▶ Example 5.   Find the equation of the curve $y = f(x)$ through the point $(2, -3)$ with slope at the point (x, y) given by $2x - 3$.

Solution.   Since the slope at any point (x, y) is given by $2x - 3$, we must have $\dfrac{dy}{dx} = f'(x) = 2x - 3$. Thus, we must find an antiderivative of $2x - 3$ such that the curve goes through the point $(2, -3)$. Thus,

$$y = \int (2x - 3)\, dx = x^2 - 3x + C.$$

To find C, we use the fact that when $x = 2$, $y = -3$. Hence, $-3 = 4 - 6 + C$ or $C = -1$ and the equation of the desired curve is given by

$$y = x^2 - 3x - 1.$$

Exercises

In Exercises 1–11, evaluate the following:

1. (a) $\displaystyle\int (x^2 - 3x + 9)\, dx.$ (b) $\displaystyle\int (x + e^x)\, dx.$

    (c) $\displaystyle\int (5 - 4x)\, dx.$ (d) $\displaystyle\int \left(4x + \frac{1}{x}\right) dx.$

2. $\displaystyle\int (3x^{1/2} + 5x^{5/2} + x^{1/19})\, dx.$ 3. $\displaystyle\int (x^{-2} + 3x^{-3})\, dx.$

4. $\displaystyle\int \left(17x^6 - x^5 + x^2 - \frac{1}{x}\right) dx.$ 5. $\displaystyle\int (-x^{-1/3} + x^{-1/6} - x^{-1/9})\, dx.$

6. $\displaystyle\int (x^{-1} - 3e^x)\, dx.$ 7. $\displaystyle\int (2e^{-x} + e^x)\, dx.$

8. $\displaystyle\int \left(4e^{-x} - \frac{5}{x}\right) dx.$ 9. $\displaystyle\int \left(\frac{1}{x} + \frac{3}{x^2} + x^{1/3} - x^{-2/3}\right) dx.$

10. $\displaystyle\int [(x + 2)(x^2 - 1)]\, dx.$ 11. $\displaystyle\int x^2(x^2 - 4)\, dx.$

12. Find a function f satisfying the conditions that
    (a) $f'(x) = x^2 - x$, $f(0) = 1$.
    (b) $f'(x) = x(x - 1)(x - 2)$, $f(2) = 4$.

(c) $f'(x) = e^x + 1/x, \quad f(1) = 1$.

(d) $f'(x) = \dfrac{1}{x} + x^{5/2}(x - 2), \quad f(1) = 0$.

13. Find the equation of the curve $y = f(x)$ through the given point, with slope at a typical point (x, y) as given.

   (a) point $(0, 0)$; slope $2x - 6x^3$.

   (b) point $(2, 1)$; slope $(3x - 1)(x - 2)$.

   (c) point $(2, 8)$; slope $8x^3 - 2x$.

14. Suppose that the marginal cost of an item is given by $(x^2 - 1)/x^2$ (see Example 4, Section 3.1), and the total cost is known to be 1 when unit 1 has been produced. Find the total cost of production, or cost function.

15. Suppose that $f''(x) = x + 1$. Find $f(x)$, given that $f'(1) = 2$ and $f(0) = 0$. [*Hint:* Remember that f' is an antiderivative of f''. Thus, you must first find an antiderivative of f'' and then an antiderivative of f'.]

5.3 More Antiderivatives and Substitution

So far we have no powerful techniques for finding antiderivatives and only a small table of known functions. To expand our repertoire we introduce Table 5.3.1. The table has been obtained in a straightforward manner from Table 3.6.1. To verify the entries, we observe that

1. $\dfrac{d}{dx}\left\{\dfrac{1}{n + 1}\,[f(x)]^{n+1} + C\right\} = \dfrac{n + 1}{n + 1}\,[f(x)]^n f'(x) = [f(x)]^n f'(x)$,

2. $\dfrac{d}{dx}\,e^{f(x)} = e^{f(x)} f'(x)$,

3. $\dfrac{d}{dx}\,\log |f(x)| = \dfrac{1}{f(x)}\,f'(x)$.

We illustrate the use of the table by several examples.

TABLE 5.3.1

---

1. $\displaystyle\int [f(x)]^n f'(x)\, dx = \dfrac{1}{n + 1}\,[f(x)]^{n+1} + C$

2. $\displaystyle\int e^{f(x)} f'(x)\, dx = e^{f(x)} + C$

3. $\displaystyle\int \dfrac{1}{f(x)}\,f'(x)\, dx = \log |f(x)| + C$

▶ **Example 1.** Find $\int 2x(x^2 + 1)^5 \, dx$.

Solution. If we study the integral for a moment we see that it has form 1 from the table, with $f(x) = x^2 + 1$, $f'(x) = 2x$, and $n = 5$.
Thus

$$\int 2x(x^2 + 1)^5 \, dx = \tfrac{1}{6}(x^2 + 1)^6 + C.$$

▶ **Example 2.** Find $\int xe^{4x^2} \, dx$.

Solution. We see that the integral almost has form 2 from the table, with $f(x) = 4x^2$. However, since $f'(x) = 8x$, not x, the formula does not quite fit. Once we can see what the difficulty is, we should also see that it is not hard to overcome. Let us write

$$\int xe^{4x^2} \, dx = \tfrac{1}{8} \int 8xe^{4x^2} \, dx$$

and now apply form 2 from the table to obtain $\tfrac{1}{8}e^{4x^2} + C$ as the antiderivative.

▶ **Example 3.** Find

$$\int \frac{x + 1}{x^2 + 2x + 3} \, dx.$$

Solution. We note that the integrand almost has form 3 from the table, with $f(x) = x^2 + 2x + 3$, although $f'(x) = 2x + 2$ not $x + 1$. Thus we write

$$\int \frac{x + 1}{x^2 + 2x + 3} \, dx = \frac{1}{2} \int \frac{2x + 2}{x^2 + 2x + 3} \, dx = \frac{1}{2} \log |x^2 + 2x + 3| + C.$$

We now wish to introduce a new technique of integration called *substitution*. To do so, it will be necessary to talk about *differentials*. In the Leibniz notation for the derivative of a function f, namely $\dfrac{df}{dx}$, we never attempted to assign a meaning to the symbols df and dx separately. Nevertheless, the chain rule suggests that we may treat these much like numbers. In particular, if $y = f(x)$, where f is differentiable, we write $dy = f'(x) \, dx$. We say that dy is the *differential of f*—often $f'(x) \, dx$ is referred to as a *first-order differential*.
Thus,

1. If $y = (4x^2 + 9)$, then $dy = 8x \, dx$.
2. If $y = \log |x| + e^{2x}$, then $dy = (1/x + 2e^{2x}) \, dx$.
3. If $y = \sqrt{1 + x^3}$, then $dy = \tfrac{3}{2}x^2(1 + x^2)^{-1/2} \, dx$.

As a mnemonic aid, write $\dfrac{dy}{dx} = f'(x)$ and then move the dx to obtain $dy = f'(x) \, dx$.

It is best to think of differentials as just a game. The rules of the game are clear enough and it is easy to play. To assuage any trepidations you might have, we do mention that:

1. Differentials can be presented in a more rigorous way and are really a disguised form of the chain rule.
2. Differentials are very useful for finding antiderivatives.

We are now ready to study substitution. This topic is best introduced by an example.

▶ **Example 4.** Find $\int 5x^4(x^5 + 1)^9 \, dx$.

Solution. We try the substitution $y = x^5 + 1$. (Hints on what to try are given later.) This yields $\int 5x^4 y^9 \, dx$. However, in substitution, we must replace every x, including dx. To this end let us find dy. Clearly, $dy = 5x^4 \, dx$. We can solve for dx, that is,

$$dx = \frac{1}{5x^4} \, dy.$$

Substituting this, we get

$$\int 5x^4 y^9 \frac{1}{5x^4} \, dy = \int y^9 \, dy = \frac{y^{10}}{10} + C.$$

(We could have saved a step by substituting $5x^4 \, dx$ for dy.)

Now, if we replace y by $(x^5 + 1)$, we obtain

$$\int 5x^4(x^5 + 1)^9 \, dx = \frac{(x^5 + 1)^{10}}{10} + C.$$

If we differentiate the right-hand side of the preceding expression, we find that

$$\frac{d}{dx}\left[\frac{(x^5 + 1)^{10}}{10} + C\right] = 5x^4(x^5 + 1)^9;$$

that is, we have obtained the right answer. Is this a stroke of luck, a miracle, or is there a procedure here that works every time? Before answering this question let us consider one more example.

▶ **Example 5.** Find $\int 3x(x + 2)^{1/2} \, dx$.

Solution. Let $y = x + 2$. When we substitute, the original integral becomes $\int 3xy^{1/2} \, dx$. We must replace all the x's including dx. It is easy to replace the term $3x$, since $x = y - 2$, and so $3x = 3(y - 2)$. The integral now becomes $\int 3(y - 2)y^{1/2} \, dx$. Since $y = x + 2$, we see that $dy = 1 \cdot dx$. Thus, our integral now has the form $\int 3(y - 2)y^{1/2} \, dy$. This is easy to handle. Indeed,

$$\int 3(y - 2)y^{1/2} \, dy = 3\int (y^{3/2} - 2y^{1/2}) \, dy = 3 \cdot \frac{y^{5/2}}{\frac{5}{2}} - 6 \cdot \frac{y^{3/2}}{\frac{3}{2}} + C.$$

If we substitute $y = x + 2$, we find

$$\int 3x(x + 2)^{1/2} \, dx = \tfrac{6}{5}(x + 2)^{5/2} - 4(x + 2)^{3/2} + C.$$

Again, if we differentiate the right-hand side of the equation we discover we have the correct answer.

By now the reader should be persuaded that substitution is a method that really works. You need only differentiate your result in order to see this. We outline the basic rules of the procedure and illustrate them by examples.

We are given an indefinite integral $\int f(x)\, dx$:

1. Choose a substitution $y = u(x)$. This choice is best made by experience. However, as a rule try differentiating the most complicated term in the integrand $f(x)$. If you get some other term in the integrand this way, you probably have a good choice for $u(x)$.
2. Replace all x's and dx's by y's and dy's. Note that $dx = [1/u'(x)]\, dy$.
3. Find $\int g(y)\, dy = G(y)$.
4. Replace $G(y)$ by $G(u(x))$; that is, replace y by $u(x)$.
5. Check your answer by differentiating.

Let us consider some additional examples.

▶ **Example 6.** Find $\int x^2 e^{x^3+2}\, dx$.

Solution. First we observe that since x^2 is closely related to the derivative of $x^3 + 2$, we should try the substitution $y = x^3 + 2$. Then $dy = 3x^2\, dx$ and hence,

$$\int x^2 e^{x^3+2}\, dx = \tfrac{1}{3} \int e^y\, dy = \tfrac{1}{3} e^y + C.$$

Substituting $y = x^3 + 2$, we obtain

$$\int x^2 e^{x^3+2}\, dx = \tfrac{1}{3} e^{x^3+2} + C.$$

▶ **Example 7.** Find

$$\int \frac{\log x}{x}\, dx.$$

Solution. We first note that $\dfrac{d}{dx} \log x = 1/x$. Thus, the substitution we try is $y = \log x$. Hence $dy = 1/x\, dx$ and

$$\int \frac{\log x}{x}\, dx = \int y\, dy = \frac{y^2}{2} + C.$$

Substituting $y = \log x$ we obtain

$$\int \frac{\log x}{x}\, dx = \frac{(\log x)^2}{2} + C.$$

▶ **Example 8.** Find

$$\int \frac{x}{\sqrt{1 - x^2}}\, e^{\sqrt{1-x^2}}\, dx.$$

Solution. This looks complicated, but if we recall that

$$\frac{d}{dx}(\sqrt{1 - x^2}) = \frac{-x}{\sqrt{1 - x^2}};$$

we see that the substitution to make is $y = \sqrt{1 - x^2}$. Then

$$dy = -\frac{x}{\sqrt{1 - x^2}}\, dx$$

and hence,

$$\int \frac{x}{\sqrt{1-x^2}} e^{\sqrt{1-x^2}}\, dx = -\int e^y\, dy = -e^y + C.$$

Substituting for y, we obtain

$$\int \frac{x}{\sqrt{1-x^2}} e^{\sqrt{1-x^2}}\, dx = -e^{\sqrt{1-x^2}} + C.$$

Let us note again what we have done in the preceding examples on substitution. In the evaluation of each antiderivative of the form above, we first looked for a quantity whose derivative appeared in the integrand. This enabled us, with the formal use of differentials, to reduce the integration to evaluating antiderivatives of the form given in Table 5.1.1. It takes a lot of practice to become acquainted with the proper substitution to make in each case.

Some of the examples above, such as 4, 6, and 8, could be done by means of Table 5.3.1. However, Example 7 (and 5) would be difficult to do by means of the table, since it might not be immediately obvious that if $f(x) = \log x$ and $f'(x) = 1/x$, then

$$\int \frac{\log x}{x}\, dx = \int f'(x)[f(x)]\, dx = \frac{[f(x)]^2}{2} + C = \frac{(\log x)^2}{2} + C.$$

As a test you might try to do Exercise 1(f) by Table 5.3.1 and then by substitution.

Exercises

1. Evaluate the following by use of Table 5.3.1:

(a) $\displaystyle\int 2x(x^2 + 1)\, dx.$

(b) $\displaystyle\int (4x^3 + 9x^2)(x^4 + 3x^3 + 1)\, dx.$

(c) $\displaystyle\int 5x^4(x^5 + 1)^{10}\, dx.$　　　　　(d) $\displaystyle\int 2xe^{x^2+1}\, dx.$

(e) $\displaystyle\int \frac{x^2 + 2x}{x^3 + 3x^2 + 1}\, dx.$　　　　(f) $\displaystyle\int \frac{1 + e^x}{x + e^x}\, dx.$

(g) $\displaystyle\int \frac{1}{x} e^{2\log x}\, dx.$　　　　　(h) $\displaystyle\int \frac{x + 1}{x^2 + 2x + 1}\, dx.$

2. Find the differentials of the following functions:

(a) $y = x^3 + e^{2x}.$　　　　　　　(b) $u = \dfrac{4}{x} - 1.$

(c) $y = e^{x^2+1}.$　　　　　　　　(d) $y = x \log x - x.$

(e) $u = \sqrt{x^2 + 1}.$　　　　　　(f) $u = \sqrt{x^3 + 4}.$

(g) $y = x^4 + 1.$　　　　　　　　(h) $u = e^x.$

(i) $y = \log x.$

Evaluate the following:

3. $\int (x^{5/3} + x^{1/6} + x^{-2})\, dx.$

4. $\int (2x + 1)^{1/2}\, dx.$

5. $\int \dfrac{(x + 1)\, dx}{(x^2 + 2x + 3)^{1/3}}.$

6. $\int \dfrac{x^3\, dx}{\sqrt{1 - x^4}}.$

7. $\int x(x + 1)^{1/2}\, dx.$

8. $\int (x + 1)(x + 3)^{3/2}\, dx.$

9. $\int x(x - 9)^{-1/2}\, dx.$

10. $\int (x - 1)(x + 1)^{100}\, dx.$

11. $\int \dfrac{(\log |x|)^2}{x}\, dx.$

12. $\int \dfrac{x}{x^2 + 1} \log (x^2 + 1)\, dx.$

13. $\int \dfrac{1}{x \log x}\, dx.$

14. $\int (x^2 + 1)e^{x^3 + 3x}\, dx.$

15. $\int \dfrac{\log x^2}{x}\, dx.$

16. $\int \dfrac{1}{x (\log |x|)^2}\, dx.$

17. $\int \dfrac{e^{\sqrt{x}}}{\sqrt{x}}\, dx.$

18. $\int x\sqrt{1 - x^2}\, dx.$

19. $\int \dfrac{x}{1 - x^2} \log \sqrt{1 - x^2}\, dx.$

20. $\int \dfrac{e^x}{1 + e^x}\, dx.$

21. $\int [x^2 e^{x^3} + 3x(x^2 + 1)^{1/2}]\, dx.$

22. $\int e^{e^x} e^x\, dx.$

23. $\int \dfrac{x + 1}{(x^2 + 2x)^9}\, dx.$

24. Find a function $f(x)$ satisfying the conditions $f'(x) = (x^2 + 1)e^{x^3 + 3x}$ and $f(0) = 1$.

25. Find a function $f(x)$ such that $f'(x) = x\sqrt{1 + x^2} + x^2$ and $f(1) = 0$.

26. Let $y = f(x)$ be a curve with slope equal to 3 at every point. Let $f(1) = -2$. Find $f(x)$.

27. If marginal revenue is known to be $x/\sqrt{9 - x}$, find the revenue function $R(x)$ if we are told that $R(0) = 1$. [See Example 4, Section 3.1.]

28. The slope of a curve at (x, y) is $Ax(x^2 - 1)$, where A is some constant. The curve crosses the x axis at $x = 3$ and it crosses the y axis at $y = 2$. Find the equation of the curve. [*Hint:* Recall that the slope of a curve at a point (x, y) is given by $\dfrac{dy}{dx}$.]

*5.4 The Riemann Integral

Most readers probably know how to find the areas of rectangles, triangles, and certain other polygons from geometry. Thus far you have not encountered techniques for determining the area of a region bounded by an arbitrary curve. In this section, we shall begin an investigation of this problem.

Let us consider the problem of determining the area under the curve from a to b in Figure 5.4.1.

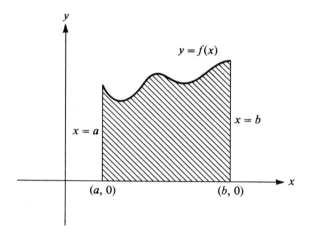

Figure 5.4.1

We have not defined what we mean by the area under a curve, but as a first step in the problem, let us rely on our intuitive notion of area and make a few estimates. We subdivide the interval between a and b into four equal pieces and then fit rectangles under the curve, as in Figure 5.4.2. You can think of the rectangles as being moved upward until they touch the curve. Next observe that the heights of the respective rectangles are $f(x_1^*)$, $f(x_2^*)$, $f(x_3^*)$, and $f(x_4^*)$ and the bases are $(x_1 - x_0)$, $(x_2 - x_1)$, $(x_3 - x_2)$, and $(x_4 - x_3)$. (The asterisks are placed on the x's as a convenient way to distinguish them from $x_1, x_2, \ldots$.)

Thus the total area of the rectangles is just equal to

$$f(x_1^*)(x_1 - x_0) + f(x_2^*)(x_2 - x_1) + f(x_3^*)(x_3 - x_2) + f(x_4^*)(x_4 - x_3).$$

Clearly the area of the rectangles is less than the area under the curve, whatever it may be. Thus this first estimate gives us a *lower* bound for the area under the curve.

We now repeat what we just did, but this time we choose the rectangles so that they *contain* the area under the curve. See Figure 5.4.3.

In this situation the heights of the rectangles are now $f(x_1')$, $f(x_2')$, $f(x_3')$, and $f(x_4')$. Thus the total area of the rectangles is just equal to

$$f(x_1')(x_1 - x_0) + f(x_2')(x_2 - x_1)f(x_3')(x_3 - x_2) + f(x_4')(x_4 - x_3).$$

---

* Page vi of the Preface contains a note to the instructor about the Riemann integral.

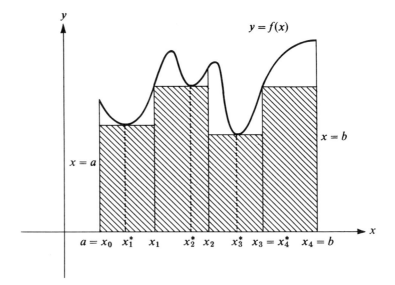

Figure 5.4.2

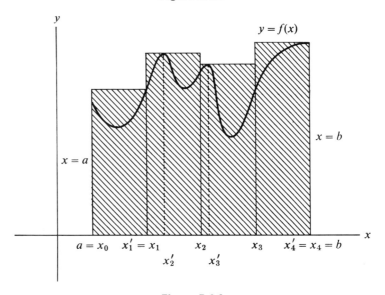

Figure 5.4.3

Clearly the area under the curve is now *less* than the area under the rectangles. Thus the area under the second set of rectangles gives us an *upper bound* for the area under the curve.

We might now try an often-used strategy; namely increase the number of subdivisions and see if our upper and lower estimates get closer together. We do this for the smaller rectangles in Figure 5.4.4 (a), (b), (c) (and a different function f). Notice that as the distance between subdivision points decreases we apparently obtain a better approximation to the area under the curve. Observe in that figure that the amount of area not counted (above the rectangles) clearly decreases.

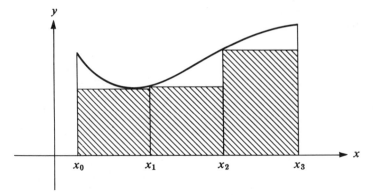

(a)   Three subdivisions

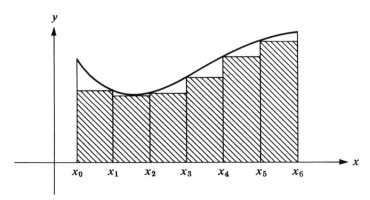

(b)   Six subdivisions

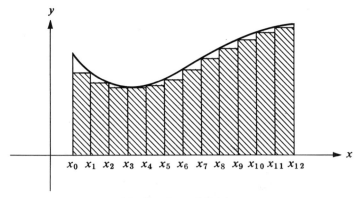

(c)   Twelve subdivisions

Figure 5.4.4

We now write down a slightly different way of implementing this strategy. Let us subdivide the interval $[a, b]$ into n subintervals, not necessarily of the same length, and let us choose an arbitrary point in each subinterval. Call x_i^* the point in the ith subinterval $[x_i, x_{i-1}]$. (See Figure 5.4.5.)

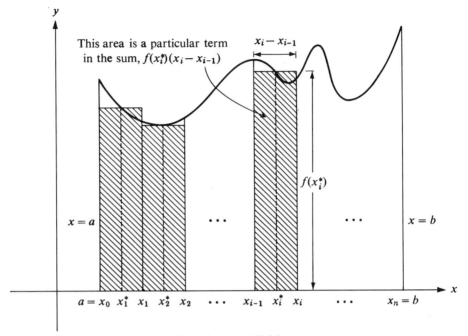

This area is a particular term in the sum, $f(x_i^*)(x_i - x_{i-1})$

Forming n subdivisions

Figure 5.4.5

Caution: The x_i^* here has no relation to the ones chosen earlier. Consider the following sum:

$$f(x_1^*)(x_1 - x_0) + f(x_2^*)(x_2 - x_1) + \cdots + f(x_n^*)(x_n - x_{n-1}).$$

Note that this sum depends on the choice of the subdivision points x_i and the choice of the x_i^*. Suppose we take a limit of such sums as the distance between subdivision points shrinks to zero. It is certainly not clear that the sums should converge to anything. Remarkably enough, in a great many cases they do. The limit of such sums is called the Riemann integral. Even if we knew the sums converged, it would appear rather difficult to calculate the limit. We shall define the area under the curve $y = f(x)$ bounded by $x = a$ on the left and $x = b$ on the right to be this limit.

We first show that the Riemann integral exists for any increasing function. Then we will introduce another integral which enables us to calculate such limits easily.

Let f be an increasing function on $[a, b]$. Let us subdivide $[a, b]$ into n subintervals by means of the points $x_0 = a, x_1, x_2, \ldots, x_n = b$. Note that the maximum value of the function f on the interval $[x_{i-1}, x_i]$ occurs at x_i and is just $f(x_i)$, while the minimum value occurs at x_{i-1} and is just $f(x_{i-1})$, since the function is increasing.

Thus for any choice whatsoever of the x_i^* in $[x_{i-1}, x_i]$ we have the following inequality:

$$L = f(x_0)(x_1 - x_0) + \cdots + f(x_{n-1})(x_n - x_{n-1})$$
$$\leq f(x_1^*)(x_1 - x_0) + \cdots + f(x_n^*)(x_n - x_{n-1})$$
$$\leq f(x_1)(x_1 - x_0) + \cdots + f(x_n)(x_n - x_{n-1}) = M.$$

(L and M are simply convenient names for the two sums.) The extreme terms in the inequality are represented geometrically in Figure 5.4.6 with the different cross-hatching representing the areas corresponding to the various sums. (Figure 5.4.6 is supposed to represent the general case when there are n subintervals. However, for

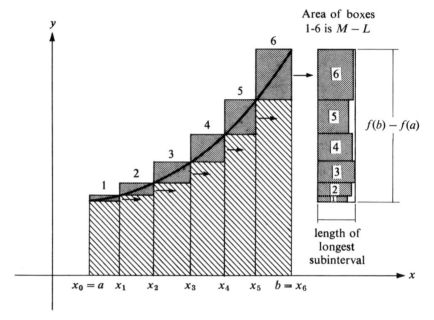

Figure 5.4.6

clarity we have depicted the situation when $n = 6$.) Let us now estimate $M - L$. By looking at Figure 5.4.6, we note that $M - L$ is just equal to the area of the boxes stacked to the right of the graph. But the area of these rectangles is no larger than the height of the stack $f(b) - f(a)$ times the base of the largest rectangle, say $[x_j - x_{j-1}]$. Thus,

$$0 \leq M - L \leq [f(b) - f(a)](x_j - x_{j-1}),$$

where $(x_j - x_{j-1})$ is the maximum distance between any two successive subdivision points. If we increase the number of subdivision points so that the successive distances tend to zero, we thus observe that the upper and lower sums approach one another. Hence the general sum

$$f(x_1^*)(x_1 - x_0) + \cdots + f(x_n^*)(x_n - x_{n-1})$$

must converge to a limit as $n \to \infty$ since it is trapped between the upper and lower sums. In this case, by converging to a limit we mean that

$$\lim_{n \to \infty} [f(x_1^*)(x_1 - x_0) + \cdots + f(x_n^*)(x_n - x_{n-1})] = S,$$

where S is a finite number. This proves every increasing function is Riemann integrable, where the last phrase simply means that the limit exists. It should be clear from the pictures that the sums are converging to the area under the curve.

Having given a somewhat sketchy proof that every increasing function is Riemann integrable, we now state a general result for continuous functions.

THEOREM 1. Let f be a continuous function on $[a, b]$. For a subdivision $x_0, x_1, \ldots,$ x_n of the interval $[a, b]$, form the sum

$$R_a^b(n) = f(x_1^*)(x_1 - x_0) + \cdots + f(x_n^*)(x_n - x_{n-1}),$$

where $x_i^* \in [x_{i-1}, x_i]$ for $i = 1, 2, \ldots, n$. Assume that as n increases the distance between the points in the subdivision shrinks to zero. Then $\lim_{n \to \infty} R_a^b(n)$ exists and its value is called the Riemann integral of f on $[a, b]$.

The last theorem tells us that the limit of a certain set of sums exists but gives us no idea how to find it. That information is contained in the next theorem.

THEOREM 2. (Fundamental Theorem of Calculus) Let f be continuous on $[a, b]$. Let F be a differentiable function on $[a, b]$, where $F'(x) = f(x)$ for all $x \in [a, b]$. Let $R_a^b(n)$ be the sum described in Theorem 1. Then

$$\lim_{n \to \infty} R_a^b(n) = F(b) - F(a).$$

It should be immediately apparent that Theorem 2 provides us with a very powerful tool for calculating an otherwise unwieldy limit. In fact, we introduce a new integral in the next section based on this result.

Exercises

1. Let $f(x) = x^2$. Assume $a = x_0 = 0, x_1 = \frac{1}{4}, x_2 = \frac{1}{2}, x_3 = \frac{3}{4}$, and $x_4 = b = 1$. Write down the corresponding Riemann sum if $x_i^* = x_{i-1}$.

2. Same as Exercise 1, but now $x_i^* = x_i$.

3. Let $f(x) = x^3$. Subdivide the interval $[2, 3]$ into 6 equal subintervals. Calculate the Riemann sum where

    (a) $x_i^* = x_{i-1}$. (b) $x_i^* = x_i$. (c) $x_i^* = \frac{1}{2}(x_{i-1} + x_i)$.

5.5 The Newton Integral

We now introduce a powerful tool for computing area, velocity, consumer surplus, and other quantities, called the Newton integral. It is as basic to calculus as the derivative. To anyone who read the previous section, the definition will come as no surprise.

DEFINITION 1. If f has an antiderivative F on the interval $[a, b]$, we say that f is Newton integrable on $[a, b]$ and the Newton integral is defined to be the number

$$\int_a^b f(x) \, dx = F(b) - F(a).$$

We also refer to $\int_a^b f(x)\,dx$ as the definite integral of f from a to b. The function f shall be referred to as the integrand; a and b are called the *limits of integration*. The definition given above is legitimate since we can easily show that the integral is independent of the particular choice of F. It depends only on a, b, and f. In order to see this, we observe that if F and G are any two antiderivatives of f on $[a, b]$, then

$$F(b) - F(a) = G(b) - G(a).$$

The proof of this is simple. If F and G are any two antiderivatives of f, then $F(x) = G(x) + C$, where C is an arbitrary constant. Thus,

$$F(b) = G(b) + C \quad \text{and} \quad F(a) = G(a) + C;$$

hence,

$$F(b) - F(a) = G(b) - G(a).$$

At the end of this chapter we shall show that if f is a continuous function, then its Riemann integral (and hence the area under the curve $y = f(x)$ bounded by $x = a$ and $x = b$) is also $F(b) - F(a)$. Thus for continuous functions the Riemann and Newton integrals are equal.

Before looking at some examples, we introduce one bit of notation.

DEFINITION 2. *Let F be a function defined on* $[a, b]$. *Then* $F(x)|_a^b = F(b) - F(a)$.

▶ **Example 1.** Find $\int_{-1}^2 x^2\,dx$.

Solution. Since $x^3/3$ is an antiderivative of x^2, we see that

$$\int_{-1}^2 x^2\,dx = \frac{8}{3} - \frac{(-1)}{3} = \frac{9}{3} = 3.$$

We could also have written

$$\int_{-1}^2 x^2\,dx = \frac{x^3}{3}\bigg|_{-1}^2 = \frac{8}{3} - \left(-\frac{1}{3}\right) = 3.$$

▶ **Example 2.** Find $\int_a^b x^n\,dx$, where n is any number except -1.

Solution. Since $x^{n+1}/(n+1)$ is an antiderivative of x^n, we see that

$$\int_a^b x^n\,dx = \frac{b^{n+1}}{n+1} - \frac{a^{n+1}}{n+1}.$$

Again we could write this as

$$\int_a^b x^n\,dx = \frac{x^{n+1}}{n+1}\bigg|_a^b = \frac{b^{n+1}}{n+1} - \frac{a^{n+1}}{n+1}.$$

▶ **Example 3.** Find $\int_0^1 e^x\,dx$.

Solution. Since e^x is an antiderivative of e^x, we see that

$$\int_0^1 e^x\,dx = e^x\bigg|_0^1 = e^1 - e^0 = e - 1.$$

▶ **Example 4.** Find $\int_1^2 1/x \, dx$.

Solution. Since $\log |x|$ is an antiderivative of $1/x$, we see that

$$\int_1^2 \frac{1}{x} \, dx = \log |x| \Big|_1^2 = \log 2 - \log 1 = \log 2.$$

Definite integrals have a number of useful properties, which we list below.

PROPERTY 1. If f is Newton integrable on $[a, b]$ and on $[b, c]$, then f is Newton integrable on $[a, c]$ and

$$\int_a^c f(x) \, dx = \int_a^b f(x) \, dx + \int_b^c f(x) \, dx.$$

Proof of Property 1: It is a simple matter to verify this result. Let F be an antiderivative of f on $[a, b]$ and G be an antiderivative of f on $[b, c]$. Let

$$H(x) = \begin{cases} F(x), & x \in [a, b], \\ G(x) + F(b) - G(b), & x \in [b, c]. \end{cases}$$

Clearly H is an antiderivative of f on $[a, c]$, since

$$H'(x) = F'(x) = f(x) \quad \text{for} \quad x \in [a, b],$$

and

$$H'(x) = G'(x) = f(x) \quad \text{for} \quad x \in [b, c].$$

Thus

$$H'(x) = f(x) \quad \text{for} \quad x \in [a, c].$$

Hence,

$$\int_a^c f(x) \, dx = H(c) - H(a).$$

But

$$\int_a^b f(x) \, dx + \int_b^c f(x) \, dx = H(b) - H(a) + H(c) - H(b) = H(c) - H(a).$$

which yields the desired result.

PROPERTY 2. If f and g are Newton integrable on $[a, b]$, then so is $f + g$ and $c \cdot f$, where c is any constant. In particular,

$$\int_a^b [f(x) + g(x)] \, dx = \int_a^b f(x) \, dx + \int_a^b g(x) \, dx$$

and

$$\int_a^b cf(x) \, dx = c \int_a^b f(x) \, dx.$$

Proof of Property 2: This is also simple to verify. Let F and G be the anti-derivatives of f and g on a, b, respectively. Then from Property 2, Section 5.2, $F + G$ is an antiderivative of $f + g$. Thus,

$$\int_a^b [f(x) + g(x)]\, dx = F(b) + G(b) - F(a) - G(a)$$

$$= [F(b) - F(a)] + [G(b) - G(a)]$$

$$= \int_a^b f(x)\, dx + \int_a^b g(x)\, dx.$$

The second part follows directly from Property 1 of Section 5.2.

We now look at some examples using these properties.

▶ **Example 5.** Find $\int_0^4 |x(x - 2)|\, dx$.

Solution. Since

$$|x(x - 2)| = x(x - 2) = x^2 - 2x \quad \text{for} \quad x \geq 2,$$

the function $|x(x - 2)|$ is Newton integrable on $[2, 4]$. Since

$$|x(x - 2)| = -x(x - 2) = -x^2 + 2x \quad \text{for} \quad 0 \leq x \leq 2,$$

the function $|x(x - 2)|$ is Newton integrable on $[0, 2]$. Thus

$$\int_0^4 |x(x - 2)|\, dx = \int_0^2 |x(x - 2)|\, dx + \int_2^4 |x(x - 2)|\, dx$$

$$= -\int_0^2 x(x - 2)\, dx + \int_2^4 x(x - 2)\, dx.$$

By Property 2,
$$= -\int_0^2 x^2\, dx + 2\int_0^2 x\, dx + \int_2^4 x^2\, dx - 2\int_2^4 x\, dx$$

$$= -\frac{x^3}{3}\bigg|_0^2 + x^2\bigg|_0^2 + \frac{x^3}{3}\bigg|_2^4 - x^2\bigg|_2^4$$

$$= -\tfrac{8}{3} + 4 + \tfrac{64}{3} - \tfrac{8}{3} - 16 + 4 = \tfrac{24}{3}.$$

Without Property 1, we could not have evaluated this integral.

PROPERTY 3. Let f be Newton integrable on a, b, and suppose that $f(x) \geq 0$ on $[a, b]$. Then, $\int_a^b f(x)\, dx \geq 0$.

Proof of Property 3: This result is also easily obtained. Let f be an antiderivative of f on $[a, b]$. Then $F'(x) = f(x) \geq 0$, and hence F is *increasing* on $[a, b]$. (Recall Section 4.2.) Thus if $b \geq a$, then $F(b) \geq F(a)$, and hence

$$\int_a^b f(x)\, dx = F(b) - F(a) \geq 0.$$

The next property, which is a direct consequence of Property 3, is useful in applications.

PROPERTY 4. Let f and g be Newton integrable on $[a, b]$, and assume that for every $x \in [a, b], f(x) \le g(x)$. Then,

$$\int_a^b f(x) \, dx \le \int_a^b g(x) \, dx.$$

Proof of Property 4: We need only realize that the function $g(x) - f(x) \ge 0$ on $[a, b]$ and apply Property 3 to obtain the desired result.

Now, let us investigate how substitutions can be used to evaluate Newton integrals. Consider the following example.

▶ Example 6. Find

$$\int_0^1 \frac{x \, dx}{\sqrt{1 + x^2}}.$$

Solution. First we must find an antiderivative of $x/\sqrt{1 + x^2}$, that is, we must find

$$\int \frac{x \, dx}{\sqrt{1 + x^2}}.$$

Let $u = 1 + x^2$; whence $du = 2x \, dx$ and

$$\int \frac{x \, dx}{\sqrt{1 + x^2}} = \frac{1}{2} \int u^{-1/2} \, du = u^{1/2} + C = \sqrt{1 + x^2} + C.$$

Hence, an antiderivative of $x/\sqrt{1 + x^2}$ is $\sqrt{1 + x^2}$. Thus,

$$\int_0^1 \frac{x \, dx}{\sqrt{1 + x^2}} = \sqrt{1 + x^2} \Big|_0^1 = \sqrt{2} - 1.$$

Now, we can actually evaluate this definite integral without first finding an antiderivative of $x/\sqrt{1 + x^2}$ as follows: Let $u(x) = 1 + x^2$. Thus when $x = 0, u(0) = 1$ and when $x = 1, u(1) = 2$. We assert that we can then write

$$\int_0^1 \frac{x \, dx}{\sqrt{1 + x^2}} = \frac{1}{2} \int_{u(0)}^{u(1)} u^{-1/2} \, du = \frac{1}{2} \int_1^2 u^{-1/2} \, du = u^{1/2} \Big|_1^2 = \sqrt{2} - 1.$$

This method avoids resubstituting $1 + x^2$ for u. The essential idea is: *when we substitute u for $1 + x^2$, we also change the limits of integration.* We will see that this can be done for most integrals. The result is stated as follows:

THEOREM 1. Let $u(x)$ be a differentiable function on $[a, b]$ and assume that $c \le u(x_0) \le d$ for $x_0 \in [a, b]$. If f is Newton integrable on $[c, d]$, then

$$\int_a^b f[u(x)]u'(x) \, dx = \int_{u(a)}^{u(b)} f(u) \, du.$$

Proof: The verification of the above theorem is straightforward. Let F be an antiderivative of f on $[u(a), u(b)]$. Then

$$\frac{d}{dx} \{F[u(x)]\} = F'[u(x)]u'(x) = f[u(x)]u'(x)$$

for all $x \in [a, b]$. Therefore, $F[u(x)]$ is an antiderivative of $f[u(x)]u'(x)$ and hence

$$\int_a^b f[u(x)]u'(x) \, dx = F[u(x)]\Big|_a^b = F[u(b)] - F[u(a)] = \int_{u(a)}^{u(b)} f(u) \, du.$$

We again repeat that this theorem enables us to evaluate many Newton integrals without explicitly producing antiderivatives. Let us look at more examples.

▶ **Example 7.** Find

$$\int_0^1 \frac{e^x}{1 + e^x} \, dx.$$

Solution. Let $u = 1 + e^x$; then $u(0) = 2$, $u(1) = 1 + e$, and $du = e^x \, dx$. Hence,

$$\int_0^1 \frac{e^x \, dx}{1 + e^x} = \int_{u(0)=2}^{u(1)=1+e} \frac{du}{u} = \log u \Big|_2^{1+e}$$

$$= \log (1 + e) - \log 2 = \log \frac{1 + e}{2}.$$

If we apply Theorem 1 to this example, then

$$f[u(x)] = \frac{1}{u(x)} = \frac{1}{1 + e^x} \quad \text{and} \quad u'(x) = e^x.$$

However, it is more convenient to use differentials than to go through this substitution each time.

▶ **Example 8.** Find $\int_0^2 (x + 1)(x^2 + 2x + 1) \, dx$.

Solution. Let $u = x^2 + 2x + 1$; then $du = 2(x + 1) \, dx$, $u(0) = 1$ and $u(2) = 9$. Therefore,

$$\int_0^2 (x + 1)(x^2 + 2x + 1) \, dx = \frac{1}{2} \int_{u(0)=1}^{u(2)=9} u \, du = \frac{u^2}{4} \Big|_1^9 = \frac{81}{4} - \frac{1}{4} = 20.$$

Note that we could integrate $(x + 1)(x^2 + 2x + 1)$ by simply multiplying it out.

Until now, we have always evaluated $\int_a^b f(x) \, dx$ for $b > a$. Suppose that $b < a$. In this case we define

$$\int_a^b f(x) \, dx = F(b) - F(a),$$

where F is any antiderivative of f on $[b, a]$. (Note that in this case we find any antiderivative on $[b, a]$ rather than on $[a, b]$.) Observe that since $a > b$,

$$\int_b^a f(x) \, dx = F(a) - F(b),$$

where F is any antiderivative of f on $[b, a]$. Hence,

$$\int_a^b f(x) \, dx = - \int_b^a f(x) \, dx.$$

It is also true that $\int_a^a f(x)\, dx = 0$, since

$$\int_a^a f(x)\, dx = F(a) - F(a) = 0.$$

We end with another illustration of how the various properties of the Newton integral can be used to evaluate them.

▶ **Example 9.*** Find $\int_{-1}^1 x^2\, dx$.

Solution. First we write

$$\int_{-1}^1 x^2\, dx = \int_{-1}^0 x^2\, dx + \int_0^1 x^2\, dx.$$

Now in the first integral let $u(x) = -x$. Then $du = -dx$, $u(-1) = 1$, $u(0) = 0$, and

$$\int_1^0 x^2\, dx = -\int_1^0 u^2\, du = \int_0^1 u^2\, du.$$

Since the variable u is really a "dummy," that is,

$$\int_0^1 u^2\, du = \int_0^1 x^2\, dx$$

(the integral *only* depends on the function in the integral and the limits; thus the variable appearing in the integrand is incidental), we see that

$$\int_{-1}^1 x^2\, dx = 2\int_0^1 x^2\, dx = 2 \cdot \frac{x^3}{3}\Big|_0^1 = \frac{2}{3}.$$

Example 9 has a useful generalization. Note that $f(x) = x^2$ is an even function, that is, $f(-x) = f(x)$. We claim that if f is an *even function*, then

$$\int_{-a}^a f(x)\, dx = 2\int_0^a f(x)\, dx.$$

This follows in exactly the same way as before. First write

$$\int_{-a}^a f(x)\, dx = \int_{-a}^0 f(x)\, dx + \int_0^a f(x)\, dx.$$

In the first integral let $u = -x$, $du = -dx$, $u(-a) = a$, and $u(0) = 0$. Thus,

$$\int_{-a}^0 f(x)\, dx = -\int_a^0 f(-u)\, du = \int_0^a f(-u)\, du = \int_0^a f(u)\, du = \int_0^a f(x)\, dx.$$

Hence,

$$\int_{-a}^a f(x)\, dx = 2\int_0^a f(x)\, dx.$$

In exactly the same way, we can show that if f is an *odd function*, that is, $f(-u) = -f(u)$, then

$$\int_{-a}^a f(x)\, dx = 0.$$

[See Exercise 4(b).]

Remark. Observe that the Newton integral is *defined* only for functions that have antiderivatives on closed intervals. It is quite useful to extend this concept of integration to functions that may not have antiderivatives. The example given in Section 5.1,

$$f(x) = \begin{cases} -1, & x \in [-1, 0), \\ 0, & x = 0, \\ 1, & x \in (0, 1], \end{cases}$$

is such a function. This leads us to the Riemann integral, which we discussed in Section 5.4. There are other functions that are Newton integrable but *not* Riemann integrable. We discuss this at greater length in the appendix.

An important theoretical question we have not settled in this chapter is: Does every continuous function have an antiderivative? The answer is yes, but we omit the proof. A proof can be found in Chapter 8 of J. W. Kitchen, *Calculus of One Variable* (Addison-Wesley, Reading, Mass., 1968). The term "Newton integral" is also due to Kitchen.

Exercises

1. Evaluate the following definite integrals:

(a) $\displaystyle\int_0^1 (1 + x)\, dx.$

(b) $\displaystyle\int_1^3 (2x - x^3)\, dx.$

(c) $\displaystyle\int_1^4 \frac{1}{x}\, dx.$

(d) $\displaystyle\int_{-1}^1 e^x\, dx.$

(e) $\displaystyle\int_{-1}^3 x^2\, dx.$

(f) $\displaystyle\int_{-1}^1 (2x^2 - x^3)\, dx.$

(g) $\displaystyle\int_1^4 \frac{dx}{2\sqrt{x}}.$

(h) $\displaystyle\int_1^2 \left(\frac{1}{x^2} - \frac{1}{x^3}\right) dx.$

(i) $\displaystyle\int_{-2}^3 e^{-(1/2)x}\, dx.$

(j) $\displaystyle\int_1^8 (1 + x^{1/3})\, dx.$

(k) $\displaystyle\int_1^2 \left(\frac{1}{x} + \frac{1}{x^2}\right) dx.$

(l) $\displaystyle\int_0^3 (x^{1/2} + x^{3/2})\, dx.$

(m) $\displaystyle\int_0^2 \frac{1}{x + 1}\, dx.$

(n) $\displaystyle\int_1^5 \frac{\log x}{x}\, dx.$

(o) $\displaystyle\int_0^2 e^{-2x}\, dx.$

2. Evaluate the following definite integrals:

(a) $\displaystyle\int_{-1}^1 |x(x - 1)|\, dx.$

(b) $\displaystyle\int_{-1}^1 |x^2(x - 1)|\, dx.$

(c) $\displaystyle\int_{-1}^2 |x^2(x - 1)|\, dx.$

(d) $\displaystyle\int_{-1}^2 \frac{dx}{|2x - 5|}.$

3. Evaluate the following:

(a) $\displaystyle\int_0^1 x^2(x^3 + 2)^2 \, dx.$

(b) $\displaystyle\int_0^1 \frac{8x^2}{(x^3 + 2)^2} \, dx.$

(c) $\displaystyle\int_0^1 \frac{x^2 \, dx}{\sqrt[4]{x^3 + 2}}.$

(d) $\displaystyle\int_{-1}^2 x\sqrt[3]{1 - x^2} \, dx.$

(e) $\displaystyle\int_0^{1/\sqrt{2}} \sqrt{x^2 - 2x^4} \, dx.$

(f) $\displaystyle\int_{-1}^0 e^x(e^x + 1)^3 \, dx.$

(g) $\displaystyle\int_1^2 \frac{(1 + x)^2}{\sqrt{x}} \, dx.$

(h) $\displaystyle\int_{-1}^2 xe^{x^2} \, dx.$

(i) $\displaystyle\int_{-4}^{-3} x(x + 6)^{1/2} \, dx.$

(j) $\displaystyle\int_1^2 \frac{x}{x^2 + 1} \log (x^2 + 1) \, dx.$

*4. (a) Using a method similar to Example 9 of this section, show that

$$\int_{-1}^1 x^3 \, dx = 0.$$

(b) Show that if f is Newton integrable on $[-a, a]$ and odd on that interval, that is, $f(-x) = -f(x)$, then

$$\int_{-a}^a f(x) \, dx = 0.$$

5. Is the function $f(x) = |x|$ Newton integrable on $[-1, 1]$? If your answer is yes, evaluate $\int_{-1}^1 f(x) \, dx$. [*Hint:* Solve the problem on $[-1, 0]$ and then $[0, 1]$ and see if you can match these at $x = 0$.]

6. Is the function $f(x) = 1/x^2$ Newton integrable on $[-1, 1]$? If not, why not? If so, evaluate $\int_{-1}^1 (1/x^2) \, dx$. (Be careful!)

*7. $\int_0^t f(x) \, dx = t^2$. Find $f(x)$ when we assume f to be Newton integrable. [*Hint:*

$$\int_0^t f(x) \, dx = F(t) - F(0)$$

where $F'(x) = f(x)$.]

5.6 Area and the Integral

How does one find the area under a parabola? The Newton integral, introduced in the previous section, provides us with a powerful computational tool for answering such questions.

Let f be a continuous function on $[a, b]$ and suppose that $f(x) \geq 0$ on $[a, b]$. (See Figure 5.6.1.) Consider the following questions.

1. What is the area of the shaded region H? (That is, the area bounded above by the curve $y = f(x)$, to the left by $x = a$, and to the right by $x = b$.)
2. Does there exist a differentiable function F on $[a, b]$ such that $F'(x) = f(x)$?

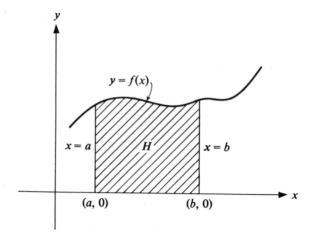

Figure 5.6.1

Off hand, the two questions do not appear to be related in any way. The surprising thing is that they are. So the reader may share our joy and astonishment at this turn of events, we state the answers in rough form now. The answers to Questions 1 and 2 are

1. Area $(H) = \int_b^a f(x)\, dx.$
2. Let $A(x_0)$ be the area under the curve bounded above by $y = f(x)$, to the left by $x = a$, and to the right by $x = x_0$. Then, $A'(x_0) = f(x_0)$. (See Figure 5.6.2.)

We now apply the result just stated to find the area in several examples. Later in the section we will justify our claim for the relation between area and the integral.

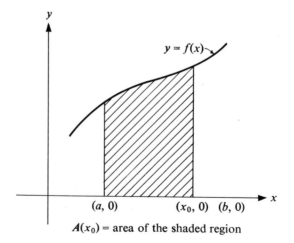

$A(x_0)$ = area of the shaded region

Figure 5.6.2

▶ Example 1.   Find the area under the curve $y = x^2$ from 0 to 1.

Solution.   First we graph the function. (See Figure 5.6.3.) We see that $f(x) = x^2$ is greater than or equal to zero on the interval $[0, 1]$. Hence,

$$\text{area} = \int_0^1 x^2 \, dx = \frac{x^3}{3}\Big|_0^1 = \frac{1}{3}.$$

We have thus found the area under part of a parabola. This problem was solved by Archimedes, a military consultant, about 250 B.C.

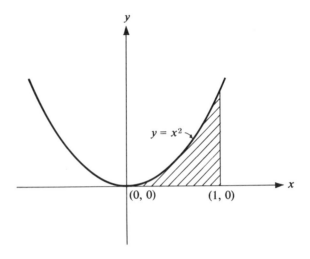

y

$y = x^2$

(0, 0)   (1, 0)

x

Figure 5.6.3

▶ Example 2.   Find the area bounded by the curve $y = 2x - x^2$ and the x axis.

Solution.   First, graph the function $2x - x^2$. Note that $f'(x) = 2(1 - x)$ and $f''(x) = -2$. Thus, $f'(x) = 0$ for $x = 1$ and since $f''(1) < 0$, f has a maximum at $x = 1$, $y = 1$. The graph is increasing for $x < 1$ and decreasing for $x > 1$, and crosses the x axis at $x = 0$ and $x = 2$ only. Thus, the area we want is that of the shaded region in Figure 5.6.4. The area is given by

$$\int_0^2 (2x - x^2) \, dx.$$

Now,

$$\int_0^2 (2x - x^2) \, dx = \left(x^2 - \frac{x^3}{3}\right)\Big|_0^2 = 4 - \frac{8}{3} = \frac{4}{3}.$$

In solving *any* area problem, it is first necessary to sketch the region in question. This is an aid in setting up the proper limits of integration.

Thus far we have restricted ourselves only to functions f such that $f(x) \geq 0$ on $[a, b]$. We now show how to extend our discussions to functions which are negative over part or all of $[a, b]$.

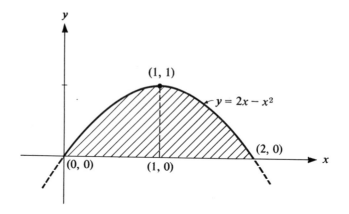

Figure 5.6.4

In Figure 5.6.5(a), we wish to find area (B), and in Figure 5.6.5(b), area $(B \cup C) =$ area (B) + area (C). In the first case [Figure 5.6.5(a)], we take

$$\text{area } (B) = -\int_a^b f(x)\, dx,$$

while in the second case,

$$\text{area } (B \cup C) = \int_a^c f(x)\, dx - \int_c^b f(x)\, dx.$$

(Note that this convention will always guarantee that the area is a positive number.)

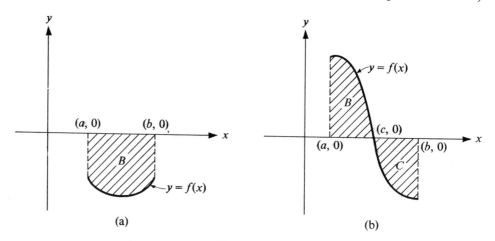

(a)

(b)

Figure 5.6.5

We proceed to illustrate this with some examples.

▶ **Example 3.** Find the area bounded by $y = x^2 - 1$, $x = 0$, and $x = 1$.

Solution. The graph of $x^2 - 1$, $x \in [0, 1]$ is shown in Figure 5.6.6.

$$\text{area } (B) = -\int_0^1 (x^2 - 1)\, dx = -\left(\frac{x^3}{3} - x\right)\Big|_0^1 = -\left(\frac{1}{3} - 1\right) = \frac{2}{3}.$$

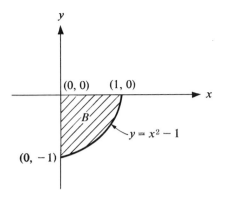

Figure 5.6.6

▶ Example 4. Find the area bounded by the curve $y = x^3$ and the lines $x = -1$ and $x = 1$.

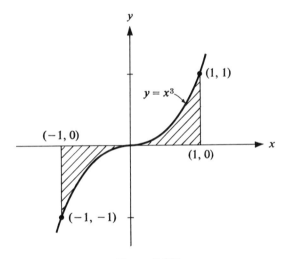

Figure 5.6.7

Solution. The graph of $y = x^3$, $x \in (-1, 1)$ is shown in Figure 5.6.7. The area in question is the area of the shaded region. The desired area is given by

$$-\int_{-1}^{0} x^3 \, dx + \int_{0}^{1} x^3 \, dx = -\frac{x^4}{4}\bigg|_{-1}^{0} + \frac{x^4}{4}\bigg|_{0}^{1} = \frac{1}{4} + \frac{1}{4} = \frac{1}{2}.$$

Recall from Exercise 4, Section 5.5, that $\int_{-1}^{1} x^3 \, dx = 0$. But the area under the curve $y = x^3$ between -1 and $+1$ is obviously not 0. It is the introduction of the minus sign for functions lying below the x axis that prevents us from obtaining 0 as the area.

Next we consider the problem of determining the area of a region bounded by the graphs of two functions f and g, both continuous on $[a, b]$, and the vertical lines $x = a$ and $x = b$. Suppose, as in Figure 5.6.8, $f(x) \geq g(x)$ for $x \in [a, b]$. We wish

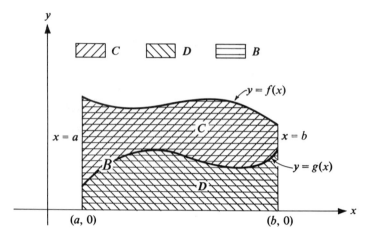

Figure 5.6.8

to determine the area of the shaded region C. Now it should be clear that if we denote the region under the curve f by B and that under g by D, then

$$\text{area } (C) = \text{area } (B) - \text{area } (D) = \int_a^b f(x)\, dx - \int_a^b g(x)\, dx$$

$$= \int_a^b [f(x) - g(x)]\, dx.$$

The above result holds *even* if f and g both take on negative values on $[a, b]$.

▶ Example 5. Find the area bounded by the curves $y = x(x - 2)$ and $y = x/2$ over the interval $[0, 2]$.

Solution. First we draw the graphs of the two functions. We must determine the area of the region B in Figure 5.6.9. Clearly, $x/2 > x(x - 2)$ on $[0, 2]$ (look at the graphs).

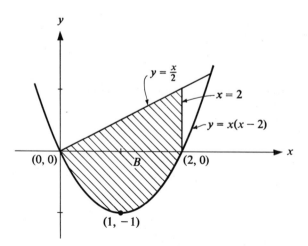

Figure 5.6.9

Hence,

$$\text{area } (B) = \int_0^2 \left[\frac{x}{2} - x(x - 2) \right] dx = \int_0^2 \left(\frac{5x}{2} - x^2 \right) dx$$

$$= \left(\frac{5}{4} x^2 - \frac{x^3}{3} \right) \Big|_0^2 = 5 - \frac{8}{3} = \frac{7}{3}.$$

▶ Example 6. Find the area bounded by the curves $y = x$ and $y = x^3$.

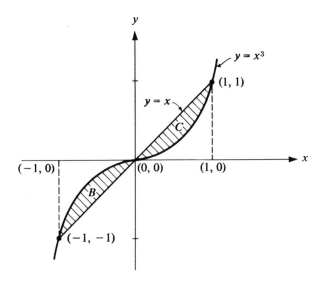

Figure 5.6.10

Solution. First we draw the graphs and find their points of intersection. Obviously the two curves intersect at points for which $x = x^3$. This occurs when $x = 0$, $x = -1$, and $x = 1$. Their graphs are shown in Figure 5.6.10. We wish to find area $(B \cup C) = $ area $(B) + $ area (C). Note that for $x \in [0, 1]$, $x \geq x^3$, while $x \leq x^3$ for $x \in [-1, 0]$. Thus,

$$\text{area } (B + C) = \int_{-1}^0 (x^3 - x) \, dx + \int_0^1 (x - x^3) \, dx$$

$$= \left(\frac{x^4}{4} - \frac{x^2}{2} \right) \Big|_{-1}^0 + \left(\frac{x^2}{2} - \frac{x^4}{4} \right) \Big|_0^1$$

$$= -\frac{1}{4} + \frac{1}{2} + \frac{1}{2} - \frac{1}{4} = \frac{1}{2}.$$

We now turn to the justification of the answers given to Questions 1 and 2 raised at the beginning of the section. In particular, we want to show that the area under the curve f from a to b really is given by $\int_a^b f(x) \, dx$. We assume first that f is continuous on $[a, b]$ and $f(x) \geq 0$ for $x \in [a, b]$. (See Figure 5.6.1.) Let us set $H = \{(x, y) \mid 0 \leq y \leq f(x) \text{ and } a \leq x \leq b\}$. Our problem is to determine the area of *H*, which we denote by area (*H*). Observe that the area is really a function that assigns to each set *H* a nonnegative number, which we call area (*H*). (We assume that there is a well-defined quantity called area.) *H* is often referred to as an *ordinate set* of *f*.

Before proceeding further, let us list three of the fundamental properties that we expect area to have. (These are intuitively plausible from geometric considerations.)

PROPERTY 1. If *H* and *K* are two ordinate sets with $H \subset K$, then area (*H*) $\leq$ area (*K*). (See Figure 5.6.11.)

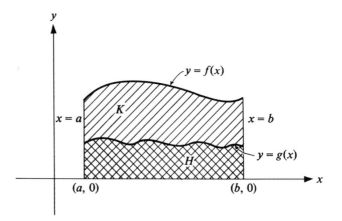

Figure 5.6.11

Suppose on $[a, b]$, $g(x) \leq f(x)$ (as in Figure 5.6.11); then $H = \{(x, y) \mid 0 \leq y \leq g(x)$ and $a \leq x \leq b\}$ and $K = \{(x, y) \mid 0 \leq y \leq f(x)$ and $a \leq x \leq b\}$. Clearly $H \subset K$. Property 1 states that area (*H*) $\leq$ area (*K*).

PROPERTY 2. If *S* is an ordinate set and the line $x = c$ divides *S* into two ordinate sets *H* and *K*, then area (*S*) = area (*H*) + area (*K*).

(Note that $S = H \cup K$. This property simply states that the whole is equal to the sum of its parts. See Figure 5.6.12.)

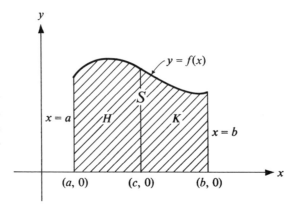

Figure 5.6.12

In Figure 5.6.12,

$$S = \{(x, y) \mid 0 \le y \le f(x), \quad a \le x \le b\},$$
$$H = \{(x, y) \mid 0 \le y \le f(x), \quad a \le x \le c\},$$
and
$$K = \{(x, y) \mid 0 \le y \le f(x), \quad c \le x \le b\}.$$

Property 2 simply states that area (S) = area (H) + area (K).

PROPERTY 3. If S is the set $\{(x, y) \mid a \le x \le b$ and $c \le y \le d\}$ (that is, a rectangle), then area $(S) = (d - c)(b - a)$. (That is, the area of a rectangle is its length times its width—see Figure 5.6.13.)

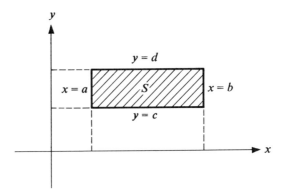

Figure 5.6.13

First, observe that $\int_a^b f(x)\,dx$ possesses Properties 1, 2, and 3, since:

1. If $f(x) \le g(x)$, then by Property 4, Section 5.5,

$$\int_a^b f(x)\,dx \le \int_a^b g(x)\,dx.$$

2. By Property 1, Section 5.5,

$$\int_a^b f(x)\,dx = \int_a^c f(x)\,dx + \int_c^b f(x)\,dx.$$

3. If $f(x) = c$, where c is a constant, then

$$\int_a^b c\,dx = c(b - a).$$

In like manner, $\int_a^b d\,dx = d(b - a)$, and thus

$$\int_a^b d\,dx - \int_a^b c\,dx = (d - c)(b - a).$$

Before starting the proofs, let us clarify our notation and hypothesis. We designate area (K) by $A(x_0)$; K is the set of points bounded above by $y = f(x)$, to the left by $x = a$, and to the right by $x = x_0$. (See Figure 5.6.14.) We assume the following:

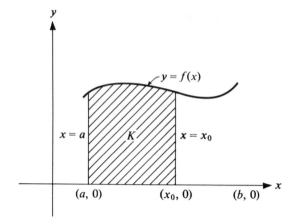

Figure 5.6.14

(a) There exists a well-defined quantity called the area under the curve f from a to x_0, for $a \leq x_0 \leq b$. (We do not give a definition of area, but assume that such a definition exists.)

(b) The area has Properties 1, 2, and 3 stated above.

In more advanced courses, it is shown that there is a definition of area with the stated properties. Indeed, part of that program is carried out in Section 5.3.

If you think about it for a moment, it is clear that $A(x)$ is a function. What is more, it is continuous. In fact, we will show that it has a derivative: Let $x_0 \geq 0$ and let x be any number "near" x_0 (see Figure 5.6.15) such that x, $x_0 \in (a, b)$ with $x > x_0$. We see that $A(x) - A(x_0)$ is the area of the shaded region in Figure 5.6.15. Now, since f is continuous on $[a, b]$ and hence on $[x_0, x]$, we know that f assumes its maximum and its minimum value on this interval at points c_x and d_x, respectively.

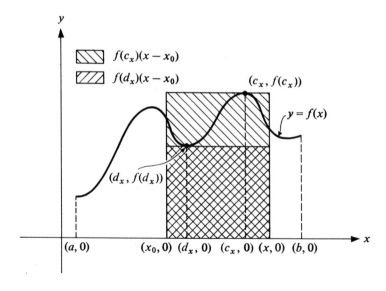

Figure 5.6.15

We use the notation c_x and d_x to indicate that the points at which f assumes its maximum and minimum values on the interval $[x_0, x]$ depend on x. One can easily see that changing x will change c_x and d_x by drawing a few pictures of different functions. All we need to know about c_x and d_x, however, is that they lie in the interval $[x_0, x]$. (This was discussed in Section 4.2.) It is clear from Figure 5.6.15 that

$$f(d_x)(x - x_0) \le A(x) - A(x_0) \le f(c_x)(x - x_0),$$

where $f(d_x)(x - x_0)$ is the area of the rectangle whose height is $f(d_x)$ and width is $x - x_0$ and $f(c_x)(x - x_0)$ is the area of a rectangle whose height is $f(c_x)$ and width is $x - x_0$. Thus,

$$f(d_x) \le \frac{A(x) - A(x_0)}{x - x_0} \le f(c_x).$$

Since f is continuous (and as $x \to x_0^+$, $d_x \to x_0$, $c_x \to x_0$),

$$\lim_{x \to x_0^+} f(d_x) = f(x_0) \quad \text{and} \quad \lim_{x \to x_0^+} f(c_x) = f(x_0).$$

Hence, by the squeezing principle for limits,

$$\lim_{x \to x_0^+} \frac{A(x) - A(x_0)}{x - x_0} = f(x_0).$$

By selecting $x < x_0$, we can easily show that

$$\lim_{x \to x_0^-} \frac{A(x) - A(x_0)}{x - x_0} = f(x_0)$$

and hence for each $x_0 \in (a, b)$,

$$A'(x_0) = \lim_{x \to x_0} \frac{A(x) - A(x_0)}{x - x_0} = f(x_0).$$

We have shown then that $A(x)$ is a differentiable function on $[a, b]$ and $A'(x) = f(x)$ for $x \in (a, b)$. This answers Question 2. Moreover,

$$\int_a^b f(x)\, dx = A(x)\Big|_a^b = A(b) - A(a) = A(b),$$

since $A(a) = 0$ by definition of the area. Since $A(b) = $ area (H), we have proved that

$$\text{area } (H) = \int_a^b f(x)\, dx.$$

(H refers to the region in Figure 5.6.1.)

Note that we answered Questions 1 and 2 simultaneously. We showed that the area function $A(x)$ satisfied the condition $A'(x) = f(x)$. Then we used this fact to show that

$$\text{area } (H) = A(b) = \int_a^b f(x)\, dx.$$

The results we have just obtained are two forms of the *fundamental theorem of calculus;* its importance, both practical and theoretical, cannot be stressed too greatly. For that reason we repeat the results here in the form of two theorems.

THEOREM 1. Let $f(x)$ be continuous on $[a, b]$ and suppose that $f(x) \geq 0$ on $[a, b]$. If area (H) denotes the area of the set of points bounded above by the curve $y = f(x)$ and on the sides by $x = a$ and $x = b$, then $\int_a^b f(x)\,dx$ exists (the Newton integral of f exists) and area $(H) = \int_a^b f(x)\,dx$.

THEOREM 2. Let $f(x)$ be continuous on $[a, b]$. Set $F(t) = \int_a^t f(x)\,dx$ for every t in $[a, b]$. Then F is differentiable on $[a, b]$, and $F'(t) = f(t)$, where $t \in [a, b]$.

One further comment is in order. We did not define the area $A(b)$ of the region bounded by the curve $y = f(x)$, $x = a$, and $x = b$, but assumed there was a definition of area that satisfied Properties 1, 2, and 3. We then proved that with this definition of area,

$$A(b) = \text{area } (H) = \int_a^b f(x)\,dx.$$

In other words, we proved that no matter what definition you choose for area, you always end up with the same number, namely $\int_a^b f(x)\,dx$ for area (H), if your area has Properties 1, 2, and 3. This result is quite remarkable. It can be interpreted as saying there is really only one way to define area, and all definitions lead to the same answer.

Exercises

1. Find the area between the curve and the x axis, where

   (a) $y = x^2 + 1$; $a = 0$, $b = 1$.   (b) $y = x^3$; $a = 1$, $b = 2$.

   (c) $y = x + 1$; $a = -1$, $b = 3$.   (d) $y = x^{1/2}$; $a = 1$, $b = 4$.

   (e) $y = e^x$; $a = 0$, $b = 1$.   (f) $y = \dfrac{1}{x + 2}$; $a = 0$, $b = 4$.

2. Find the area of the region lying above the x axis and under the parabola $y = 4x - x^2$.

3. Find the area of the region bounded by $y = 1/x$, $x = 1$ and $x = 3$, and the x axis.

4. Find the area of the region bounded by $y = -x^2 - 2x + 15$ and the x axis between $x = 0$ and $x = 5$.

5. Find the area of the region bounded by the parabolas $y = 6x - x^2$ and $y = x^2 - 2x$.

6. Find the area of the region bounded by the curve $y = e^x$ and the lines $y = 1 - x$ and $x = 1$. Note that $e^0 = 1$.

7. Find the area of the region bounded by the curve $y = e^{-x}$ and the lines $y = 2x - 1$, $x = -1$ on the left, and $x = 0$ on the right.

8. (a) Find the area of the region bounded by the curves $y = x$ and $y = x^3/4$ over the interval $[-1, 2]$.

   (b) Find the area of the region bounded by the curves $y = x$ and $y = x^3/4$.

9. Find the area of the region inside the closed curve $y^2 = x^2 - x^4$.

10. Find the area of the region bounded by $y = |x|$ and $y = x^2 - 1$ over the interval $[-1, 1]$.

11. Let $f(x) = x - x^2$ and $g(x) = bx$. Determine b so that the region above the graph of g and below the graph of f has area $\frac{9}{2}$.

12. Show that the area of the region that lies between the parabolas $y^2 = 2ax$ and $x^2 = 2by$ $(a, b > 0)$ is $4(ab/3)$.

13. The graph of $y = \sqrt{a^2 - x^2}$ over $-a \leq x \leq a$ is a semicircle of radius a.

    (a) Using this fact, explain why it is true that

    $$\int_{-a}^{a} \sqrt{a^2 - x^2}\, dx = \tfrac{1}{2}\pi a^2.$$

    (b) Evaluate

    $$\int_{0}^{a} \sqrt{a^2 - x^2}\, dx.$$

14. The area bounded by the curve $y = x^2$ and the line $y = 4$ is divided into two equal portions by the line $y = c$. Find c.

15. (a) If $F(x) = \int_0^x \sqrt{1 - t^2}\, dt$, $0 < x < 1$, what is $F'(x)$? (Use Theorem 2 of this section.)

    (b) Suppose $F(x) = \int_0^{x^2} \sqrt{1 - t^2}\, dt$, $0 < x < 1$. How could you find $F'(x)$? Find it.

*16. The area bounded by the x axis, the curve $y = f(x)$, and the lines $x = a$, $x = b$ is equal to $\sqrt{b^2 - a^2}$ for all $b > a$. Find $f(x)$.

5.7 Other Applications of Integration

In this section, we discuss some applications of integration to the social and biological sciences.

▶ Example 1. (*Cost Analysis*) Lightning Bug Electric, a power company, is considering the purchase of a new switching facility. The cost of the machine is $25,000. The management believes that, after a short period of installation and adjustment, cost savings will be realized through increased efficiency. The rate of cost savings over 6 years is thought to be $c(x) = 2,000x$, where x represents years and $c(x)$ represents dollars per year savings at any given time. [*Note:* $c(x)$ does *not* represent the number of dollars saved after x years but rather the *rate* at which dollars are being saved after x years.] Would the machine pay for itself in 4 years? If not, in how many years of operation would the machine pay for itself?

Solution. Since the rate of cost savings is given by $c(x) = 2,000x$, we see that the actual savings after n years, $S(n)$, must be given by

$$S(n) = \int_{0}^{n} 2,000x\, dx.$$

Thus the total savings during the first four years is given by

$$S(4) = \int_0^4 2,000x \, dx = 2,000 \left.\frac{x^2}{2}\right|_0^4 = 16,000.$$

Clearly in four years the machine would not pay for itself, since the savings are only $16,000. To determine in how many years the machine would pay for itself, we must find a number n so that

$$\int_0^n 2,000x \, dx = \left.\frac{2,000x^2}{2}\right|_0^n = \frac{2,000n^2}{2} = 25,000.$$

Thus $n^2 = 25$ or $n = 5$. At the indicated rate of saving, it would take 5 years for this new machine to pay for itself.

▶ Example 2. (*Consumer and Producer Surplus*) Recall from Chapter 4 that a demand function indicates the relationship between the quantity of a commodity demanded and other variables such as the price of the commodity. Let us consider an ideal situation where quantity and demand are the only two variables. If Q is the demand and p is the price then $Q(p)$ is the amount demanded at price p (i.e., people are willing to buy Q units of the commodity at price p). Usually as price increases the demand for the product decreases; it has become too expensive for some. Similarly as the price drops the demand increases; more people can now afford the product (or the same people will buy more).

Let us denote the quantity demanded by x and the price by p. The relationship between x and p may be expressed in three ways: (1) the price p may be given directly in terms of x, that is, $p = f(x)$; or (2) x may be given directly in terms of p, that is, $x = g(p)$; or (3) a relationship exists between x and p such as $x^2 + p^2 - 9 = 0$, or, in general, a relationship of the form $h(x, p) = 0$. Any of these three cases is called the demand law. We remark at this point that the demand is assumed to be a *monotonically decreasing function* of the price—that is, the lower the price the greater the demand; the higher the price the lower the demand. Traditionally economists have preferred formulation (1), although (2) seems somewhat more natural to a mathematician. (See Figure 5.7.1.)

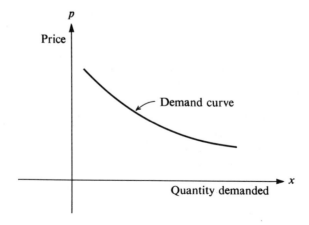

Figure 5.7.1

A supply function for a specific product represents the relationship between the price and the amount manufacturers are willing to produce. Thus if S is the supply and p is the price, then $S(p)$ is the amount manufacturers are willing to supply at price p. As the price increases, additional manufacturers are induced to produce the item. As the price drops and profits dwindle, fewer manufacturers are willing to produce the item. As above, there are several ways of expressing the supply function. Suppose x is the quantity supplied and p the price of one unit of x. Then the relationship between x and p can be expressed as (1) $p = F(x)$, or (2) $x = G(p)$, or (3) $H(p, x) = 0$. We assume that the *price* is a monotonically increasing function of the quantity supplied. Again, economists have traditionally preferred formulation (1). (See Figure 5.7.2.)

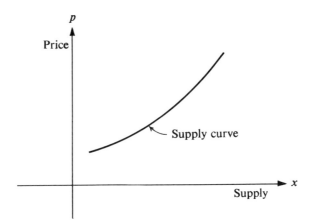

Figure 5.7.2

We now can state a fundamental rule of supply and demand. Market equilibrium is said to occur under pure competition if the demand is equal to the supply. The equilibrium price and equilibrium supply are defined to be the coordinates of the point of intersection of the supply and demand curves. The equilibrium amount and price can be found algebraically by solving the two equations simultaneously. (See Figure 5.7.3.) Note in Figure 5.7.3 that both supply and demand appear on the x axis.

If you stare at Figure 5.7.3 for a minute you can see why the price will approach the equilibrium price p under pure competition. If the price is above p, then more of the item will be produced than is demanded. Inventories will build up and merchants will lower prices to get the goods off their shelves. Suppose, on the other hand, the price is below p. Then more of the product is demanded than is being produced. Merchants will not be able to keep the item in stocks, so great will be the demand, and it will occur to them to raise their prices. Only at price p will there be a balance between supply and demand.

If a demand curve is given and the market demand x_0 and the corresponding price p_0 are determined in some way, then consumers who would have been willing to pay more than the market price p_0 have gained by the setting of the price at p_0 rather than at the maximum price they would have been willing to pay. If a supply

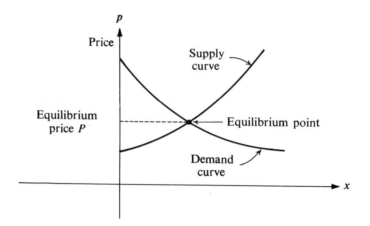

Figure 5.7.3

curve were given and x_0, p_0 were determined by pure competition, then x_0, p_0 would be the equilibrium supply and price. Under certain economic assumptions the total consumer gain, as a result of this fixed price p_0, called the *consumer's surplus*, is represented (Figure 5.7.4) by the area above the line $p = p_0$ and below the *demand* curve. In Figure 5.7.4, we consider the case of pure competition and let $p = f(x)$ be the demand law and $p = F(x)$, the supply law. Then, $p = p_0$ is the market equilibrium price.

It is clear from Figure 5.7.4 that the consumer's surplus C_s is given by

$$C_s = \int_0^{x_0} [f(x) - p_0] \, dx = \int_0^{x_0} f(x) \, dx - p_0 x_0.$$

Note that $p_0 x_0$ is just the area of the rectangle adjacent to the x and y axes. As a concrete example of consumer surplus in action, consider the commodity water. You would probably be willing to pay a dollar a gallon for drinking water while you

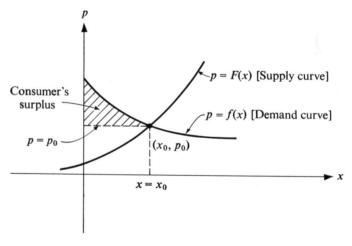

Figure 5.7.4

would be willing to pay only a few cents a gallon for water to water lawns or wash cars. The equilibrium price for water is determined mainly by consumer demand for such purposes as watering lawns and washing cars, which have low priority. Hence the price of water is low. (The argument here is a bit delicate and involves marginal utility, which we shall avoid.) Since you obtain drinking water, for which you would be willing to pay dearly, at the same low price, this is an example of consumer surplus.

If a supply curve is given and if the amount supplied x_0 and the corresponding price p_0 are determined in some way, then producers who would have been willing to supply the commodity below the price p_0 have gained by the setting of the price at p_0. The total producer's gain, called the producer's surplus, is represented by the area below the line $p = p_0$ and above the supply curve. (See Figure 5.7.5.)

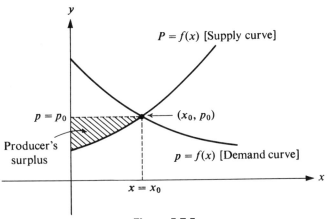

Figure 5.7.5

In Figure 5.7.5, x_0 and p_0 represent the market equilibrium amount and price, respectively. The producer's surplus P_s is given by

$$P_s = \int_0^{x_0} [p_0 - F(x)] \, dx = p_0 x_0 - \int_0^{x_0} F(x) \, dx.$$

For example, if the demand and supply curves are

$$p = F(x) = 14 + x \quad \text{and} \quad p = f(x) = (6 - x)^2, \qquad 0 \le x \le 6,$$

we find the consumer and producer's surplus, where the demand and price are determined under pure competition. First, we construct the two curves and find their point of intersection in order to determine the market equilibrium supply x_0 and price p_0. (See Figure 5.7.6.)

Algebraically, the equilibrium supply is determined by setting the supply equal to the demand (since we assume pure competition). Thus,

$$(6 - x)^2 = 14 + x$$

or

$$36 - 12x + x^2 = 14 + x$$

whence

$$x^2 - 13x + 22 = 0,$$

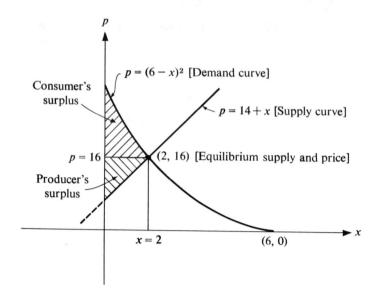

p = (6 − x)² [Demand curve]

Consumer's surplus

p = 14 + x [Supply curve]

p = 16

(2, 16) [Equilibrium supply and price]

Producer's surplus

x = 2

(6, 0)

Figure 5.7.6

and factoring the last expression we obtain $(x - 11)(x - 2) = 0$. Since $0 \le x \le 6$, $x \ne 11$, and hence $x = 2$. Thus, $x_0 = 2$ and $p_0 = 14 + x_0 = 16$. The consumer's surplus is then

$$C_s = \int_0^2 (6 - x)^2 \, dx - 2 \cdot 16 = -\frac{(6 - x)^3}{3}\Big|_0^2 - 32$$

$$= -\tfrac{64}{3} + \tfrac{216}{3} - 32 = \tfrac{56}{3}$$

and the producer's surplus is

$$P_s = 32 - \int_0^2 (14 + x) \, dx = 32 - \left(14x + \frac{x^2}{2}\right)\Big|_0^2$$

$$= 32 - 28 + 2 = 6.$$

Hence, the consumer's surplus is $18.67 and the producer's surplus is $6.00.

▶ Example 3. The process of finding the average value of a finite number of data is familiar to most students. For example, if $a_1, a_2, \ldots, a_n$ were the grades of a class of students on a certain hour exam, then the class average of the test is

$$y_{av} = \frac{a_1 + a_2 + \cdots + a_n}{n}. \tag{5.7.1}$$

When the number of the data is not finite, the above equation is not feasible. For example, if the data y are given by a continuous function $y = f(x)$, $a \le x \le b$, it is not possible to use Equation (5.7.1). In this case, we define the average value of y with respect to x by

$$[y_{av}]_x = \frac{1}{b - a} \int_a^b f(x) \, dx. \tag{5.7.2}$$

Thus the average value of the function $y = x^2$ for $x \in [1, 3]$ is given by

$$[y_{av}]_x = \tfrac{1}{2} \int_1^3 x^2 \, dx = \tfrac{1}{2}(\tfrac{27}{3} - \tfrac{1}{3}) = \tfrac{13}{3}.$$

If we multiply both sides of Equation (5.7.2) by $(b - a)$, we see that

$$(b - a)[y_{av}]_x = \int_a^b f(x) \, dx.$$

The left-hand side is simply the area of a rectangle of height $[y_{av}]_x$ and width $b - a$. (See Figure 5.7.7.) Thus, $[y_{av}]_x$ is that ordinate of the curve $y = f(x)$ that should be used as the altitude if one wishes to construct a rectangle whose base is the interval $[a, b]$ and whose area is the area under the region bounded by $y = f(x)$, the x axis, $x = a$, and $x = b$.

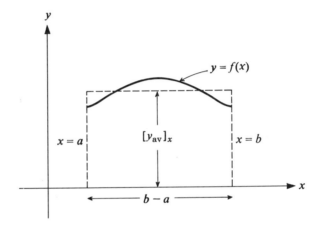

Figure 5.7.7

We now consider an example of some biological phenomena.

▶ **Example 4.** A growing culture of bacteria weighs $5e^{0.3t}$ grams at time t. What is its average rate of growth during the first three hours—that is, from $t = 0$ to time $t = 3$?

Solution. Letting $y = 5e^{0.3t}$, we see that the rate of growth of the culture at any time t is given by $dy/dt = 1.5e^{0.3t}$ grams per hour. [Let $y = f(t)$ be the amount of a substance at time t. Then the rate of growth of the substance at time t is $dy/dt = f'(t)$. We take up this topic at considerable length in Chapter 6.] The average during the first three hours of the growth rate, considered as a function of time t, is defined as

$$\frac{1}{3} \int_0^3 (1.5)e^{0.3t} \, dt = \left[\left(\frac{1.5}{3}\right) \cdot \frac{1}{3} \, e^{0.3t} \right]\Big|_0^3 = \frac{5}{3}(e^{0.9} - 1).$$

Thus the average growth rate in question is $\tfrac{5}{3}(e^{0.9} - 1)$, or about 2.43 grams per hour.

In the preceding example we asked for the average growth rate with respect to time. Suppose we ask for the average growth rate with respect to the amount of bacteria present. In this case, the growth rate at any time t is still $dy/dt = 1.5e^{0.3t}$, where $y = 5e^{0.3t}$. As a function of weight y, this growth rate is $dy/dt = 0.3y$. Thus, we wish to find the average of $0.3y$ for the first three hours. This is the average of $0.3y$ in the interval $5e^{0.3(0)} = 5$ to $5e^{0.3(3)} = 5e^{0.9}$. The average growth rate with respect to weight y is

$$\frac{\int_5^{5e^{0.9}} 0.3y \, dy}{5(e^{0.9} - 1)} = \frac{0.3}{10(e^{0.9} - 1)} \cdot y^2 \Big|_5^{5e^{0.9}} = \frac{0.15[(5e^{0.9})^2 - 5^2]}{5e^{0.9} - 5}.$$

This works out to about 2.59 grams per hour.

▶ Example 5. (*Interest Rates*) Banks compute interest over varying lengths of time: yearly, semiannually, monthly, and even daily. (The latter practice sometimes stems from an effort to avoid certain regulations of the Federal Reserve Board.) It is possible to compute interest instantaneously. If P_0 is the initial amount of money and if interest is computed instantaneously at the rate r, then the accrued interest I over a period of time T is given by the formula

$$I = \int_0^T P_0 r e^{rt} \, dt.$$

Thus if $P(T)$ represents the principal (amount of money) at time T, then

$$P(T) = P_0 + \int_0^T P_0 r e^{rt} \, dt.$$

(It is not very difficult to derive these formulas, but we will not do so here. See Chapter 8 for the derivation.) Let us now consider some specific applications of the above.

▶ Problem 1. How much interest does $1,000 draw at a rate of $5\% = \frac{5}{100}$ compounded instantaneously for 1 year?

Solution. Here $T = 1$, $P_0 = 1,000$, and $r = 0.05$. Thus,

$$I = \int_0^1 P_0 r e^{rt} \, dt = P_0 e^{rt} \Big|_0^1 = P_0 e^r - P_0 = P_0(e^r - 1).$$

We now substitute the known values (it is easier to integrate before substituting) to obtain $I = 1,000(e^{0.05} - 1)$. By checking Table A1, we find

$$I = 1,000(1.0513 - 1) = 51.30.$$

Hence the interest accrued is $51.30. The interest rate in this problem is thus equivalent to 5.13% computed annually.

▶ Problem 2. How long will it take an amount of money P to double if interest is computed instantaneously at a rate of 10%?

Solution. The unknown in this problem is T. Our equation becomes

$$P(T) = 2P = P + \int_0^T Pre^{rt}\, dt$$

or

$$2P = P + Pe^{rt}\Big|_0^T = Pe^{rT} \quad \text{and thus} \quad 2 = e^{rT}.$$

We now substitute $r = \frac{1}{10}$ to obtain $2 = e^{T/10}$. To determine T, we first take logarithms of both sides of the equation and find that

$$\log 2 = \log e^{T/10} = T/10 \quad \text{or} \quad T = 10 \log 2.$$

Checking Table A2 in the back of the text, we find

$$T = 10(0.6931) = 6.931.$$

Thus the money will double in slightly less than seven years.

Exercises

1. The Lightning Bug Electric Company is considering the purchase of a new computer. The cost of the machine is $50,000. The director of the computing center believes that after a short period of adjustment, savings from increased efficiency of various departments utilizing the computer will offset the cost. It is thought that the rate of cost savings will be $C(x) = 6,000x + 2,000$ over 10 years, where x represents years and $C(x)$ is dollars per year savings at any given time. Will the machine pay for itself during its 10-year life? At what time would the break-even point come?

2. In Exercise 1, suppose another machine costing $100,000 is being considered with an expected life of 20 years. Assume that the cost savings function $C(x)$ is the same as in Exercise 1. Would the machine pay for itself in 10 years? Where would the break-even point be? Suppose that the Lightning Bug Electric Company would consider the $100,000 machine against two consecutive $50,000 machines (with installation and procedural change costs for each). Which would be less expensive for a 20-year period?

3. If the demand law is $p = 36 - 3x^2$, find the consumer's surplus (a) if $x_0 = 2$ and (b) if the commodity is free—that is, $p_0 = 0$. Draw the appropriate diagram.

4. The quantity demanded and the corresponding price, under pure competition, are determined by the demand and supply laws

$$p = 36 - x^2 \quad \text{and} \quad p = 6 + \tfrac{1}{4}t^2,$$

where x is the supply and t is the demand. Determine the corresponding consumer's surplus and producer's surplus. Draw the appropriate diagram.

5. If the supply curve is $p = 4e^{x/3}$, and $x_0 = 3$, find the producer's surplus.

6. (a) Find the average value of $f(x) = x^3$ for $x \in [1, 3]$.

   (b) How does this average compare to $f(2)$? To the average of $f(1)$ and $f(3)$?

7. The amount of a quantity y present is a function of time t. Show that the average of the growth rate $\dfrac{dy}{dt}$, considered as a function of time, is simply "the change in y divided by the length of time considered."

8. A growing culture of bacteria weighs 10^t grams at time t. What is its average weight of growth during the first three hours (a) with respect to time t, and (b) with respect to the amount of bacteria present?

9. Find the average value of y with respect to x for that part of the curve $y = \sqrt{ax}$ between $x = a$ and $x = 3a$.

10. If \$200 draws interest at a rate of 4% compounded instantaneously, how much interest will accrue in 3 years?

11. How long will it take for an amount P to triple at a rate of 20% compounded instantaneously?

12. An amount of money P draws interest at the rate r compounded instantaneously. If the amount doubles in 15 years, what is r?

13. An amount of money P draws interest at 5% compounded instantaneously. At the end of 10 years the total amount is \$2,000. What was the initial amount P?

14. If interest is compounded instantaneously at the rate of 10%, how much money must you start with to obtain \$1,000 in interest at the end of one year?

15. The Denver Dairy Trust is considering the purchase of a new slush machine. There are two models available, the Economy and the Delux. The Economy costs \$1,000 and its rate of cost savings function is $C(x) = 400x$ over a 10-year period. The Delux costs \$2,000 and its rate of cost savings function is $C(x) = 300x^2$ over a 10-year period.

   Compute the amount each machine saves over a period of 1, 2, 3, 4, 5 years. Note the saving is negative for the first few years since the cost of the machine must be taken into account. How many years does it take for the Delux machine to out-perform the Economy model? (Use your previous calculations to answer this question.)

5.8 Integration by Parts

In Section 5.2 we remarked that there was no rule analogous to the rule for the differentiation of the product of two functions for antiderivatives, and hence Newton integrals. The integration by parts formula which we now introduce comes closest to filling this gap. It also allows us to integrate a wide variety of rather complicated-looking integrals. In Chapter 7, Section 7.2, we shall show how one could use tables of integrals for other problems.

This formula is based upon the rule for differentiating products,

$$\frac{d}{dx}[f(x)g(x)] = f(x)\frac{dg(x)}{dx} + g(x)\frac{df(x)}{dx}.$$

We take antiderivatives of both sides, which yields (apart from the arbitrary constant),

$$\int \frac{d}{dx}[f(x)g(x)]\,dx = \int f(x)\frac{dg(x)}{dx}\,dx + \int g(x)\frac{df(x)}{dx}\,dx$$

and thus obtain

$$f(x)g(x) = \int f(x)\frac{dg(x)}{dx}\,dx + \int g(x)\frac{df(x)}{dx}\,dx$$

or equivalently,

$$\int f(x)\frac{dg(x)}{dx}\,dx = f(x)g(x) - \int g(x)\frac{df(x)}{dx}\,dx.$$

If we let $u = f(x)$, $v = g(x)$, $du = f'(x)\,dx$, and $dv = g'(x)\,dx$, then simply rewriting the last equation yields

$$\int u\,dv = uv - \int v\,du.$$

It is a good idea to commit this formula to memory. We observe that for definite integrals, the formula reads

$$\int_a^b u\,dv = u(x)v(x)\Big|_a^b - \int_a^b v\,du.$$

When using this technique, one tries to choose u and v in such a way that the integral appearing on the right is easier to evaluate than that on the left. It takes a good deal of practice to learn to choose the u and v properly.

Let us look at some examples.

▶ **Example 1.** Find $\int xe^x\,dx$.

Solution. In this case we choose $u = x$ and $dv = e^x\,dx$. Then, $du = dx$ and $v = e^x$. (To find v, of course, we must also find an antiderivative.) Thus,

$$\int xe^x\,dx = \int u\,dv = uv - \int v\,du$$

$$= xe^x - \int e^x\,dx = xe^x - e^x + C = (x - 1)e^x + C.$$

▶ **Example 2.** Find $\int \log x\,dx$.

Solution. In this case we are not left with much choice. Choose $u = \log x$ and $dv = dx$. Obviously $v = x$ and $du = (1/x)\,dx$. Thus,

$$\int \log x\,dx = \int u\,dv = uv - \int v\,du$$

$$= x\log x - \int x\cdot\frac{1}{x}\,dx = x\log x - \int dx = x\log x - x + C.$$

Note that we really use integration by parts to evaluate integrals of form $\int f(x)g(x)\,dx$. We choose $u = f(x)$ and $dv = g(x)\,dx$ if (1) $v\,du$ is easier to compute than $u\,dv$, and

(2) it is a simple matter to find an antiderivative of $g(x)$. If this is not the case, then we choose $u = g(x)$ and $dv = f(x)\, dx$.

The next examples show that we must be careful how we choose u and v.

▶ **Example 3.** Let us attempt to evaluate $\int x^{-1}\, dx$ by parts.

Solution. We would probably set $u = x^{-1}$, $dv = dx$, and, thus, $du = -x^{-2}\, dx$ and $v = x$. This would yield

$$\int x^{-1}\, dx = 1 + \int x^{-2}x\, dx = 1 + \int x^{-1}\, dx.$$

Obviously this is false. (Otherwise $0 = 1$.) It is clear in this case that an *arbitrary* constant *must* be included; that is,

$$\int x^{-1}\, dx = 1 + \int x^{-1}\, dx + C,$$

which implies $C = -1$. (For integration by parts with a definite integral, this sort of thing cannot happen; see Exercise 7.) However, we were *not* able to evaluate $\int x^{-1}\, dx$ in this manner. We already know that $\int x^{-1}\, dx = \log |x| + C$. Thus integration by parts is not an appropriate technique in this problem.

▶ **Example 4.** Evaluate $\int x^2 e^x\, dx$.

Solution. Suppose we begin by letting $u = e^x$, $du = e^x\, dx$, $dv = x^2\, dx$, and $v = x^3/3$. Then,

$$\int x^2 e^x\, dx = uv - \int v\, du = \frac{x^3 e^x}{3} - \frac{1}{3}\int x^3 e^x\, dx.$$

Clearly we are worse off than when we started: $\int x^3 e^x\, dx$ is more difficult to evaluate than $\int x^2 e^x\, dx$. Now try $u = x^2$ and $dv = e^x\, dx$; whence $du = 2x\, dx$ and $v = e^x$. Thus

$$\int x^2 e^x\, dx = x^2 e^x - 2\int x e^x\, dx.$$

By Example 1 (that is, another application of integration by parts on $\int x e^x\, dx$),

$$\int x e^x\, dx = (x - 1)e^x + C.$$

Thus,

$$\int x^2 e^x\, dx = x^2 e^x - 2(x - 1)e^x + C.$$

This shows that one must use some insight in choosing u and dv.

▶ **Example 5.** Evaluate $\int_0^1 x(1 + x)^{1/2}\, dx$.

Solution. It is best in this example if we choose $u = x$ and $dv = (1 + x)^{1/2}\, dx$. (Otherwise, $\int v\, du$ becomes extremely complicated.) In this case, $du = dx$ and

$$v = \int (1 + x)^{1/2}\, dx = \tfrac{2}{3}(x + 1)^{3/2}.$$

Then

$$\int_0^1 x(1 + x)^{1/2}\, dx = u(x)v(x)\Big|_0^1 - \int_0^1 v\, du$$

$$= \tfrac{2}{3}x(1 + x)^{3/2}\Big|_0^1 - \tfrac{2}{3}\int_0^1 (1 + x)^{3/2}\, dx$$

$$= \tfrac{2}{3}2^{3/2} - \tfrac{2}{3}\int_0^1 (1 + x)^{3/2}\, dx.$$

We must evaluate $\int_0^1 (1 + x)^{3/2}\, dx$. This could be done by making the substitution $w = 1 + x$. But by inspection we see that

$$\int_0^1 (1 + x)^{3/2}\, dx = \tfrac{2}{5}(1 + x)^{5/2}\Big|_0^1 = \tfrac{2}{5}2^{5/2} - \tfrac{2}{5}.$$

Thus

$$\int_0^1 x(1 + x)^{1/2}\, dx = \tfrac{2}{3}2^{3/2} - \tfrac{4}{15}2^{5/2} + \tfrac{4}{15}.$$

This problem could also have been done by substitution. We have done it by integration by parts to illustrate the techniques of the section.

Exercises

1. Find, the following antiderivatives, using integration by parts:

   (a) $\int 2xe^x\, dx.$ (b) $\int xe^{-x}\, dx.$ (c) $\int x \log x\, dx.$

   (d) $\int x(1 + x)^{3/2}\, dx.$ (e) $\int x^3 e^x\, dx.$ (f) $\int x^2 \log x\, dx.$

   (g) $\int x^{1/2} \log x\, dx.$ (h) $\int xe^{4x}\, dx.$

2. Find the following Newton integrals, using integration by parts:

   (a) $\int_0^1 2xe^{-x}\, dx.$ (b) $\int_1^e x \log x\, dx.$

   (c) $\int_0^4 x^2\sqrt{4 - x}\, dx.$ (d) $\int_0^1 \dfrac{x}{\sqrt{1 + x}}\, dx.$

3. Show that the integrals $\int_a^b f(x)g'(x)\, dx$ and $\int_a^b f'(x)g(x)\, dx$ have as their sum $f(b)g(b) - f(a)g(a)$. [*Hint:* Integrate one of them by parts.]

4. Use integration by parts to show that

   $$\int \sqrt{1 - x^2}\, dx = x\sqrt{1 - x^2} + \int \frac{x^2\, dx}{\sqrt{1 - x^2}}.$$

   Write $x^2 = (x^2 - 1) + 1$ in the second integral and deduce the formula

   $$\int \sqrt{1 - x^2}\, dx = \tfrac{1}{2}x\sqrt{1 - x^2} + \tfrac{1}{2}\int \frac{dx}{\sqrt{1 - x^2}}.$$

5. Show that the formula

$$\int_a^b xg'(x)\,dx = xg(x)\Big|_a^b - \int_a^b g(x)\,dx$$

is a special case of the integration by parts formula for definite integrals.

6. Evaluate $\int_0^1 x(1 + x)^{1/2}\,dx$ by making the substitution $w = x + 1$ initially.

7. Evaluate $\int_a^b x^{-1}\,dx$ as in Example 3 by integration by parts. (Assume $a > 0$, $b > 0$.) Note that the right- and left-hand sides are equal. Why?

5.9 Chapter 5 Summary

1. A function F, differentiable on an interval I (open or closed), is said to be an *antiderivative* of f if

$$F'(x) = f(x) \qquad \text{for} \quad x \in I.$$

We also write $F(x) = \int f(x)\,dx + C$.

2. Antiderivatives of common functions. (In all parts, C is an arbitrary constant.)

   (a) $\int K\,dx = Kx + C$, where K is a constant.

   (b) $\int x^n\,dx = x^{n+1}/(n + 1) + C$ for $x \neq -1$.

   (c) $\int x^{-1}\,dx = \log |x| + C$ for $x \neq 0$.

   (d) $\int e^x\,dx = e^x + C$.

3. Properties of antiderivatives

   (a) $\int [f(x) \pm g(x)]\,dx = \int f(x)\,dx \pm \int g(x)\,dx$.

   (b) $\int kf(x)\,dx = k \int f(x)\,dx$, where k is a constant.

4. A function f is said to be *Newton integrable* on $[a, b]$ if f has an antiderivative F on $[a, b]$. We define the Newton integral as

$$\int_a^b f(x)\,dx = F(x)\Big|_a^b = F(b) - F(a).$$

5. Substitution

   Let $u(x)$ be a differentiable function on $[a, b]$. Then

$$\int f[u(x)]u'(x)\,dx = \int f(u)\,du + C,$$

   For antiderivatives,

$$\int_a^b f[u(x)]u'(x)\,dx = \int_{u(a)}^{u(b)} f(u)\,du.$$

   where C is an arbitrary constant.

6. Integration by parts

$$\int_a^b u \, dv = uv \Big|_a^b - \int_a^b v \, du$$

or for antiderivatives,

$$\int u \, dv = uv - \int v \, du \qquad \text{(where } u = u(x), \, v = v(x)\text{),}$$

where C is an arbitrary constant.

7. The area of the region bounded above by the curve $y = f(x)$, below by the x axis, on the left by $x = a$, and on the right by $x = b$ (see Figure 5.9.1) is given by

$$\text{area } (H) = \int_a^b f(x) \, dx.$$

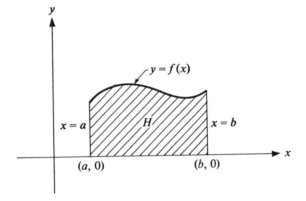

Figure 5.9.1

(a) If the curve $y = f(x)$ lies *below* the x axis, then

$$\text{area } (H) = -\int_a^b f(x) \, dx.$$

(b) In Figure 5.9.2,

$$\text{area } (H) + \text{area } (I) = \int_a^c f(x) \, dx - \int_c^b f(x) \, dx.$$

(c) In Figure 5.9.3, $f(x) \le g(x)$, $x \in [a, b]$. The area between the two curves, area (I), is given by

$$\text{area } (I) = \text{area } (B) - \text{area } (A) = \int_a^b g(x) \, dx - \int_a^b f(x) \, dx$$

$$= \int_a^b [g(x) - f(x)] \, dx.$$

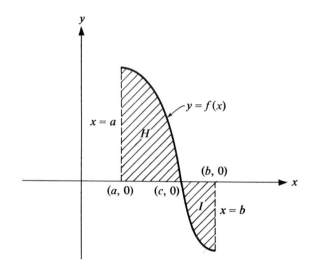

Figure 5.9.2

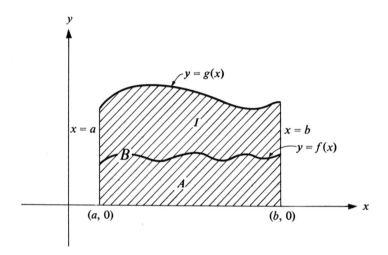

Figure 5.9.3

Review Exercises

1. Evaluate the following:

(a) $\int (5x^2 - 9x^{-10} + 10x^{-5/3})\, dx.$

(b) $\int \left(104e^x - \dfrac{1}{x} + 3\right) dx.$

(c) $\int_{-1}^{1} e^{10x+1}\, dx.$

(d) $\int \dfrac{\log (2x)}{x}\, dx.$

(e) $\int \dfrac{x}{\sqrt{1 - x^2}}\, dx.$

(f) $\int_{0}^{1} \dfrac{\log (x + 1)}{x + 1}\, dx.$

(g) $\displaystyle\int x(4x^2 - 9)^5 \, dx.$

(h) $\displaystyle\int_{-1}^{0} x^2 e^{x^3+1} \, dx.$

(i) $\displaystyle\int x(x - 3)^{1/2} \, dx.$

(j) $\displaystyle\int x^2 e^x \, dx.$

(k) $\displaystyle\int_{1}^{2} x \log x \, dx.$

(l) $\displaystyle\int \frac{(\log x)^5}{x} \, dx.$

(m) $\displaystyle\int \frac{\log x}{x^4} \, dx.$

(n) $\displaystyle\int_{1}^{2} \frac{\log (x + 1)}{x^2} \, dx. \left[Hint: \frac{1}{x(x + 1)} = \frac{1}{x} - \frac{1}{x + 1}. \right]$

(o) $\displaystyle\int_{3}^{5} xe^{x+1} \, dx.$

(p) $\displaystyle\int_{1}^{10} \left(\frac{1}{x} + x\right) dx.$

(q) $\displaystyle\int_{1}^{2} \frac{x}{\sqrt{3x^2 - 1}} \, dx.$

(r) $\displaystyle\int_{0}^{1} x(2x + 1)^{1/3} \, dx.$

2. Find the area enclosed by the curves $y = x^2$ and $y = x^5$.

3. Find the area enclosed by the curves $y = x^2$ and $y = x^4$. (Draw a picture.)

4. Find the area bounded by the curve $f(x) = e^x - 2$, the x axis, and the lines $x = 0$ and $x = 1$.

5. The amount of money in the Monopoly Bank over a two-year period is given by the function $f(x) = 2x^2 - 3x + 9$ for $0 \le x \le 2$. What was the average amount of money in the bank over this period?

*6. The area bounded by the x axis, the curve $y = f(x)$, and the lines $x = 1$ and $x = t$ is equal to $(t^2 + 1)^{1/2} - 2^{1/2}$ for all $t > 1$. Find $f(x)$.

7. The rate of change of sales s with respect to advertising expenditure x is given by

$$\frac{ds}{dx} = 5 + \frac{10}{x} + x \log x \qquad \text{for} \quad 1 \le x \le 5.$$

Determine the total change in sales as advertising expenditure is increased from 1 to 5.

8. The marginal cost of producing Supergoodie Cake Mix is given by

$$m(x) = 225 - 1{,}200e^{-2x},$$

where x, the number of packages produced, is in dozens, and the cost is in cents. Find a formula for $C(x)$, the total cost of producing x dozen packages of the cake mix, given that 2 dozen can be produced at a cost of $8.00. How much will it cost to produce 4 dozen packages of the mix?

9. Evaluate the Newton integral $\int_{-1}^{1} (x^3 - x) \, dx$ and explain the result by drawing the region whose area the definite integral is supposed to represent.

10. Let $y = x^2 + 1$ be the supply curve for a certain product (y is the supply and x is the price). Let $y = -x + 7$ be the demand curve for the same product (y is the demand and x is the price). Assuming pure competition, find the equilibrium price and supply. Find the consumer surplus and the producer surplus.

11. If \$200 draws interest at the rate of 5% compounded instantaneously, find the total amount at the end of 10 years.

12. How long will it take for an amount of money P to double if it draws 1% interest compounded instantaneously?

13. A piece of real estate worth \$20 billion in 1970 is alleged to have been worth \$20 in 1640. What rate of interest, compounded instantaneously, would yield this increase in the same time?

14. Find all differentiable functions f which satisfy the equation

$$[f(x)]^2 = \int_0^x f(t)\, dt.$$

[*Hint:* First differentiate both sides of the equation with respect to x, the left-hand side by the chain rule, the right by the fundamental theorem of calculus. Then, perform the obvious cancellation. Now, integrate both sides of the equation and use the fact that $f(0) = 0$.]

A P P E N D I X   1

5.10   The Newton and Riemann Integrals Revisited

In this section we compare the Newton and Riemann integrals and answer several questions about them. We begin by recalling the definition of the Riemann integral given in Section 5.4.

Let f be any function bounded on $[a, b]$. (A function f is bounded on $[a, b]$ if there exists a number M such that $|f(x)| \le M$, $x \in [a, b]$.) If n is a positive integer, we choose points $x_0, x_1, x_2, \ldots, x_n$ such that $a = x_0 \le x_1 \le x_2 \le \cdots \le x_n = b$. We then choose points $c_1, c_2, \ldots, c_n$ such that $c_i \in [x_{i-1}, x_i]$. Finally, we consider the sum

$$R_a^b(n) = f(c_1)(x_1 - x_0) + f(c_2)(x_2 - x_1) + \cdots + f(c_n)(x_n - x_{n-1}). \quad (5.10.1)$$

This sum is called a Riemann sum. We regard the function f and the interval $[a, b]$ as fixed, but the integer n and the points x_i, c_i may be chosen in various ways. In most cases, the number $R_a^b(n)$ obtained in this manner will vary as we vary n and the choice of the x_i's and c_i's.

If we increase n and space the points $x_0, x_1, \ldots, x_n$ in such a way that the maximum of the distance between consecutive points (that is, $|x_i - x_{i-1}|$) approaches 0 as $n \to \infty$, then $R_a^b(n)$ may or may not approach a limit. If it does, we say the function f is Riemann integrable. Thus, if

$$\lim_{n \to \infty} R_a^b(n) = \lim_{\substack{\max|x_i - x_{i-1}| \to 0 \\ n \to \infty}} [f(c_1)(x_1 - x_0) + \cdots + f(c_n)(x_n - x_{n-1})]$$

exists, we say f is Riemann integrable.

We denote the limit by $I_a^b(f)$ to indicate that it depends only on the function f and the interval $[a, b]$. If f is continuous on $[a, b]$, then f is Riemann integrable on $[a, b]$, although this fact is not simple to prove. At this point, it is natural to ask about the relation between the Riemann and the Newton integral. More specifically, (1) if a function is Riemann integrable, is it Newton integrable and vice versa? (2) if a function is both Riemann and Newton integrable, do the integrals have the same value?

For the first question, the answer is no in both directions. For example, the function

$$f(x) = \begin{cases} 0 & \text{for } 0 \le x < 1, \\ 2 & \text{for } 1 \le x \le 2, \end{cases}$$

is Riemann integrable, but is not Newton integrable since f does not have an anti-derivative on $[0, 2]$. This is not too hard to show. To produce a function that is Newton integrable but not Riemann integrable is more difficult and we press the point no further.

However, the answer to the second question is yes. The question is not as difficult as it seems, so let us write down the answer with a proof.

THEOREM. *Let f be both Riemann and Newton integrable on $[a, b]$. Then*

$$I_a^b(f) = \int_a^b f(x)\, dx = F(b) - F(a),$$

where $F'(x) = f(x)$ for $x \in [a, b]$.

Proof: Since f is Newton integrable, we know that a function $F(x)$ exists with the property that $F'(x) = f(x)$. Since f is Riemann integrable, we know that

$$\lim_{\substack{\max|x_i - x_{i-1}| \to 0 \\ n \to \infty}} [f(c_1)(x_1 - x_0) + \cdots + f(c_n)(x_n - x_{n-1})]$$

exists for any choice of x_i and c_i as long as $c_i \in [x_{i-1}, x_i]$ and $\max |x_i - x_{i-1}| \to 0$. Since the limit exists for any choice of the c_i's, let us choose them to our advantage. Consider $x_0, x_1, \ldots, x_n$ to be fixed. By the mean value theorem, we know that

$$\frac{F(x_i) - F(x_{i-1})}{x_i - x_{i-1}} = F'(p_i)$$

for some $p_i \in (x_{i-1}, x_i)$. Thus,

$$F(x_i) - F(x_{i-1}) = F'(p_i)(x_i - x_{i-1}) = f(p_i)(x_i - x_{i-1}).$$

We will choose $c_i = p_i$. Then

$$\begin{aligned} R_a^b(n) &= [f(p_1)(x_1 - x_0) + \cdots + f(p_n)(x_n - x_{n-1})] \\ &= [(F(x_1) - F(x_0)) + (F(x_2) - F(x_1)) + \cdots + (F(x_n) - F(x_{n-1}))] \\ &= F(x_n) - F(x_0) = F(b) - F(a). \end{aligned}$$

(The second from the last line reduces to the last line if we perform all available cancellations.) Thus,

$$\lim_{n \to \infty} R_a^b(n) = I_a^b(f) = F(b) - F(a) = \int_a^b f(x)\, dx.$$

The first equality is just the definition of the Riemann integral. The last equality is just the definition of the Newton integral.

As a special case of this theorem, we note that if f is continuous on $[a, b]$, then f is both Newton and Riemann integrable (with the same value, of course). This case is intimately related to the fundamental theorem of calculus.

While the Riemann integral is a bit more flexible, the Newton integral suffices for most applications to the social and biological sciences.

For a more detailed discussion of the ideas presented in this section, the reader is referred to either J. W. Kitchen, *Calculus of One Variable* (Addison-Wesley, Reading, Mass., 1968), or T. Apostol, *Calculus* (Xerox, Lexington, Mass., 1967), 2nd ed., Vol. I.

A P P E N D I X   2

5.11   Volumes

Integrals can also be used to calculate the volume of various solids. We first discuss the volume of a solid of known cross-sectional area. Many solids are conveniently described in terms of three-dimensional Cartesian coordinates. We introduce such a system by first selecting three mutually perpendicular lines called the coordinate axes. Their common point of intersection is called the origin of the coordinate system. Using a common unit of length, we introduce coordinates onto each of the coordinate axes (called the x, y, and z axes). (See Figure 5.11.1.) We do this in such a way that if the index and middle finger of the right hand is pointed in the direction of the positive x and y axes, then the right thumb is pointed in the direction of the positive z axis. The three planes determined by pairs of coordinate axes are called the *coordinate planes*. These planes are the xy plane formed by the x and y axes, the yz plane formed by the y and z axes, and the xz plane formed by the x and z axes. Just as we were able to establish a one-to-one correspondence between points in the two-dimensional coordinate system and ordered pairs of numbers, we could do

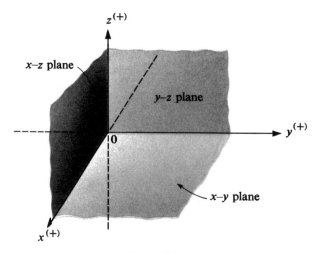

Figure 5.11.1

exactly the same between points in the three-dimensional system and ordered triples of numbers. Given any point P_0 in space, we let x_0, y_0, and z_0 be the perpendicular projections of P onto the x, y, and z axes. We then write for P_0 the ordered triple (x_0, y_0, z_0). (See Figure 5.11.2.)

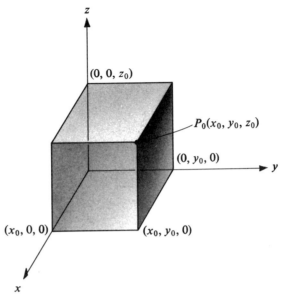

Figure 5.11.2

We now proceed to show that if B is any solid whose cross-sectional area is given by $A(x)$, $a \leq x \leq b$, then the volume of H is

$$\text{volume } (H) = \int_a^b A(x)\, dx.$$

Although this is true more generally, we will assume that $A(x)$ is continuous.

Let us consider the solid illustrated in Figure 5.11.3. We wish to determine the volume of the solid H, whose cross-sectional area is known for any $a \leq x \leq b$, to be $A(x)$. (*Caution:* Do not confuse the present notation with that of the previous section.) In order to show that volume $(H) = \int_a^b A(x)\, dx$, we proceed as we did in determining areas. Imagine that the solid is generated by the continuous expansion of the set $A(a)$, indicated in Figure 5.11.3. Thus, corresponding to x_0, we get the solid S_{x_0}, which is that part of H that lies between the planes $x = a$ and $x = x_0$, indicated in Figure 5.11.4. We denote the volume of S_x by $V(x)$ and the volume of S_{x_0} by $V(x_0)$. We study the rate at which the volume increases. In particular, we show that for any $x_0 \in [a, b]$, $V'(x_0) = A(x_0)$. First observe that the volume $[V(x) - V(x_0)]$ is the volume of the slice shown in Figure 5.11.5. Now if we consider the solid cylinder C_{x_0}, whose cross-sectional area is $A(x_0)$ and height is $x - x_0$ (see Figure 5.11.6), and the solid cylinder C_x, whose cross-sectional area is $A(x)$ and height is $(x - x_0)$ (see Figure 5.11.7), then the solid slice $K \subset C_x$ and $C_{x_0} \subset K$. (See Figure 5.11.8.) Thus, we see that geometrically

$$\text{volume } (C_{x_0}) \leq V(x) - V(x_0) \leq \text{volume } (C_x);$$

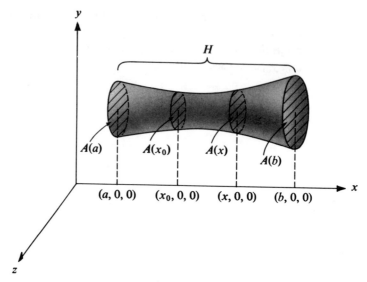

Figure 5.11.3

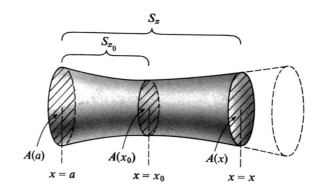

Figure 5.11.4

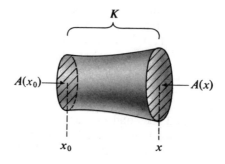

Figure 5.11.5

$A(x_0)$

The cylinder C_{x_0}

Figure 5.11.6

$A(x)$

The cylinder C_x

Figure 5.11.7

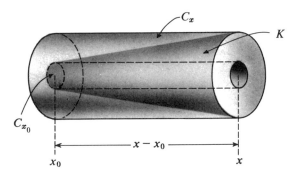

C_x

K

C_{x_0}

$x - x_0$

x_0

x

Figure 5.11.8

but the volume of the cylinder is simply the area of its base times its height. Hence, volume $(C_x) = A(x)(x - x_0)$, volume $(C_{x_0}) = A(x_0)(x - x_0)$ and

$$A(x_0)(x - x_0) \le V(x) - V(x_0) \le A(x)(x - x_0)$$

or

$$A(x_0) \le \frac{V(x) - V(x_0)}{x - x_0} \le A(x).$$

Since A is continuous on $[a, b]$, $\lim_{x \to x_0^+} A(x) = A(x_0)$ and by the squeezing principle,

$$\lim_{x \to x_0^+} \frac{V(x) - V(x_0)}{x - x_0} = A(x_0).$$

If we take $x < x_0$, then we find

$$\lim_{x \to x_0^-} \frac{V(x) - V(x_0)}{x - x_0} = A(x_0),$$

thus yielding

$$V'(x_0) = \lim_{x \to x_0} \frac{V(x) - V(x_0)}{x - x_0} = A(x_0).$$

We have shown, then, that for any $x \in (a, b)$, $V'(x) = A(x)$ and hence

$$\text{volume } (H) = V(b) - V(a) = \int_a^b A(x)\, dx,$$

where A is the cross-sectional area of the solid.

▶ Example 1. A solid has a circular base of radius 4. Find the volume of the solid if every plane section perpendicular to a fixed diameter is an equilateral triangle.

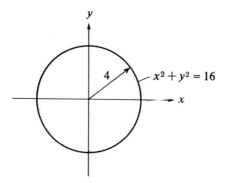

Figure 5.11.9

Solution. Since the solid has a circular base of radius 4, we first sketch the base in the xy plane. (See Figure 5.11.9.) Next, we use the fact that every plane section perpendicular to a fixed diameter is an equilateral triangle. Hence, the solid must appear as shown in Figure 5.11.10, where $y = \sqrt{16 - x^2}$. Hence,

$$A(x) = \tfrac{1}{2}(2y)\sqrt{3}y = \sqrt{3}y^2 = \sqrt{3}(16 - x^2).$$

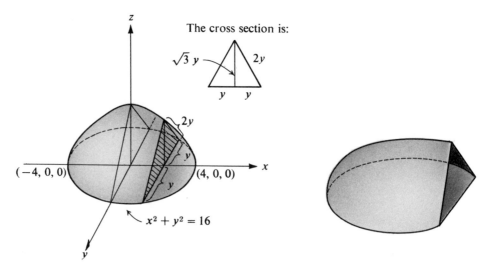

Figure 5.11.10

Thus, the volume V is given by

$$V = \sqrt{3} \int_{-4}^{4} (16 - x^2)\, dx.$$

Since $16 - x^2$ is an *even* function,

$$V = 2\sqrt{3} \int_{0}^{4} (16 - x^2)\, dx = 2\sqrt{3}\left(16x - \frac{x^3}{3}\right)\Big|_{0}^{4} = 2\sqrt{3}\left(64 - \frac{64}{3}\right) = \sqrt{3} \cdot \frac{256}{3}.$$

▶ Example 2. The base of a certain solid is an equilateral triangle of side s with one vertex at the origin and an altitude along the x axis. Each plane section perpendicular to the x axis is a square, one side of which lies in the base of the solid. Find the volume of the solid.

Solution. The base is shown in Figure 5.11.11, and the solid is shown in Figure 5.11.12. The volume is given by $V = \int_{0}^{h} A(x)\, dx$. (We call h the length of the

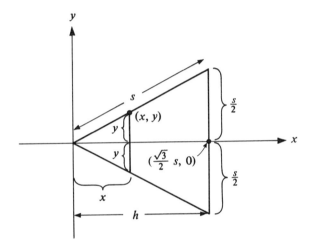

Figure 5.11.11

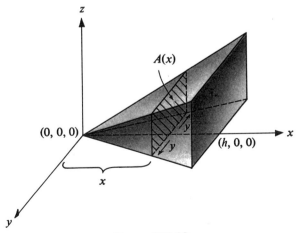

Figure 5.11.12

altitude.) We must compute the cross-sectional area $A(x) = 4y^2$. Thus, we must find y as a function of x. If we let h be the altitude of the triangle in Figure 5.11.11, then we note that

$$h^2 = s^2 - \frac{s^2}{4} = \frac{3s^2}{4} \quad \text{or} \quad h = \frac{\sqrt{3}}{2}s.$$

Now, using similar triangles,

$$\frac{y}{x} = \frac{s/2}{h} = \frac{s/2}{\sqrt{3}s/2} = \frac{1}{\sqrt{3}}$$

and thus $A(x) = \frac{4}{3}x^2$. Hence,

$$V = \int_0^h A(x)\, dx = \int_0^{\sqrt{3}s/2} \frac{4}{3}x^2\, dx = \frac{4}{9}x^3 \Big|_0^{\sqrt{3}s/2} = \frac{4}{9} \cdot \frac{3\sqrt{3}}{8} \cdot s^3 = \frac{\sqrt{3}s^3}{6}.$$

Solids can also be generated by revolving ordinate sets, that is, a set

$$S = \{(x, y) \mid a \le x \le b \quad \text{and} \quad 0 \le y \le f(x)\}$$

about the x axis. These solids are called solids of revolution. Consider a function f that is continuous and nonnegative on $[a, b]$. (See Figure 5.11.13.) Let B be the

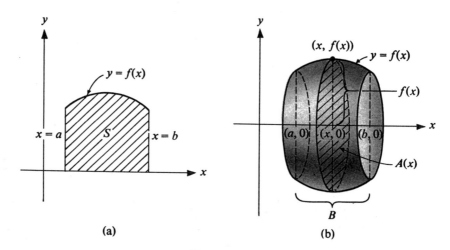

(a) (b)

Figure 5.11.13

solid generated by revolving S about the x axis. The cross-sectional area of B at x, $A(x) = \pi[f(x)]^2$, since the cross section of B corresponding to x is a circle of radius $f(x)$. Hence, the volume of B, $V(B)$, is given by

$$V(B) = \pi \int_a^b [f(x)]^2\, dx.$$

▶ **Example 3.** Find the volume of a sphere of radius r by rotating the ordinate set of the function

$$f(x) = \sqrt{r^2 - x^2} \quad \text{for} \quad -r \le x \le r.$$

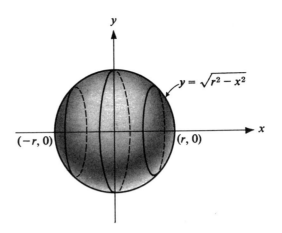

Figure 5.11.14

Solution. (See Figure 5.11.14.) By the foregoing result,

$$V = \pi \int_{-r}^{r} (r^2 - x^2)\, dx = 2\pi \int_{0}^{r} (r^2 - x^2)\, dx = 2\pi \left(r^2 x - \frac{x^3}{3} \right) \Big|_{0}^{r} = \frac{4\pi r^3}{3}.$$

▶ **Example 4.** Find the volume generated by revolving about the x axis the area bounded by the curve $y = e^x$, $x = 1$, and $x = 2$.

Solution. First sketch the region in question. (See Figure 5.11.15.)

$$V = \pi \int_{1}^{2} (e^x)^2\, dx = \pi \int_{1}^{2} e^{2x}\, dx.$$

Let $u = 2x$, $du = 2\, dx$; then $u(1) = 2$, $u(2) = 4$, and

$$\int_{1}^{2} e^{2x}\, dx = \frac{1}{2} \int_{2}^{4} e^u\, du = \frac{e^4 - e^2}{2}.$$

Thus,

$$V = \pi \cdot \frac{e^4 - e^2}{2}.$$

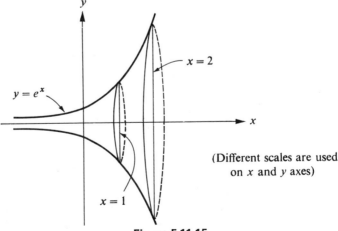

(Different scales are used on x and y axes)

Figure 5.11.15

Exercises

In all exercises, first sketch the solid whose volume is to be found.

1. The base of a certain solid is the circle $x^2 + y^2 = a^2$. Each plane section of the solid cut out by a plane perpendicular to the x axis is a square with one edge of the square in the base of the solid. Find the volume of the solid.

2. The base of a solid is the figure bounded by the parabola $y^2 = x$ and the line $x = 1$. Every cross section perpendicular to the x axis is a square. Find the volume of the solid.

3. Find the volume in Exercise 2 if the cross sections perpendicular to the x axis are equilateral triangles.

4. A tower is 60 ft high and 30 ft square at the base, and every cross section parallel to the base is square. (See Figure 5.11.16.) The side, s feet, of any of these cross sections is given by the formula $s = (15x^2 - x^3/6)/600$, where x is the distance in feet from the peak. Find the volume of the tower.

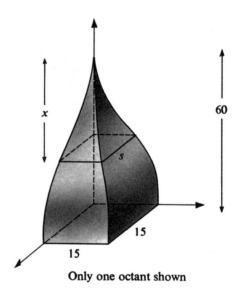

Only one octant shown

Figure 5.11.16

5. Find the volume of the solid generated by revolving about the x axis the region bounded by the graphs of

   (a) $f(x) = \sqrt{x}$, the x axis, $x = 0$, and $x = 1$.

   (b) $f(x) = x + 1/x$, the x axis, $x = 1$, and $x = 4$.

   (c) $f(x) = \sqrt{x}(x^2 + 1)^{1/4}$, the x axis, $x = 0$, and $x = 1$.

   (d) $f(x) = xe^{\sqrt{x}}$, the x axis, $x = 1$, and $x = 2$.

   (e) $f(x) = \log x$, the x axis, $x = 1$, and $x = 2$.

   (f) $f(x) = x(x^3 + 5)^{20}$, the x axis, $x = 0$, and $x = 2$.

6. Find the volume of the solid generated by revolving about the x axis the region bounded by the parabola $y = 9 - x^2$ and the straight line $y = 8$. [*Hint:* Find two volumes and subtract.]

7. Find the volume of the solid generated by revolving about the x axis the region bounded by $y = x^2$ and $y = x^3$.

8. Suppose the region in Exercise 7 was revolved about the y axis. Find the volume of the solid so generated.

9. Find the volume generated by revolving the region bounded by $f(x) = 1 - x^2$ and the x axis about the line $y = -1$.

Miscellaneous Topics. I. Differentiation

6.1 Implicit Differentiation

Consider the two functions $f(x) = \sqrt{1 - x^2}$ and $f(x) = -\sqrt{1 - x^2}$ whose graphs are shown in Figure 6.1.1. Their graphs are the top and bottom halves of a circle,

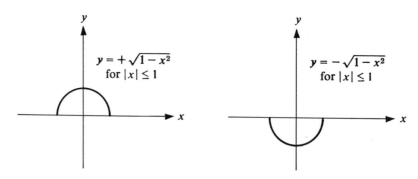

$$y = +\sqrt{1 - x^2}$$
$$\text{for } |x| \le 1$$

$$y = -\sqrt{1 - x^2}$$
$$\text{for } |x| \le 1$$

Figure 6.1.1

respectively. It is convenient to combine both in one equation: $x^2 + y^2 - 1 = 0$, which is the equation of a circle. We can also write this equation as $F(x, y) = 0$ where $F(x, y) = x^2 + y^2 - 1$.

If we are given a "function" of the form $F(x, y) = 0$, is it possible to find $\dfrac{dy}{dx}$

without first solving for y? If so, this would save considerable time and effort. (In fact, it may *not* be possible to solve for y in terms of x.)

Let us first rewrite the equation of a circle as

$$x^2 + [f(x)]^2 = 1.$$

Now formally differentiate the equation. Thus,

$$\frac{d}{dx}(x^2 + [f(x)]^2) = \frac{d}{dx}(1).$$

By our rules for sums this becomes

$$\frac{d}{dx}(x^2) + \frac{d}{dx}[f(x)]^2 = 0.$$

The first term is easy, $\dfrac{d}{dx}(x^2) = 2x$; but how about $\dfrac{d}{dx}[f(x)]^2$? At first glance, one might think we have no way of handling such expressions, but recall the chain rule. In particular,

$$\frac{d}{dx}[u(x)]^n = n[u(x)]^{n-1}\frac{du(x)}{dx}.$$

Thus,

$$\frac{d}{dx}[f(x)]^2 = 2f(x)\frac{df}{dx}.$$

Our formally differentiated equation thus becomes

$$2x + 2f(x)\frac{df}{dx} = 0 \quad \text{or} \quad \frac{df}{dx} = -\frac{x}{f(x)}.$$

We can, if we wish, write this as $\dfrac{dy}{dx} = -\dfrac{x}{y}$, where $y = f(x)$. The process we have just indulged in is referred to as implicit differentiation. Before going further, a comment is in order. First, although we shall make no attempt to prove it, this formal differentiation is valid (it yields the correct value for the derivative). Second, since $x^2 + y^2 = 1$ represents not *one* but *two* functions (see Figure 6.1.1), how can the expression $\dfrac{dy}{dx} = -\dfrac{x}{y}$ give the correct derivative for both? The answer lies in the y or $f(x)$ term in $-x/y$ or $-x/f(x)$. It is this term which automatically makes the necessary adjustments in the derivative $\dfrac{dy}{dx} = -\dfrac{x}{y}$.

RULE. To differentiate implicitly a function of the form $F(x, y) = 0$, simply differentiate formally, using the chain rule on terms containing y. [We assume, of course, that y is a function of x, say $y = f(x)$.]

Let us consider another example.

▶ **Example 1.** If $x^2y + e^y + y^5 + \log xy = 0$, find $\dfrac{dy}{dx}$.

Solution. This time we will not replace y with f. Differentiating both sides formally with respect to x, we obtain

$$\frac{d}{dx}(x^2y) + \frac{d}{dx}(e^y) + \frac{d}{dx}(y^5) + \frac{d}{dx}\log xy = 0,$$

and hence

$$y\frac{d}{dx}(x^2) + x^2\frac{d}{dx}(y) + e^y\frac{dy}{dx} + 5y^4\frac{dy}{dx} + \frac{1}{xy}\frac{d}{dx}(xy) = 0.$$

(We have used the chain rule on the second and third terms.) Continuing on, we obtain

$$2yx + x^2\frac{dy}{dx} + e^y\frac{dy}{dx} + 5y^4\frac{dy}{dx} + \frac{1}{xy}\left(y + x\frac{dy}{dx}\right) = 0.$$

If we wish, we can now solve for $\dfrac{dy}{dx}$. Thus,

$$\frac{dy}{dx}\left(x^2 + e^y + 5y^4 + \frac{x}{xy}\right) + 2yx + \frac{y}{xy} = 0$$

and so

$$\frac{dy}{dx} = \frac{-(2yx + (1/x))}{x^2 + e^y + 5y^4 + (1/y)}.$$

We give one more example to illustrate the method.

▶ **Example 2.** If $e^{5x^3y^2} = 1$, find $\dfrac{dy}{dx}$.

Solution. Differentiating both sides we obtain

$$\frac{d}{dx}(e^{5x^3y^2}) = 0.$$

Since $5x^3y^2$ can be considered a function of x, by the chain rule we have

$$e^{5x^3y^2}\frac{d}{dx}(5x^3y^2) = 0.$$

Thus,

$$e^{5x^3y^2}\left[y^2\frac{d}{dx}(5x^3) + 5x^3\frac{d}{dx}(y^2)\right] = 0$$

and so,

$$e^{5x^3y^2}\left[15y^2x^2 + 10x^3y\frac{dy}{dx}\right] = 0.$$

We will not solve for $\dfrac{dy}{dx}$. Note though, that no matter how complicated the equation $F(x, y) = 0$, one can solve for $\dfrac{dy}{dx}$ and, in fact, one need only solve an equation of the form $A \cdot \dfrac{dy}{dx} + B = 0$ to obtain $\dfrac{dy}{dx}$. (In this case A and B are themselves functions of x and y.)

Actually, implicit differentiation is a special case of a more general situation. Given any equation $F(x, y, u, v, \ldots) = 0$, where $y, u, v, \ldots$ are functions of x, we can differentiate it formally to obtain a relation between $x, y, u, v, \ldots$ and their derivatives. This last statement is imposingly abstract; let us consider an example. Let $e^u + w^2 + uv = 0$ when u, v, w are functions of x. We differentiate with respect to x and obtain

$$\frac{d}{dx}(e^u) + \frac{d}{dx}(w^2) + \frac{d}{dx}(uv) = 0.$$

By the chain rule, this yields

$$e^u \frac{du}{dx} + 2w \frac{dw}{dx} + u \frac{dv}{dx} + v \frac{du}{dx} = 0.$$

This is really all there is to it.

Two other examples will clearly illustrate the technique.

▶ **Example 3.** If $V = \frac{4}{3}\pi r^3$ when V and r are functions of t, find dV/dt.

Solution. Differentiating both sides with respect to t, we obtain

$$\frac{dV}{dt} = \frac{d}{dt}\left(\frac{4\pi}{3} \cdot r^3\right) = \frac{4\pi}{3} \frac{d}{dt}(r^3) = \frac{4\pi}{3} \cdot 3r^2 \frac{dr}{dt} = 4\pi r^2 \frac{dr}{dt}$$

by the chain rule.

▶ **Example 4.** Differentiate the expression $u + u^2 + \log uv^3$ with respect to x when u, v are functions of x.

Solution.

$$\frac{d}{dx}(u + u^2 + \log uv^3) = \frac{du}{dx} + \frac{d}{dx}(u^2) + \frac{d}{dx}(\log uv^3)$$

$$= \frac{du}{dx} + 2u \frac{du}{dx} + \frac{1}{uv^3} \frac{d}{dx}(uv^3)$$

$$= \frac{du}{dx} + 2u \frac{du}{dx} + \frac{1}{uv^3}\left(v^3 \frac{du}{dx} + 3v^2 u \frac{dv}{dx}\right).$$

Next we give an example to demonstrate how implicit differentiation can be useful in solving certain maxima and minima problems.

▶ **Example 5.** A piece of ground has the shape of a right triangle. The length of the two legs are 600 and 400 yards. A rectangular building is to be set in the lot as shown in Figure 6.1.2. Find the maximum possible area of the base of the building.

Solution. The area of the inscribed rectangle is given by

$$A = xy.$$

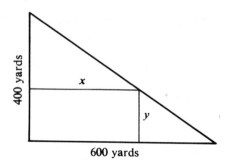

Figure 6.1.2

If we treat y as a function of x (without eliminating y) and differentiate implicitly, we obtain

$$A'(x) = y + xy'.$$

Setting $A'(x) = 0$, we see that

$$y' = -(y/x). \tag{6.1.1}$$

From Figure 6.1.2 it is clear that

$$\frac{x}{600} = \frac{400 - y}{400},$$

and hence

$$x = \tfrac{3}{2}(400 - y) \quad \text{or} \quad y = 400 - \tfrac{2}{3}x, \tag{6.1.2}$$

while

$$y' = -\tfrac{2}{3}.$$

Thus (6.1.1) yields the relation

$$3y = 2x.$$

Using this together with (6.1.2), we see that $y = 200$, $x = 300$ yields a critical value. To determine whether this critical value gives maximum or minimum area, we find $A''(x)$.

$$A''(x) = xy'' + 2y'$$

and when $A'(x) = 0$, then $y' = -\tfrac{2}{3}$ and $y'' = 0$. Hence,

$$A''(x) = -\tfrac{4}{3} < 0$$

and thus A attains its maximum value when $x = 300$ yards.

Frequently the procedure illustrated above is far more practical than direct elimination of one of the unknowns.

Exercises

1. Find $\dfrac{dy}{dx}$ for each of the following equations:

   (a) $xy = 1.$

   (b) $3x^2y + 4xy^2 + 5 = 0.$

   (c) $e^x + \log y = 0.$

   (d) $e^y \log x + 1 = 0.$

(e) $x^5 \log 2y + y^3 \log x = 1$.

(f) $xy^3 + 2x^2y^2 + x^3y = 9$.

(g) $e^{x^2}y = 1$.

(h) $\frac{1}{4}x^2 + \frac{1}{9}y^2 = 1$.

(i) $b^2x^2 + a^2y^2 = a^2b^2$ (a, b are constants).

(j) $y^2 - 2xy + b^2 = 0$ (b is constant).

(k) $x^3 + y^3 - 3axy = 0$ (a is constant).

2. (a) If $x^2 + y^2 = 4$, find $\dfrac{dy}{dx}$ at $(1, \sqrt{3})$.

(b) If $e^{xy^2} = e$, find $\dfrac{dy}{dx}$ at $(1, -1)$.

(c) If $\frac{1}{4}x^2 - \frac{1}{9}y^2 = -3$, find $\dfrac{dy}{dx}$ at $(2, 6)$.

(d) If $2x^2 - 3xy + 2y^2 = 2$, find $\dfrac{dy}{dx}$ at $(-1, -\frac{3}{2})$.

(e) If $x^2 + x + 2 - y^2 = 0$, find $\dfrac{dy}{dx}$ at $(1, 2)$.

(f) If $x + \log xy^3 = 8$, find $\dfrac{dy}{dx}$ at $(8, \frac{1}{2})$.

3. If A, B, C are functions of x, formally differentiate the following equation with respect to x:
$$A^2B + B^2C + C^2A = 0.$$

4. Given that U and V are functions of t, find $\dfrac{dV}{dt}$ when $V^3 + 2 \log U = 0$.

5. Find $\dfrac{d}{dx} [e^{u(x)} + \log v(x)]$.

6. If $A = \pi r^2$ when A and r are functions of t, find $\dfrac{dr}{dt}$.

7. If u, v, w are functions of x, find $\dfrac{d}{dx} [2uv^2 + \log (u - 2w) + e^{w^2+u}]$.

8. Find the equation of the line tangent to the curve
$$x^3 - 2x^2y^2 + 3y = x - 2$$
at $(1, 2)$.

9. Find the equation of the line tangent to the curve
$$x = 2a^3/(2a^2 - y^2)$$
at $(2a, -a)$.

10. Find the value of $\dfrac{d^2y}{dx^2}$ at the points where $\dfrac{dy}{dx} = 0$, of the following:

(a) $2\sqrt{x} + 3\sqrt{y} = 8$.

(b) $x^2 = y^4 - y^2$.

11. A rectangle is to be inscribed in a circle of radius 1. (See Figure 6.1.3.) Determine the dimensions of the rectangle of maximum area using implicit differentiation.

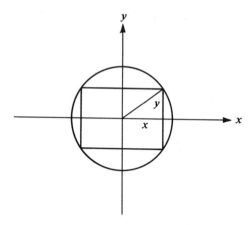

Figure 6.1.3

12. A lot adjacent to a river is to be fenced off. No fence is needed along the river and there is to be a 15-yd opening in front. If the fencing along the front costs $4 per yard, while that along the side costs $1 per yard, find the largest area that can be fenced off for $300. Use implicit differentiation.

6.2 Derivatives of Inverse Functions

Recall that in Section 1.6 we noted that if f is a one-to-one function, then its inverse f^{-1} exists and

$$f^{-1}[f(x)] = f[f^{-1}(x)] = x. \tag{6.2.1}$$

We now investigate the problem of determining the derivative of f^{-1}, written $[f^{-1}]'$. It turns out that we can actually calculate the derivative of the inverse function with only a knowledge of the derivative of the original function. One need only apply the chain rule in a straightforward manner.

Let $f^{-1} = g$. Thus $g[f(x)] = x$ and, we see from the chain rule,

$$\frac{d}{dx}[g[f(x)]] = g'[f(x)]f'(x) = 1.$$

Hence, we observe that if $a \in$ domain of f and $b = f(a)$, then

$$g'(b) = (f^{-1})'(b) = \frac{1}{f'(a)}. \tag{6.2.2}$$

In Leibniz notation, this relation becomes

$$\frac{d}{dx}f^{-1}(x)\Big|_{x=b} = \frac{1}{\dfrac{d}{dx}f(x)\Big|_{x=a}}, \qquad \text{where } b = f(a).$$

Another useful result of the differentiation of the inverse function is the following.

Suppose $y = g(x)$. Then $\dfrac{dy}{dx} = g'(x)$. Now assume g is one-to-one and $x = g^{-1}(y)$

Then,

$$\frac{dx}{dy} = (g^{-1})'(y) = \frac{1}{g'(x)} = \frac{1}{\dfrac{dy}{dx}}.$$

Thus we can write

$$\frac{dx}{dy} = \frac{1}{\dfrac{dy}{dx}}. \tag{6.2.3}$$

We must be careful to note that $\dfrac{d^2x}{dy^2} \neq \dfrac{1}{\dfrac{d^2y}{dx^2}}.$

We now apply the foregoing to some examples.

▶ **Example 1.** Suppose $f(x) = x^3$. Find $(f^{-1})'(2)$.

Solution. First we do the problem by means of the formula. Let $g = f^{-1}$. Then

$$g'(2) = \frac{1}{f'(2^{1/3})}$$

since $f(2^{1/3}) = 2$. But $f(x) = x^3$,

$$f'(x) = 3x^2 \quad \text{and} \quad f'(2^{1/3}) = 3 \cdot 2^{2/3}.$$

Thus,

$$g'(2) = \frac{1}{3 \cdot 2^{2/3}} = (f^{-1})'(2).$$

This problem can be done without recourse to the formula. We start with the equation $f[g(x)] = x$, where $g = f^{-1}$. This equation can be written as

$$[g(x)]^3 = x \quad \text{since} \quad f(x) = x^3.$$

Now differentiating by the chain rule, we find that

$$3[g(x)]^2 g'(x) = 1.$$

Hence,

$$g'(x) = \frac{1}{3[g(x)]^2}.$$

Finally,

$$g'(2) = \frac{1}{3 \cdot 2^{2/3}}$$

since $g(2) = 2^{1/3}$. Of course, in this case, since we know that $f^{-1}(x) = x^{1/3}$, it is just as easy to find f^{-1} by differentiating $x^{1/3}$.

Let us look at a different sort of example.

▶ **Example 2.** Consider the function $g(x) = x^{1/n}$, for $x \geq 0$ and n a positive integer. Using the fact that $\dfrac{d}{dx} \cdot x^n = nx^{n-1}$, verify that

$$g'(x) = \frac{1}{n} x^{(1/n)-1}.$$

Solution. Let $f(x) = x^n$, $x \geq 0$. It should be clear that an inverse exists and $f^{-1}(x) = g(x) = x^{1/n}$. Now, $f[g(x)] = x$ and thus $[g(x)]^n = x$. Using the chain rule, we see that

$$n[g(x)]^{n-1}g'(x) = 1$$

and thus

$$g'(x) = \frac{1}{n[g(x)]^{n-1}} = \frac{1}{n[x^{1/n}]^{n-1}}$$

$$= \frac{1}{n} \frac{1}{x^{1-(1/n)}} = \frac{1}{n} x^{(1/n)-1}.$$

▶ **Example 3.** Use the fact that $\dfrac{d}{dx}(e^x) = e^x$ to show that $\dfrac{d}{dy}(\log y) = \dfrac{1}{y}$.

Solution. Set $y = e^x$. Then, $x = f^{-1}(y) = \log y$. Observe that we wish to find $\dfrac{dx}{dy}$. From (6.2.3),

$$\frac{dx}{dy} = \frac{1}{\dfrac{dy}{dx}} = \frac{1}{e^x} = \frac{1}{y}.$$

Hence,

$$\frac{d}{dy}(\log y) = \frac{1}{y}.$$

▶ **Example 4.** Suppose that $y = e^{2x}$ expresses x as a function of y. (In fact, we know that $x = \frac{1}{2}\log y$.) We wish to determine $\dfrac{dx}{dy}$ directly.

Solution. $\dfrac{dy}{dx} = 2e^{2x}$ and *hence*

$$\frac{dx}{dy} = \frac{1}{\dfrac{dy}{dx}} = \frac{1}{2e^{2x}} = \frac{1}{2y}.$$

This example provides further verification of the fact that the derivative of $\frac{1}{2}\log y$ is $1/2y$.

Exercises

1. Let $y = x^2 + 3$.

   (a) Verify that $\dfrac{dy}{dx} = \dfrac{1}{\dfrac{dx}{dy}}$.

(b) Find $\dfrac{d^2x}{dy^2}$. Does it follow that $\dfrac{d^2x}{dy^2} = \dfrac{1}{\dfrac{d^2y}{dx^2}}$? Why?

2. Find $\dfrac{dx}{dy}$ if

(a) $y = 3x^2 + 9$.

(b) $y = xe^x$.

(c) $y = x^2 + e^{2x}$.

(d) $y = \log x + x^3$.

3. (a) Find $\dfrac{dx}{dy}$ if $y = \sqrt{x + 1}$.

(b) Find $\dfrac{dy}{dx}$ at the point $(27, 7)$ if $x = (y + 2)^{3/2}$.

4. If $f(x) = \dfrac{x - 1}{x + 1}$, find $f^{-1}(x)$ and determine, in two ways, $[f^{-1}]'(x)$.

5. If f is defined by $f(x) = \sqrt{1 - x^2}, 0 \le x \le 1$, determine whether or not f^{-1} exists. If f^{-1} exists, determine, in two ways, $[f^{-1}]'(x)$.

6. Let $g(x) = x^4 + 13$ and let h be the inverse of g. First compute $h'(14)$ by finding a formula for h and then by using the fact that

$$\frac{d}{dx}\,[f^{-1}](x) = \frac{1}{f'[f^{-1}(x)]}.$$

7. Find $h'(19)$ if h is the inverse of the function $g(x) = x^7 + x^5 + 17$.

8. Let $f(x) = x^5 + x^3 + 1$. Does f^{-1} exist? [*Hint:* Consider f'.] Find $(f^{-1})'(1), (f^{-1})'(-1)$, and $(f^{-1})'(41)$. What must you do in order to find $(f^{-1})'(0)$?

9. Let F be the function defined by

$$F(x) = \int_0^x \frac{dt}{1 + t^2}.$$

Show that if $y = F^{-1}(x)$, then y satisfies the equation $y' = 1 + y^2$.

10. (a) If h is the inverse of a function g, where g'' exists, find h'' in terms of the derivatives for g.

(b) Let $g(x) = (x - 1)/(x + 1)$. Apply (a) to finding h'', where h is the inverse of g.

11. Find $\dfrac{dy}{dx}$ if y is implicitly defined as a function of x by

$$ye^x - xe^y = \frac{xy}{\log (xy)}.$$

*12. If $g[f(x)] = 3x^2 + 1$, find g' in terms of f and f'.

6.3 Rate, Change, and Velocity

The world of experience offers no shortage of quantities which are a function of time. The gross national product, batting averages, the position of a rocket, the number of cars in Chicago, the Dow Jones industrial average, the price of copper, the water level in Lake Michigan, and the number of people in the United States are all functions of time. For at least some of these quantities there is considerable interest—emotional, financial, and otherwise—in the rate at which the quantity is changing. It is a remarkable and extremely useful fact that the derivative enables us to determine this rate of change for such a diverse range of phenomena. We have observed some of this in various examples in Chapter 3. Let us now consider another example in some detail.

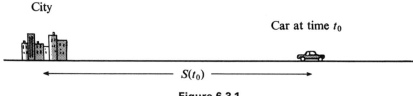

Figure 6.3.1

A car moves along a straight road R. Its distance from a city C is given as a function of time by $S(t)$. Suppose we determine the position of the car at time t_1 and its position a short time later at time t_2. (See Figures 6.3.1 and 6.3.2.) Then the quantity $S(t_2) - S(t_1)$ represents the distance traveled in the time interval between t_1 and t_2. The expression

$$\frac{S(t_2) - S(t_1)}{t_2 - t_1}$$

represents the average speed over this time period.

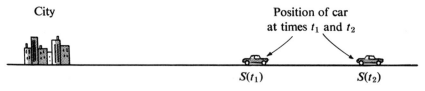

Figure 6.3.2

Certainly this last calculation is familiar to everyone and is used in everyday life to calculate the average speed of cars, boats, bicycles, etc. Of course, this average speed does not tell us how fast the car was going at t_1. To make a good estimate of the speed at t_1, one might choose t_2 close to t_1; that is, make the time interval for the average small. Let us carry the last step a bit further by letting t approach t_1, as we compute our averages. Consider

$$\lim_{t \to t_1} \frac{S(t) - S(t_1)}{t - t_1}.$$

On the one hand, this is the limit of the average speed of the car over the time interval from t_1 to t; thus we would expect it to approach the speed of the car at $t = t_1$. On the other hand, the expression is just $S'(t_1)$ [if $S(t)$ is differentiable at t_1].

This discussion was intended to make plausible our next statement.

DEFINITION. If $S(t)$ represents the position of a particle at time t, then the velocity *of the particle at time t_0 is $S'(t_0)$, if S is differentiable at t_0.*

This definition only applies to a particle which travels in a straight line, and we restrict ourselves to this case.

GENERAL DEFINITION. If $g(t)$ represents some quantity, then the rate of change *of $g(t)$ at t_0 equals $g'(t_0)$, if the derivative exists.*

We should observe that we defined the *velocity*, not the *speed*, to be $S'(t_0)$. The number $S'(t_1)$ may be either plus or minus, and the sign may be interpreted as saying the particle is either coming or going, moving up or down, etc. The speed, on the other hand, is *always* a positive quantity. The relation between them is simple: speed $=$ |velocity|. This is amply illustrated in the examples.

▶ Example 1. A particle moves along the x axis in such a way that its position (x coordinate) at time t is $5t^3 - 9t + 3$. Find the velocity of the particle when $t = 7$.

Solution. Let $S(t) = 5t^3 - 9t + 3$ represent the position of the particle at time t. Then $S'(t) = 15t^2 - 9$. The velocity of the particle at $t = 7$ equals $S'(7) = 15(7^2) - 9 = 726$.

▶ Example 2. A ball is dropped from the top of a building which is 1024 ft high. Its distance $S(t)$ from the top of the building is $16t^2$ at time t (assuming it was dropped at $t = 0$). What is the speed of the ball when it hits the ground?

Solution. First we draw a picture to illustrate the problem. (See Figure 6.3.3.) To determine when the ball will hit the ground, we set $16t^2 = 1024$. Thus, $t^2 = 64$ or $t = 8$. Since $S(t) = 16t^2$, $S'(t) = 32t$. The velocity at $t = 8$ is $S'(8) = 32 \cdot 8 = 256$. Thus the speed of the ball when it hits the ground is 256 ft/sec. (About 180 mph.)

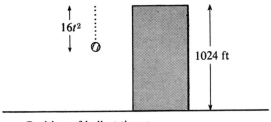

Position of ball at time t

Figure 6.3.3

▶ Example 3.   A ball is thrown straight upward at time $t = 0$. Its distance (in feet), $S(t)$, above the ground is $288t - 16t^2$ at time t (in seconds). (See Figure 6.3.4.)

(a)   What is its velocity at the moment it leaves the ground?
(b)   How high does it get before it comes down (the high point is just when the velocity is zero)?
(c)   When does it reach the ground?
(d)   What is its velocity when it reaches the ground?

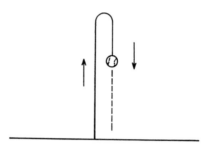

Figure 6.3.4

Solution.   As before, we set $S(t) = 288t - 16t^2$, where $S(t)$ represents the distance from the ground to the ball. To answer (a), we find $S'(t) = 288 - 32t$. Thus the velocity when the ball leaves the ground is just $S'(0) = 288$ ft/sec. To find the high point, we set $0 = S'(t) = 288 - 32t$. Thus, $t = 288/32 = 9$. At $t = 9$, $S(9) = 288 \cdot 9 - 16 \cdot 9^2 = 1296$ ft. To find when the ball reaches the ground, we set $S(t) = 0 = 288t - 16t^2$. Thus, $16t(18 - t) = 0$, and so the ball returns to the ground after 18 sec. (The other root of the equation $t = 0$ simply represents the time when the ball left the ground.) The velocity of the ball when it strikes the ground is

$$S'(18) = 288 - 32 \cdot 18 = -288.$$

This is just the velocity with which it left the ground, except for the sign. This is to be expected if there is no air resistance. Note that $S'(0) = 288$, while $S'(18) = -288$. The velocity is positive when the ball is going up and negative when it comes down, since it is going in opposite directions.

We will now consider another type of "rate" problem.

*▶ Example 4.   A rock is thrown into a pond and a ripple moves out at a constant velocity of 5 ft/sec. What is the rate of change of the surface area encompassed by the ripple when the ripple is 10 ft from the place where the rock struck? (See Figure 6.3.5.)

Solution.   Note that the ripple moves out in a straight line from the place where the rock strikes as shown by the arrows. At first this may seem like a very different sort of rate problem and one beyond the scope of present techniques. As we shall see, it is not that difficult. Let r be the distance of the ripple from the center. The area surrounded by the ripple is $A = \pi r^2$. Certainly A and r are functions of time. What we do now is simply differentiate implicitly the equation $A = \pi r^2$ with respect

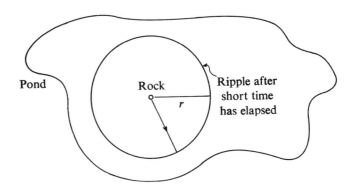

Figure 6.3.5

to t, obtaining $\dfrac{dA}{dt} = 2\pi r \dfrac{dr}{dt}$. Now $\dfrac{dA}{dt}$ is the rate of change of the surface area, which is what we are looking for. The quantity π is familiar and easy to handle. Finally, $\dfrac{dr}{dt}$ is just the rate of change of the radius, or the velocity of the ripple, and that was given. More precisely, $\dfrac{dr}{dt} = 5$. Thus the rate of change of the surface area where the ripple is 10 ft from the center is

$$\frac{dA}{dt} = 2 \cdot \pi \cdot 10 \cdot 5 = 100\pi \text{ (sq ft/sec).}$$

▶ **Example 5.** A water tank in the shape of an inverted cone is being filled with water at a rate of 5 cu ft/min. (See Figure 6.3.6 for the dimensions of the cone.) How fast is the surface of the water rising when the water is 10 ft deep? [The volume of the cone in Figure 6.3.6 is $\frac{1}{3}\pi r^2 h$.]

Solution. If the water is h feet deep, the volume of the water is $V = \frac{1}{3}\pi r^2 h$. First, observe that, since triangle OPQ is similar to triangle ORS, $r/h = \frac{40}{80}$, so that

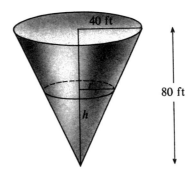

Figure 6.3.6

$r = \frac{1}{2}h$. Thus, $V = \frac{1}{12}\pi h^3$. Now both the volume and the height of the water are functions of time. Differentiating both sides with respect to t, we obtain

$$\frac{dV}{dt} = \frac{\pi}{12}\, 3h^2\, \frac{dh}{dt}.$$

But $\dfrac{dV}{dt} = 5$, since the tank is being filled at a rate of 5 cu ft/min. Hence,

$$5 = \frac{\pi}{4}\, h^2\, \frac{dh}{dt} \quad \text{or} \quad \frac{dh}{dt} = \frac{20}{\pi h^2}.$$

Thus when $h = 10$,

$$\frac{dh}{dt} = \frac{20}{\pi 10^2} = \frac{1}{5\pi}\ \text{ft/sec.}$$

Note that $\dfrac{dh}{dt}$, the "velocity" of the surface level, decreases as h increases. Why is this obvious on physical grounds?

Rather than do another rate problem, we will give a general scheme for handling such problems.

1. Draw a picture of the situation and label it clearly.
2. Write an equation relating the quantities known and unknown in the problem (it may be necessary to use auxiliary equations to do this).
3. Differentiate the equation with respect to time.
4. Substitute the known quantities to obtain an answer.

It is also useful to discuss the converse to the problems treated earlier in this section. That is, given the rate of change of a function g, find the function. From Chapter 5, we know that this is simply a problem in integration. Let us just consider a few examples.

▶ Example 6. The velocity at time t of a moving body is given by gt, where g is a constant. If the body's position at time $t = 0$ is given by s_0, find the distance s as a function of time t.

Solution. We know that the velocity is given by $\dfrac{ds}{dt}$. Hence, $\dfrac{ds}{dt} = gt$ and $s(t) = \frac{1}{2}gt^2 + c_1$, where c_1 is an arbitrary constant. However, using the fact that $s(0) = s_0$, we see that $c_1 = s_0$ and thus $s(t)$ is given by

$$s(t) = \tfrac{1}{2}gt^2 + s_0.$$

▶ Example 7. Acceleration, a, is simply the rate of change of velocity, v. That is, if s is the distance as a function of time, then $v = \dfrac{ds}{dt}$ and

$$a = \frac{dv}{dt} = \frac{d^2s}{dt^2}.$$

A stone was thrown straight down from a stationary balloon 10,000 ft above the ground with a speed of 48 ft/sec. Assuming that the acceleration is 32 ft/sec^2, locate the stone and find its speed 20 sec later.

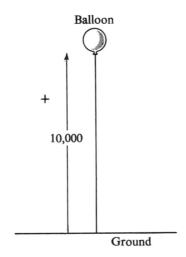

Figure 6.3.7

Solution. Let us assume that positive distance and velocity are directed upward. See Figure 6.3.7. When the stone leaves the balloon, it is subject to the law $a = \dfrac{dv}{dt} = -32$ ft/sec^2, and hence $v = -32t + c_1$, where c_1 is a constant. However, we are told that when $t = 0$, $v = -48$ ft/sec. Hence, $c_1 = -48$ and $v(t) = \dfrac{ds}{dt} = -32t - 48$. (The minus sign occurs since we said that positive distance and acceleration are upward, and the stone is thrown downward.) Thus,

$$s(t) = -32\frac{t^2}{2} - 48t + c_2,$$

where c_2 is a constant. However, at time $t = 0$, $s(0) = 10{,}000$. Thus, $c_2 = 10{,}000$ and

$$s(t) = -16t^2 - 48t + 10{,}000.$$

The stone's position 20 sec after being dropped is $s(20) = -16(20)^2 - 48(20) + 10{,}000 = 2{,}640$ ft above the ground and, since $v(20) = -32(20) - 48 = -688$, its speed is 688 ft/sec.

Exercises

1. A particle moves along the x axis in such a way that its x coordinate is equal to $(t^3 - e^{2t})$ at time t.

    (a) What is the velocity of the particle at $t = 2$?

    (b) When, if ever, is the velocity zero?

2. A particle moves along the positive x axis in such a way that its distance from the origin is $[4t^2 + \log(1 + t^2)]$ at time t.

    (a) What is the velocity of the particle when $t = 0$?

    (b) What is the velocity of the particle when $t = 1$?

3. Let the expression below represent the x coordinate of a particle at time t. Find the velocity at the time indicated.

(a) $t^5 - 6t^2 + 9$, $t = -2$.

(b) $e^{4t^2} - t$, $t = 3$.

(c) $t^6 - 4t^4 + t^2$, $t = 1$.

(d) $2t + 4 \log t$ for $t > 0$, $t = 9$.

(e) $t\sqrt{1 + t^2}$, $t = -1$.

(f) $\dfrac{1 - t^2}{\sqrt{1 + t}}$, $t = 3$.

4. In Exercise 3(f), discuss what happens at $t = -1$.

5. A rocket is fired in a vertical direction at $t = 0$. The equation of its distance from the ground is $h(t) = 4{,}800t - 16t^2$.

(a) What is its velocity when it leaves the ground? 4800

(b) What is its velocity 10 sec after it is launched? 4480 $V=$

(c) What is its maximum altitude? 360,000

(d) At what time does it return to the ground? 300 sec

(e) What is its velocity when it reaches the ground? 4800

6. A bowling ball is dropped from the top of a building 512 ft high at time $t = 0$. Its distance from the ground is $512 - 16t^2$ at time t.

(a) What is its initial velocity (i.e., at $t = 0$)?

(b) When does it strike the ground?

(c) What is its velocity when it strikes the ground?

7. A tank in the shape of a cone is filled with water. (See Figure 6.3.8.) The water flows out at a rate of 12 cu ft/min. How fast is the surface level dropping when the release valve is first opened? When the water is 30 ft deep? [*Hint:* Recall that the volume of a cone is $\frac{1}{3}\pi r^2 h$.]

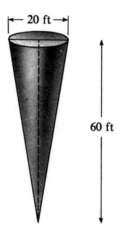

├── 20 ft ──┤

60 ft

Figure 6.3.8

8. A ladder 40 ft long is standing against a building. The base of the ladder is moved away from the building at the rate of 2 ft/sec. How fast is the top of the ladder moving down the wall when the base of the ladder is 0 ft, 20 ft, 30 ft, from the foot of the building? Refer to Figure 6.3.9.

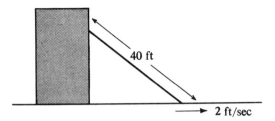

40 ft

2 ft/sec

Figure 6.3.9

9. A rubber balloon is being inflated with air at a constant rate of 6 cu ft/sec. The volume V and area A are related by the equation $A^3 = 36\pi V^2$. What is the rate of change of the area when the volume is 20 cu ft?

10. As in Exercise 9, a balloon is being inflated at a constant rate of 6 cu ft/sec. The volume is $V = \frac{4}{3}\pi r^3$, where r is the radius. How fast is the radius changing when the radius is 10 ft? 20 ft? What is the rate of change of the radius when the radius is zero? Does this seem reasonable?

11. A ball is rolled over a level lawn with an initial velocity of 25 ft/sec. Due to friction, the velocity decreases at the rate of 6 ft/sec². How far will the ball roll?

12. The acceleration at time t seconds is known to be $(12t + 2)$ ft/sec². At time $t = 3$ sec, the moving object is at $s = 5$ ft and its velocity is -20 ft/sec. Find its position at $t = 5$ sec.

13. The acceleration at time t of an object moving on a straight line is $12t^2$ ft/sec². Its position at time $t = 2$ sec is known to be 28 ft, and time $t = 3$ sec, it is 98 ft. Find its position at time t.

14. A ball was dropped from a balloon 640 ft above the ground. If the balloon was rising at the rate of 48 ft/sec, find

   (a) the greatest distance above the ground attained by the ball.

   (b) the time the ball was in the air.

   (c) the speed of the ball when it struck the ground.

   [*Hint*: Assume the acceleration a is governed by $a(t) = -32$ ft/sec².]

*15. Oil is escaping from a well in the floor of the Santa Barbara Channel at the rate of 100 gallons per minute. Assume that one gallon of oil forms a uniform film covering one square yard of ocean surface. Assume further that the oil covers a circular area as it spreads. How fast is the radius of the oil slick changing five minutes after the oil began escaping? one hour after it began escaping? (Assume no time loss for the oil in traveling from the ocean floor to the surface.)

6.4 L'Hôpital's Rule

Suppose we wish to find $\lim_{x \to a} f(x)/g(x)$ when f, g are continuous at x. In most cases we can apply Property 6 of Section 2.4 to obtain

$$\lim_{x \to a} \frac{f(x)}{g(x)} = \frac{\lim_{x \to a} f(x)}{\lim_{x \to a} g(x)} = \frac{f(a)}{g(a)}.$$

But what if $f(a) = g(a) = 0$? Although the foregoing method does not apply, we will now present a technique for handling this case.

THEOREM. (L'Hôpital's Rule) Let f and g be differentiable near a but not necessarily at a. Assume

$$\lim_{x \to a} f(x) = \lim_{x \to a} g(x) = 0, \text{ and } g'(x) \neq 0 \text{ near } a$$

Then

$$\lim_{x \to a} \frac{f(x)}{g(x)} = \lim_{x \to a} \frac{f'(x)}{g'(x)}$$

if the latter limit exists.

The proof of this theorem uses an extended version of the mean value theorem. We omit the proof here and refer the reader to any of the standard advanced calculus texts. In spite of the many hypotheses, this theorem is quite easy to apply in practice.

▶ **Example 1.** Find $\lim_{x \to 0} \dfrac{e^x - 1}{2x}$.

Solution. Note that both the numerator and denominator are 0 at $x = 0$. Since $f(x) = e^x - 1, f'(x) = e^x$. Since $g(x) = 2x, g'(x) = 2$. Thus,

$$\lim_{x \to 0} \frac{e^x - 1}{2x} = \lim_{x \to 0} \frac{e^x}{2} = \frac{1}{2}.$$

▶ **Example 2.** Find $\lim_{x \to 0} \dfrac{1 + x - e^x}{x^2}$.

Solution. If $f(x) = 1 + x - e^x$ and $g(x) = x^2$, then $f(0) = g(0) = 0$. By L'Hôpital's rule,

$$\lim_{x \to 0} \frac{1 + x - e^x}{x^2} = \lim_{x \to 0} \frac{1 - e^x}{2x}.$$

But again both the numerator and denominator vanish at $x = 0$, so that it is not clear what $\lim_{x \to 0} (1 - e^x)/2x$ is. However, we can simply apply L'Hôpital's rule again. Thus,

$$\lim_{x \to 0} \frac{1 - e^x}{2x} = \lim_{x \to 0} \frac{-e^x}{2} = -\frac{1}{2}.$$

▶ **Example 3.** Find $\lim_{x \to 1} \dfrac{\log x}{1 - x}$.

Solution. By L'Hôpital's rule,

$$\lim_{x \to 1} \frac{\log x}{1 - x} = \lim_{x \to 1} \frac{1/x}{-1} = -1.$$

Remark. In Example 2, we applied L'Hôpital's rule twice to obtain the answer. One might think that it is permissible to apply the rule as many times as one wished to a quotient,

$$\lim_{x \to a} \frac{f(x)}{g(x)} = \lim_{x \to a} \frac{f'(x)}{g'(x)} = \lim_{x \to a} \frac{f''(x)}{g''(x)} = \cdots .$$

This is true as long as the hypotheses are satisfied and in particular both numerator and denominator vanish at x. But note, in Example 3, if we had continued another step we would have obtained

$$\lim_{x \to 1} \frac{\log x}{1 - x} = \lim_{x \to 1} \frac{1/x}{-1} \overset{?}{=} \lim_{x \to 1} \frac{1/x^2}{0} = \frac{-1}{0} = -\infty .$$

Since we know the term before the question mark equals -1, this is clearly not valid.

Exercises

1. Find

   (a) $\displaystyle \lim_{x \to 2} \frac{x - 2}{e^{x-2} - 1}$.

   (b) $\displaystyle \lim_{x \to 1} \frac{1 - e^{1-x}}{\log x}$.

   (c) $\displaystyle \lim_{x \to 1} \frac{x \log x}{\sqrt{1 - x^2}}$.

   (d) $\displaystyle \lim_{x \to 0} \frac{xe^x - x}{\log (1 - x)}$.

   (e) $\displaystyle \lim_{x \to 2} \frac{\sqrt{x^2 - 4}}{x - 2}$.

2. There is a version of L'Hôpital's rule which is valid for the case when $\lim_{x \to a} f(x) = \pm\infty = \lim_{x \to a} g(x)$. (Note that in this situation the expression

$$\frac{\lim_{x \to a} f(x)}{\lim_{x \to a} g(x)}$$

is ambiguous. As before,

$$\lim_{x \to a} \frac{f(x)}{g(x)} = \lim_{x \to a} \frac{f'(x)}{g'(x)}$$

if this limit exists. For example, let us find $\lim_{x \to 0} \log x/x^{-2}$. Note that

$$\lim_{x \to 0} \log x = -\infty \quad \text{and} \quad \lim_{x \to 0} x^{-2} = \infty .$$

By the second version of L'Hôpital's rule,

$$\lim_{x \to 0} \frac{\log x}{x^{-2}} = \lim_{x \to 0} \frac{1/x}{-2x^{-3}} = \lim_{x \to 0} \frac{x^3}{-2x} = \lim_{x \to 0} \frac{3x^2}{-2} = 0 .$$

(Why is the penultimate step valid?) It also follows that L'Hôpital's rule is valid at ∞, as is seen in the following exercises.

Evaluate the following limits:

   (i) $\displaystyle \lim_{x \to \infty} \frac{x^3}{e^x}$.

Solution. Applying the above,

$$\lim_{x \to \infty} \frac{x^3}{e^x} = \lim_{x \to \infty} \frac{3x^2}{e^x} = \lim_{x \to \infty} \frac{6x}{e^x} = \lim_{x \to \infty} \frac{6}{e^x} = 0.$$

(ii) $\displaystyle\lim_{x \to \infty} \frac{x^4}{e^x}$.

(iii) $\displaystyle\lim_{x \to \infty} \frac{e^x}{\log x}$.

(iv) $\displaystyle\lim_{x \to \infty} \frac{e^x + 1}{e^x + x}$.

3. Find $\lim_{x \to 0} x \log x$. [*Hint:* Rewrite the initial expression as either $\dfrac{\log x}{1/x}$ or $\dfrac{x}{1/\log x}$ and apply Exercise 2. *Caution:* One rewritten expression works but the other does not.]

4. Find $\displaystyle\lim_{x \to 0} \sqrt{x} \log x$.

5. Find $\lim_{x \to 0} x^x$ for $x > 0$. [*Hint:* First consider $\log x^x = x \log x$.]

*6.5 Trigonometric Functions

In this section we introduce the derivatives of the well-known trigonometric functions. We discussed the functions sine, cosine, and tangent in Section 1.9. As a reminder, we present their graphs in Figures 6.5.1, 6.5.2, and 6.5.3. Also, recall that $\tan x = \sin x / \cos x$.

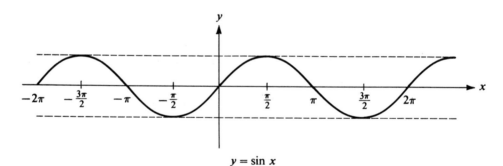

$y = \sin x$

Figure 6.5.1

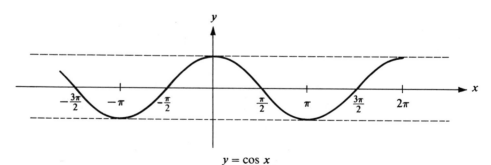

$y = \cos x$

Figure 6.5.2

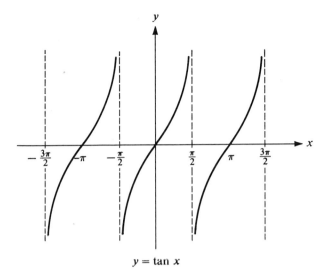

$$y = \tan x$$

Figure 6.5.3

We now introduce the derivatives of these functions in Table 6.5.1. Using the chain rule for composite functions, one can easily obtain the formulas in Table 6.5.2.

TABLE 6.5.1	$f(x)$	$f'(x)$
	$\sin x$	$\cos x$
	$\cos x$	$-\sin x$
	$\tan x$	$\dfrac{1}{(\cos x)^2} = (\cos x)^{-2} = \sec^2 x$

TABLE 6.5.2

1. $\dfrac{d}{dx} \sin u(x) = (\cos u(x)) \dfrac{du}{dx}$

2. $\dfrac{d}{dx} \cos u(x) = (-\sin u(x)) \dfrac{du}{dx}$

3. $\dfrac{d}{dx} \tan u(x) = \dfrac{1}{(\cos u(x))^2} \dfrac{du}{dx} = [\sec^2 u(x)] \dfrac{du}{dx}$

We shall not derive the formulas in Tables 6.5.1 and 6.5.2, but do point out that once you know that $\dfrac{d}{dx} \sin x = \cos x$, the rest can be obtained without much trouble. This is illustrated both in the exercises at the end of the section and in the next example.

▶ **Example 1.** Show that $\dfrac{d}{dx} \cos x = -\sin x$.

Solution. Recall the well-known formula $(\sin x)^2 + (\cos x)^2 = 1$. We differentiate this implicitly.

$$\frac{d}{dx} 1 = \frac{d}{dx} [(\sin x)^2 + (\cos x)^2]$$

Thus,

$$0 = \frac{d}{dx} (\sin x)^2 + \frac{d}{dx} (\cos x)^2$$

$$= 2 \sin x \frac{d}{dx} \sin x + 2 \cos x \frac{d}{dx} \cos x$$

(this last step is just the chain rule)

$$= 2 \sin x \cos x + 2 \cos x \frac{d}{dx} \cos x.$$

Of course, we don't know what $\dfrac{d}{dx} \cos x$ is; this is what we are trying to determine. Solving this equation, we obtain

$$2 \cos x \frac{d}{dx} \cos x = -2 \cos x \sin x.$$

Thus,

$$\frac{d}{dx} \cos x = -\sin x,$$

at least when $\cos x \neq 0$. (Why?) We stop at this point.

▶ **Example 2.** In a study of certain beetles, biologists have found that the average angle of orientation β at which the beetle climbed an inclined plane was related to v, the angle of inclination of the plane, according to the function

$$\beta(v) = \Gamma_1 \sin v + \Gamma_2, \qquad \Gamma_1 > 0$$

where Γ_1, Γ_2 are constants. The rate of change of β with respect to v is given by

$$\beta'(v) = \Gamma_1 \cos v.$$

▶ **Example 3.** If $f(x) = \sin (\cos x^2)$, find $f'(x)$.

Solution.

$$\frac{d}{dx} [\sin (\cos x^2)] = [\cos \cos x^2] \frac{d}{dx} (\cos x^2) \qquad \text{(by Table 6.5.2)}$$

$$= [\cos \cos x^2](-\sin x^2) \frac{d}{dx} x^2 \quad \text{(by Table 6.5.2)}$$

$$= -(\cos \cos x^2)(\sin x^2)2x.$$

We now introduce the inverse trigonometric functions. Observe from Figures 6.5.1, 6.5.2, and 6.5.3 that since the trigonometric functions are periodic, they do not have the property that if $x_1 \neq x_2$, then $f(x_1) \neq f(x_2)$. In fact, just draw any line

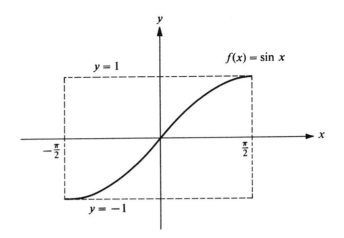

Figure 6.5.4

parallel to the x axis through the graphs and it is easy to see that there are many points $a_1, a_2, \ldots$ such that $f(a_1) = f(a_2) = \cdots$ etc. However, these functions may be restricted to intervals on which they are increasing, and in this manner partial inverses may be defined.

Let us begin with the function $f(x) = \sin x$, where $-\frac{1}{2}\pi \leq x \leq \frac{1}{2}\pi$. (See Figure 6.5.4.) Observe that for $-\frac{1}{2}\pi \leq x \leq \frac{1}{2}\pi$, $f'(x) = \cos x$ and $\cos x \geq 0$ on this interval. Thus, by Exercise 2 in Section 6.2, we see that on $[-\frac{1}{2}\pi, \frac{1}{2}\pi]$ the sine function has an inverse, $f^{-1}(x)$. We call $f^{-1}(x)$ the arc sine function and write $f^{-1}(x) =$ arc sin x. Thus, if $y =$ arc sin x, then $x = \sin y$ by definition of inverse. The domain of arc sin x must be $[-1, 1]$. Its graph is shown in Figure 6.5.5.

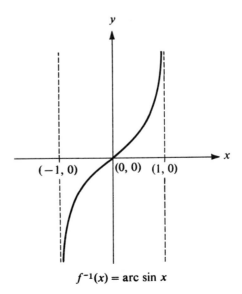

$f^{-1}(x) =$ arc sin x

Figure 6.5.5

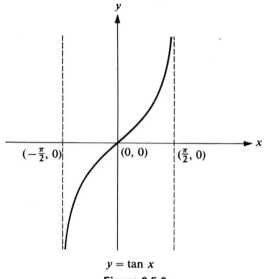

$$y = \tan x$$

Figure 6.5.6

Partial inverses can be defined for all of the trigonometric functions in exactly the same way as for the sine function. We consider only one more case, that of the tangent function. It is clear from the graph in Figure 6.5.3 that the tangent does not have an inverse function. However, consider the function $g(x) = \tan x$, $-\frac{1}{2}\pi \le x \le \frac{1}{2}\pi$. Its graph is shown in Figure 6.5.6. It is clear that in the interval $[\frac{1}{2}\pi, \frac{1}{2}\pi]$ the function is increasing; that is, $f'(x) > 0$, and hence an inverse $g^{-1}(x)$ exists. We call this inverse the *arc tangent function* and write $g^{-1}(x) = $ arc tan x. The domain must be the set of *all* real numbers. Its graph, obtained by reflecting the tangent function about the line $y = x$, is shown in Figure 6.5.7.

It is a simple matter to obtain the derivatives of the arc sin and arc tan functions.

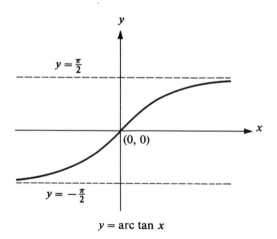

$$y = \text{arc tan } x$$

Figure 6.5.7

Surprisingly, their derivatives are simple algebraic functions. We begin with the arc sin function. Let $y = \text{arc sin } x$, $-1 \leq x \leq 1$. It follows that $x = \sin y$, and hence by implicit differentiation (and Table 6.5.1), $1 = \cos y \dfrac{dy}{dx}$. Thus,

$$\frac{dy}{dx} = \frac{d}{dx} (\text{arc sin } x) = \frac{1}{\cos y} = \frac{1}{\sqrt{1 - x^2}}$$

(since $\cos y = \sqrt{1 - \sin^2 y} = \sqrt{1 - x^2}$). In like manner, if $y = \text{arc tan } x$, then $\tan y = x$ and

$$\sec^2 y \frac{dy}{dx} = 1 \quad \text{or} \quad \frac{dy}{dx} = \frac{1}{\sec^2 y} = \frac{1}{1 + \tan^2 y} = \frac{1}{1 + x^2}.$$

We summarize our results in Tables 6.5.3 and 6.5.4. The results in Table 6.5.4 are verified in Exercise 9 of this section.

TABLE 6.5.3	$f(x)$	$f'(x)$
	arc sin x	$\dfrac{1}{\sqrt{1 - x^2}}$
	arc tan x	$\dfrac{1}{1 + x^2}$

TABLE 6.5.4	$f(x)$	$f'(x)$
	arc sin $u(x)$	$\dfrac{1}{\sqrt{1 - [u(x)]^2}} \dfrac{du}{dx}$
	arc tan $u(x)$	$\dfrac{1}{1 + [u(x)]^2} \dfrac{du}{dx}$

*Exercises

1. Find the derivative of each of the following functions:

    (a) $\cos 5x$.

    (b) $\sin \sqrt{1 - x}$.

    (c) $\sqrt{\sin (1 - x)}$.

    (d) $e^{\sin 3x}$.

    (e) $\log (\cos 2x)$.

    (f) $\tan (x + 5)^3$.

    (g) $\tan e^x$.

    (h) $\log \left(\tan \dfrac{1 + x}{1 - x} \right)$.

    (i) $\log [\sin (\cos x)]$.

    (j) $\dfrac{e^x}{(\sin x)^2}$.

(k) tan (log x).

(l) sin [sin (sin x)].

(m) $\dfrac{e^{\tan x}}{\sqrt{1 + \sin x}}$.

2. Find the following limits (use L'Hôpital's rule):

(a) $\lim\limits_{x\to 0} \dfrac{\sin x}{x}$.

(b) $\lim\limits_{x\to 0} \dfrac{1 - \cos x}{x^2}$.

(c) $\lim\limits_{x\to 0} \dfrac{1 - \cos x}{x}$.

(d) $\lim\limits_{x\to 1} \dfrac{\log x}{\cos (\frac{1}{2}\pi x)}$.

(e) $\lim\limits_{x\to 0} \dfrac{\tan x}{x}$.

3. Using the fact that $\cos x = \sin (x + \frac{1}{2}\pi)$ and $\cos (x + \frac{1}{2}\pi) = -\sin x$; show that $\dfrac{d}{dx} \cos x = -\sin x$. [You may use $\dfrac{d}{dx} \sin x = \cos x$.]

4. Show that $\dfrac{d}{dx} \tan x = 1/(\cos x)^2$. [*Hint:* $\tan x = \sin x/\cos x$.]

5. Derive the expressions in Table 6.5.2.

6. How would you define the inverse function, arc cos x, viewing the cosine only on the interval $[0, \pi]$?

7. What is the derivative of arc cos?

8. Differentiate the following:

(a) arc sin $(x^2 - 1)$.

(b) arc cos $(2x + 5)$.

(c) arc tan $\sqrt{x}$.

(d) x arc sin x.

(e) arc sin x + arc tan x.

(f) arc tan $(\sin x^2)$.

9. Derive the formulas in Table 6.5.4.

10. A man whose eyes are 6 ft above the ground is walking toward a lamp post 10 ft tall at a rate of 5 ft/sec. (See Figure 6.5.8.) He keeps his eyes fixed on the

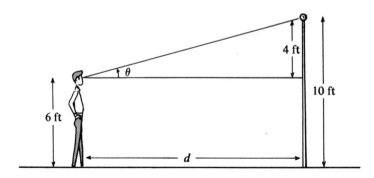

Figure 6.5.8

light. How fast is his head rotating when he is 24 ft from the lamp post? [*Remark:* If θ is the angle indicated, then $\dfrac{d\theta}{dt}$ is the rate at which his head is rotating—it is also referred to as the angular velocity.] [*Hint:* Obtain an equation relating θ and d. Differentiate it with respect to time.]

11. A searchlight turns at the constant rate of two revolutions per minute. (It throws a narrow beam of light which we will consider to be a straight line.) The beam is parallel to the ground. There is a wall 100 ft from the light. (See Figure 6.5.9.) How fast is the beam moving along the wall when it is 100 ft from point A, the point nearest the light? [*Hint:* Two revolutions per minute means the light has an angular velocity (rate of rotation) of 4π radians per minute.]

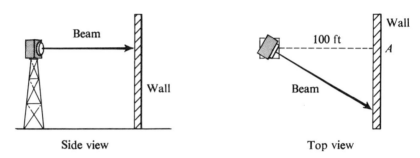

Beam

Wall

Side view

Wall

100 ft

A

Beam

Top view

Figure 6.5.9

6.6 Chapter 6 Summary

1. If we are given a function $y = g(x)$ in the form $F(x, y) = 0$, it is possible to find $\dfrac{dy}{dx}$ without first solving for y. Simply differentiate formally using the chain rule on terms containing y.

2. (a) If f is a function where $x_1 \neq x_2$ implies that $f(x_1) \neq f(x_2)$ for $x_1, x_2 \in$ domain of f, then f has associated with it an inverse function f^{-1} defined by
$$f[f^{-1}(x)] = x \quad \text{and} \quad f^{-1}[f(x)] = x.$$

   (b) If $f'(x) > 0$ for $x \in$ domain of f, then f has an inverse function f^{-1} defined as in part (a).

   (c) If f and f^{-1} are differentiable, then implicit differentiation yields that
$$(f^{-1})'(x) = \frac{1}{f'[f^{-1}(x)]}.$$

3. If $s(t)$ represents the position of a particle at time t, the *velocity* of the particle at time $t = t_0$ is given by $s'(t_0)$, if s is differentiable at t_0 and the acceleration at time $t = t_0$ is given by $s''(t_0)$ (provided that s'' exists). Given the velocity and acceleration with certain initial conditions, it is possible to find the distance function s.

4. *L'Hôpital's Rule.* Let f and g be differentiable near a. If $\lim_{x \to a} f(x) = \lim_{x \to a} g(x) = 0$ and $g'(x) \neq 0$ near a, then

$$\lim_{x \to a} \frac{f(x)}{g(x)} = \lim_{x \to a} \frac{f'(x)}{g'(x)},$$

provided the latter limit exists.

Review Exercises

1. Find $\dfrac{dy}{dx}$ if y is implicitly defined as a function of x by

   (a) $2xy + y^{1/2} = x + y$.

   (b) $x^2 y^2 = x^2 + y^2$.

   (c) $(x + y)^4 + (x - y)^4 = x^5 + y^5$.

   (d) $e^{x^2 y} + x \log xy = x^3$.

   (e) $x^3 - xy + y^3 = 10$.

   (f) $\dfrac{1}{y} + \dfrac{1}{x} = 1$.

   (g) $e^y \log xy = y^2$.

2. Find the equation of the line tangent to the following curves at the indicated points:

   (a) $x^2 + y^2 = 25$ at $(3, -4)$.

   (b) $x^2 y^2 = 9$ at $(-1, 3)$.

   (c) $(y - x)^2 = 2x + 4$ at $(6, 2)$.

3. The total surface area of a right circular cone of height h and radius of base r is $S = \pi(r^2 + r\sqrt{r^2 + h^2})$. If S is constant, find $\dfrac{dr}{dh}$ when $r = 3$ and $h = 4$.

4. A ladder 15 ft long rests against a house. It slides down, the lower end slipping along the level ground at the rate of 2 ft/sec. How fast is the upper end of the ladder sliding down the wall when it is 12 ft from the ground?

5. Two airplanes fly eastward on parallel courses 12 miles apart. One flies at 240 mph, the other at 300 mph. How fast is the distance between the planes changing when the slower plane is 5 miles farther east than the faster plane?

6. Suppose that a particle moves along a straight line in such a way that its distance from some fixed point on the line at any time t is given by $s(t) = t^2 - 2 \log (t + 1)$, where t is measured in seconds and s in feet.

   (a) Compute the velocity and acceleration at the end of 2 sec and at the end of 3 sec.

   (b) What happens to the velocity and acceleration as t becomes very large?

7. A ball, rolling up a certain incline, is slowing down at the rate of 9 ft/sec². If the ball is moving 12 ft/sec when it passes a certain point, how far does it roll before it stops and begins to roll down. [*Hint:* Use the fact that acceleration

$$a = \frac{dv}{dt} = \frac{d^2 s}{d^2 t} = \frac{dv}{ds} \frac{ds}{dt} = v \frac{dv}{ds}.]$$

8. Evaluate the following limits:

    (a) $\lim\limits_{h \to 0} \dfrac{\sqrt{4 + h} - 2}{h}$.

    (b) $\lim\limits_{x \to 0} \dfrac{e^x - (1 + x)}{x^2}$.

    (c) $\lim\limits_{x \to 1^+} \left(\dfrac{1}{x - 1} - \dfrac{1}{\sqrt{x - 1}} \right)$.

    (d) $\lim\limits_{x \to 0^+} x \log x$.

    (e) $\lim\limits_{x \to 0^+} x^x$. [*Hint*: $x^x = e^{x \log x}$.]

    (f) $\lim\limits_{x \to 1} x^{1/(1 - x)}$. [*Hint*: $x^{1/(1 - x)} = e^{\log x/(1 - x)}$.]

9. Determine whether or not an inverse exists for each of the following functions. If so, explicitly find the inverse and its derivatives in two different ways.

    (a) $f(x) = 1/x$, $x \neq 0$.

    (b) $f(x) = 3x - 2$.

    (c) $f(x) = x^4$.

    (d) $f(x) = x^4$, $x > 0$.

    (e) $f(x) = |x|$.

10. An object moves along the curve $y = x^3$. At what points on the curve will the x and y coordinates be changing at the same rate?

11. Assume that the brakes on an automobile produce a constant deceleration of p ft/sec^2.

    (a) Determine what p must be to bring an automobile traveling 60 mph (88 ft/sec) to rest in a distance of 100 ft from the point where the brakes are applied.

    (b) With the same p, how far would a car traveling 30 mph travel before being brought to a stop?

Miscellaneous Topics. II. Integration

In this section we first discuss antiderivatives for the basic trigonometric functions, sin, cos, and tan. We recall from Chapter 4 that

$$\frac{d}{dx} \sin x = \cos x \quad \text{and} \quad \frac{d}{dx} \cos x = -\sin x.$$

Hence, we immediately conclude that

$$\int \sin x \, dx = -\cos x + C \tag{7.1.1}$$

and

$$\int \cos x \, dx = \sin x + C, \tag{7.1.2}$$

where C is any arbitrary constant.

In order to find an antiderivative of the tangent function, we first note that

$$\int \tan x \, dx = \int \frac{\sin x}{\cos x} \, dx.$$

Let us try a substitution. If in the second integral we let $u = \cos x$ and $du = -\sin x \, dx$, then

$$\int \frac{\sin x}{\cos x} \, dx = -\int \frac{du}{u} = -\log |u| + C = -\log |\cos x| + C = \log |\cos x|^{-1} + C$$

$$= \log \left| \frac{1}{\cos x} \right| + C = \log |\sec x| + C,$$

where C is an arbitrary constant. We also note that

$$\frac{d}{dx} \tan x = \frac{d}{dx}\left(\frac{\sin x}{\cos x}\right) = \frac{\cos^2 x + \sin^2 x}{\cos^2 x} = \frac{1}{\cos^2 x} = \sec^2 x.$$

Hence,

$$\int \sec^2 x \, dx = \tan x + C,$$

where C is an arbitrary constant. We summarize the results in Table 7.1.1.

TABLE 7.1.1	$f(x)$	$\int f(x) \, dx$
1.	$\sin x$	$-\cos x + C$
2.	$\cos x$	$\sin x + C$
3.	$\tan x$	$\log |\sec x| + C$
4.	$\sec^2 x$ $\csc^2 x$	$\tan x + C$ $-\cot x + C$

Let us now consider the following examples.

▶ Example 1.   Evaluate the following definite integral:

$$\int_0^{\pi/2} 2 \cos 2x \, dx.$$

Solution.   Let $u = 2x$; $du = 2 \, dx$; then

$$\int_0^{\pi/2} 2 \cos 2x \, dx = \int_0^{\pi} \cos u \, du = \sin u \Big|_0^{\pi} = 0.$$

▶ Example 2.   Evaluate

$$\int \cos x e^{\sin x} \, dx.$$

Solution.   If we try a substitution, $u = \sin x$, then $du = \cos x \, dx$ and

$$\int \cos x e^{\sin x} \, dx = \int e^u \, du = e^u + C = e^{\sin x} + C.$$

▶ Example 3.   Evaluate $\int_0^{\pi} x \cos x \, dx$.

Solution.   It is clear that substitution will not work in this case. Let us try integration by parts with $u = x$ and $dv = \cos x \, dx$, yielding $v = \sin x$ and $du = dx$. Then,

$$\int_0^{\pi} x \cos x \, dx = x \sin x \Big|_0^{\pi} - \int_0^{\pi} \sin x \, dx = 0 + \cos x \Big|_0^{\pi} = -2.$$

▶ **Example 4.** Find the area bounded by the curves $y = \sin x$, $y = \cos x$, the y axis, and the first point where these curves intersect for $x > 0$.

Solution. We first draw a graph of the two curves and indicate the region in question. (See Figure 7.1.1.) The area of the region in question is given by

$$\int_0^{\pi/4} (\cos x - \sin x) \, dx = (\sin x + \cos x)\Big|_0^{\pi/4} = \sqrt{2} - 1.$$

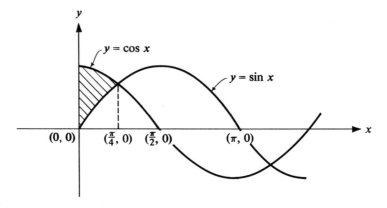

Figure 7.1.1

▶ **Example 5.** Find the volume of the solid generated by revolving the curve $y = \cos x$ between $x = 0$ and $x = \frac{1}{4}\pi$ about the x axis.

Solution. The volume of the solid given in Figure 7.1.2 is

$$V = \pi \int_0^{\pi/4} \cos^2 x \, dx.$$

Now, in order to find an antiderivative of $\cos^2 x$, it is necessary for us to use the fact that

$$\cos^2 x = \frac{1 + \cos 2x}{2}$$

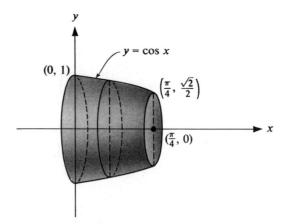

Figure 7.1.2

since we observe from Table 7.1.1 that there is apparently no simple way of calculating an antiderivative of $\cos^2 x$. [*Note:* If the integrand were $\sin^2 x$, then we would use the fact that $\sin^2 x = (1 - \cos 2x)/2$.] Thus

$$\int_0^{\pi/4} \cos^2 x \, dx = \int_0^{\pi/4} \frac{1 + \cos 2x}{2} \, dx$$

$$= \frac{1}{2} \int_0^{\pi/4} dx + \frac{1}{2} \int_0^{\pi/4} \cos 2x \, dx.$$

In the second integral we let $u = 2x$, obtaining $du = 2 \, dx$ and hence

$$\int_0^{\pi/4} \cos 2x \, dx = \frac{1}{2} \int_0^{\pi/2} \cos u \, du = \frac{1}{2} \sin u \Big|_0^{\pi/2} = \frac{1}{2},$$

$$\int_0^{\pi/4} \cos^2 x \, dx = \frac{1}{8} \pi + \frac{1}{4},$$

and

$$V = \tfrac{1}{8}\pi^2 + \tfrac{1}{4}\pi.$$

Exercises

1. Evaluate the following:

   (a) $\displaystyle\int \cos 5x \, dx.$

   (b) $\displaystyle\int x \sin x^2 \, dx.$

   (c) $\displaystyle\int \tan 2x \, dx.$

   (d) $\displaystyle\int \sin x \cos x \, dx.$

   (e) $\displaystyle\int \cos x \sin^2 x \, dx.$

   (f) $\displaystyle\int \sec^2 x e^{\tan x} \, dx.$

   (g) $\displaystyle\int \sin x \log |\cos x| \, dx.$

   (h) $\displaystyle\int x \sin 2x \, dx.$

2. Evaluate the following definite integrals:

   (a) $\displaystyle\int_{\pi/2}^{3\pi/4} \sin 2x \, dx.$

   (b) $\displaystyle\int_0^{\pi/2} \sin^2 x \cos x \, dx.$

   (c) $\displaystyle\int_{\pi/6}^{\pi/3} \tan x \sec^2 x \, dx.$

   (d) $\displaystyle\int_0^{\pi/2} \sqrt{1 - \sin x} \cos x \, dx.$

   (e) $\displaystyle\int_0^{\pi/2} \frac{\cos x}{1 + \sin x} \, dx.$

3. Find the area under the curve $y = \cos x$ and above the x axis from $x = 0$ to $x = \tfrac{1}{2}\pi$.

4. Find the area bounded by the curve $y = \sin x$, the y axis, and the line $y = 1$.

5. Find the area bounded by $y = \sin x$, $y = \cos x$, and the x axis, for $0 \le x \le \tfrac{1}{2}\pi$.

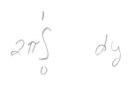

6. Find the volume of the solid generated by revolving the curve $y = \sin x$, $0 \le x \le \frac{1}{2}\pi$ about the x axis.

7. Find the volume of the solid generated by revolving the curve $y = \sec x$, $0 \le x \le \frac{1}{4}\pi$ about the x axis.

8. Find the area bounded by the curve $y = \tan x \sec^2 x$, the x axis, and the line $x = \frac{1}{3}\pi$.

7.2 Use of Tables of Integrals

The integration techniques introduced in Chapter 5, integration by parts and substitution, are adequate for many of the functions needed in applications. However, there still are many important types of functions for which neither of the above techniques will apply. For example, consider the evaluation of the integral

$$\int \frac{2x + 1}{(3x^2 + 2x + 1)^3} \, dx.$$

In order to evaluate such an integral, we would be required to use techniques involving trigonometric functions and an algebraic device called partial fraction decomposition. Fortunately, a large number of integrals involving complicated techniques appear in tables of integrals such as Table A3 in the Appendix. Let us begin by demonstrating how to use the table of integrals effectively.

▶ Example 1. Evaluate

$$\int \frac{dx}{\sqrt{9x^2 + 1}}.$$

We see from Table A3 that this is almost of the form of entry 14, namely

$$\int \frac{dx}{\sqrt{x^2 + a^2}} = \log (x + \sqrt{x^2 + a^2}) + C$$

To make our integral look more like the integral in the table, we factor the coefficient of x^2, namely 9, out of the radical sign. Thus, we first write

$$\int \frac{dx}{\sqrt{9x^2 + 1}} = \frac{1}{3} \int \frac{dx}{\sqrt{x^2 + \frac{1}{9}}}.$$

We then use entry 14 with $a = \frac{1}{3}$. Hence,

$$\int \frac{dx}{\sqrt{9x^2 + 1}} = \frac{1}{3} \log x + \sqrt{x^2 + \frac{1}{9}} + C$$

where C is an arbitrary constant.

▶ Example 2. Evaluate

$$\int \sqrt{\frac{x + 1}{x - 1}} \, dx.$$

We first note that this is not in a form suitable for the tables. However, if we rationalize the expression $\sqrt{(x + 1)/(x - 1)}$ by multiplying numerator and denominator by $\sqrt{x + 1}$ to obtain

$$\sqrt{\frac{x + 1}{x - 1}} = \frac{x + 1}{\sqrt{x^2 - 1}},$$

we see that

$$\int \sqrt{\frac{x + 1}{x - 1}} \, dx = \int \frac{x \, dx}{\sqrt{x^2 - 1}} + \int \frac{dx}{\sqrt{x^2 - 1}}.$$

The first integral can be evaluated by setting $u = x^2 - 1$, so that

$$\int \frac{x \, dx}{\sqrt{x^2 - 1}} = \frac{1}{2} \int \frac{du}{\sqrt{u}} = \sqrt{u} + C_1 = \sqrt{x^2 - 1} + C_1.$$

From entry 15 in Table A3, with $a = 1$, we obtain

$$\int \frac{dx}{\sqrt{x^2 - 1}} = \log (x + \sqrt{x^2 - 1}) + C_2,$$

and thus

$$\int \sqrt{\frac{x + 1}{x - 1}} \, dx = \sqrt{x^2 - 1} + \log (x + \sqrt{x^2 - 1}) + C,$$

where C is the sum of the two arbitrary constants C_1 and C_2.

▶ **Example 3.** Evaluate $\int x^4 e^{2x} \, dx$.

Using entry 34 in Table A3 with $n = 4$ and $a = 2$, we obtain

$$\int x^4 e^{2x} \, dx = \frac{x^4 e^{2x}}{2} - 2 \int x^3 e^{2x} \, dx + C.$$

We again use entry 34 repeatedly to obtain

$$\int x^3 e^{2x} \, dx = \frac{x^3 e^{2x}}{2} - \frac{3}{2} \int x^2 e^{2x} \, dx = \frac{x^3 e^{2x}}{2} - \frac{3}{2} \left(\frac{x^2 e^{2x}}{2} - \int x e^{2x} \, dx \right)$$

$$= \frac{x^3 e^{2x}}{2} - \frac{3}{2} \left(\frac{x^2 e^{2x}}{2} - \frac{x e^{2x}}{2} + \frac{1}{2} \int e^{2x} \, dx \right)$$

$$= \frac{x^3 e^{2x}}{2} - \frac{3}{2} \left(\frac{x^2 e^{2x}}{2} - \frac{x e^{2x}}{2} + \frac{1}{4} e^{2x} \right) + C_1.$$

Hence,

$$\int x^4 e^{2x} \, dx = \frac{x^4 e^{2x}}{2} - x^3 e^{2x} + \tfrac{3}{2} x^2 e^{2x} - \tfrac{3}{2} x e^{2x} + \tfrac{3}{4} e^{2x} + C,$$

where C is an arbitrary constant.

In Example 3, we could have repeatedly used the integration by parts formula, but that is certainly more cumbersome than using the integral tables. It should be noted that the tables were prepared using techniques we have already learned, such as integration by parts.

▶ Example 4. Evaluate

$$\int_0^{1/2} \frac{dx}{1 - x^2}.$$

We first observe that an antiderivative of $1/(1 - x^2)$ can be found by means of entry 16 in Table A3 with $a = 1$. Thus,

$$\int_0^{1/2} \frac{dx}{1 - x^2} = \tfrac{1}{2} \log \left(\frac{1 + x}{1 - x}\right)\Big|_0^{1/2} = \tfrac{1}{2} \log \left(\frac{1 + \tfrac{1}{2}}{1 - \tfrac{1}{2}}\right) - \tfrac{1}{2} \log 1 = \tfrac{1}{2} \log 3.$$

In evaluating integrals by use of Table A3, first see if the integral is of the form of one of the entries in the table. If not, try to get the integral into one of these forms by either the method of substitution or integration by parts.

As a final example, we show how useful the technique of integration by parts can be in using Table A3.

▶ Example 5. Evaluate

$$\int \frac{x^4 \, dx}{\sqrt{1 - x^2}}.$$

As we scan Table A3, we do not find an entry of the form of this integral. Suppose we rewrite the above integral as

$$\int \frac{x^4 \, dx}{\sqrt{1 - x^2}} = \int x^3 \cdot \frac{x \, dx}{\sqrt{1 - x^2}}.$$

Now, if we perform an integration by parts with $u = x^3$ and

$$dv = \frac{x}{\sqrt{1 - x^2}} \, dx,$$

yielding $v = -\sqrt{1 - x^2}$, we note that

$$\int x^3 \cdot \frac{x \, dx}{\sqrt{1 - x^2}} = -x^3 \sqrt{1 - x^2} + 3 \int x^2 \sqrt{1 - x^2} \, dx.$$

Now from entry 26 with $a = 1$, we find that

$$\int x^2 \sqrt{1 - x^2} \, dx = -\tfrac{1}{4} x (1 - x^2)^{3/2} + \tfrac{1}{8} x \sqrt{1 - x^2} + \tfrac{1}{8} \arcsin x + C_1,$$

where C_1 is an arbitrary constant. Thus,

$$\int \frac{x^4 \, dx}{\sqrt{1 - x^2}} = -x^3 \sqrt{1 - x^2} - \tfrac{3}{4} x (1 - x^2)^{3/2} + \tfrac{3}{8} x \sqrt{1 - x^2} + \tfrac{3}{8} \arcsin x + C,$$

where C is an arbitrary constant. Note that the example we have just worked out could itself be added to the table of integrals.

Exercises

1. By the use of tables, evaluate the following integrals:

(a) $\displaystyle\int \sqrt{4x^2 + 9}\ dx.$

(b) $\displaystyle\int \frac{dx}{x\sqrt{9 - 2x^2}}.$

(c) $\displaystyle\int \sqrt{\frac{x - 1}{x + 2}}\ dx.$

(d) $\displaystyle\int x^3 e^{8x}\ dx.$

(e) $\displaystyle\int \frac{e^x\ dx}{3e^{2x} + 5}.$

(f) $\displaystyle\int x^4 e^{2x}\ dx.$

(g) $\displaystyle\int \sqrt{\frac{x - 3}{x + 3}}\ dx.$

(h) $\displaystyle\int \frac{dx}{x\sqrt{9 - x^2}}.$

(i) $\displaystyle\int \frac{x\ dx}{5x + 4}.$

(j) $\displaystyle\int x^2 \sqrt{25 - x^2}\ dx.$

2. Evaluate the following:

(a) $\displaystyle\int_3^4 \frac{dx}{x^2 - 2}.$

(b) $\displaystyle\int_0^1 x^4 e^{2x}\ dx.$

(c) $\displaystyle\int_0^2 \frac{x\ dx}{2x + 3}.$

(d) $\displaystyle\int_1^2 x^3 \log x\ dx.$

*(e) $\displaystyle\int_0^1 e^x (\sin e^x)^2\ dx.$

3. By first reducing to a form suitable for using Table A3, evaluate the following:

(a) $\displaystyle\int \frac{x^2\ dx}{\sqrt{2x^2 + 3}}.$

(b) $\displaystyle\int_{\log 3/2}^{\log 2} \frac{e^x\ dx}{e^{2x} - 1}.$

(c) $\displaystyle\int \frac{x^9\ dx}{\sqrt{1 - x^4}}.$

4. Find the average value of the function $f(x) = x^3 e^x$ on the interval $[0, 1]$.

5. Find the volume of the solid generated by revolving about the x axis the region bounded by $y = x^2 e^x$, $x = 0$, and $x = 1$.

7.3 Additional Comments on Integration. Improper Integrals

In Chapter 5, we always assumed that, in order to evaluate definite integrals on an interval $[a, b]$, the function in the integrand had to have an antiderivative on $[a, b]$, or had to be continuous on $[a, b]$. (See Section 5.1.) Actually, this is really more than

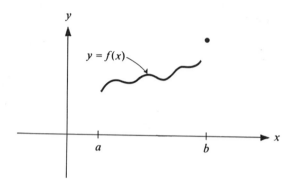

Figure 7.3.1

one needs. It turns out that if f is bounded on $[a, b]$, that is, if there exists a number M such that $|f(x)| \leq M$ for $x \in [a, b]$, and if f is continuous except at a finite number of points, say $c_1, c_2, \ldots, c_n$ (with $a \leq c_1 \leq c_2 \leq \cdots \leq b$), we can still define $\int_a^b f(x)\, dx$ in a reasonable manner.

Let us begin by considering the following example. (See Figure 7.3.1.) The function f is defined on $[a, b]$ and has an antiderivative $F(x)$ on $[a, b)$. We also assume that f is bounded on $[a, b]$. Thus f is Newton integrable on $[a, b - h]$ for $h > 0$ and

$$\int_a^{b-h} f(x)\, dx = F(b - h) - F(a).$$

It is an obvious and attractive strategy to define

$$\int_a^b f(x)\, dx = \lim_{h \to 0^+} \int_a^{b-h} f(x)\, dx = \lim_{h \to 0^+} (F(b - h) - F(a)) = \lim_{h \to 0^+} F(b - h) - F(a).$$

Clearly this definition works only if $\lim_{h \to 0^+} F(b - h)$ is defined. It is our good fortune that it always is. To repeat, if f is bounded on $[a, b]$ and has an antiderivative F on $[a, b)$, then the $\lim_{h \to 0^+} F(b - h)$ *always* exists.

We now expand this method slightly to cover a number of situations that arise frequently in practice.

We first consider a function f, bounded on $[a, b]$ and having a discontinuity at one point, say $c_1 \in [a, b]$. In this case one can prove that $\int_a^b f(x)\, dx$ exists and has the value

$$\int_a^b f(x)\, dx = \lim_{h \to 0^+} \left[\int_a^{c_1 - h} f(x)\, dx + \int_{c_1 + h}^b f(x)\, dx \right].$$

We are really saying that if f is continuous except at c_1 and bounded on $[a, b]$, then the limit on the right exists. Note that f is certainly Newton integrable on $[a, c_1 - h]$ and $[c_1 + h, b]$, but not on $[a, b]$. Thus,

$$\int_a^{c_1 - h} f(x)\, dx = F(c_1 - h) - F(a) \quad \text{and} \quad \int_{c_1 + h}^b f(x)\, dx = G(b) - G(c_1 + h),$$

where F is any antiderivative of f on $[a, c_1 - h]$ and G any antiderivative of f on $[c_1 + h, b]$. In the event that the discontinuities occur at a finite number of points $c_1, c_2, \ldots, c_m$, the value of the integral is given by

$$\int_a^b f(x)\, dx = \lim_{h \to 0^+} \left[\int_a^{c_1 - h} f(x)\, dx + \int_{c_1 + h}^{c_2 - h} f(x)\, dx + \cdots \right.$$

$$\left. + \int_{c_{m-1} + h}^{c_m - h} f(x)\, dx + \int_{c_m + h}^b f(x)\, dx \right].$$

(See Example 3 to follow.) One can always show that if $|f(x)| \le M$, a constant, for $x \in [a, b]$, then the limit on the right exists. Let us consider some examples.

▶ Example 1. Suppose f is defined on $[0, 2]$ as follows:

$$f(x) = \begin{cases} x, & 0 \le x < 1, \\ 2, & 1 \le x \le 2. \end{cases}$$

The graph of f is shown in Figure 7.3.2. Find $\int_0^2 f(x)\, dx$.

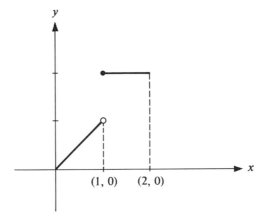

Figure 7.3.2

Solution. It is clear that on $[0, 2]$, $|f(x)| \le 2$ and the only discontinuity occurs at $x = 1$. Hence, $\int_0^2 f(x)\, dx$ exists and is given by

$$\int_0^2 f(x)\, dx = \lim_{h \to 0^+} \left[\int_0^{1-h} f(x)\, dx + \int_{1+h}^2 f(x)\, dx \right]$$

$$= \lim_{h \to 0^+} \left[\int_0^{1-h} x\, dx + \int_{1+h}^2 2\, dx \right]$$

$$= \lim_{h \to 0^+} \left[\frac{x^2}{2} \Big|_0^{1-h} + 2x \Big|_{1+h}^2 \right]$$

$$= \lim_{h \to 0^+} \left[\frac{(1 - h)^2}{2} + 4 - 2(1 + h) \right]$$

$$= \tfrac{1}{2} + 4 - 2 = \tfrac{5}{2}.$$

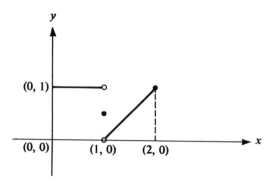

Figure 7.3.3

▶ Example 2. Consider the function defined on $[0, 2]$ by

$$f(x) = \begin{cases} 1, & 0 \le x < 1, \\ \frac{1}{2}, & x = 1, \\ x - 1, & 1 < x \le 2. \end{cases}$$

The graph of f is shown in Figure 7.3.3. Find $\int_0^2 f(x)\, dx$.

Solution. We immediately observe that $|f(x)| \le 1$ for $x \in [0, 2]$ and the only point of discontinuity occurs at $x = 1$. Hence,

$$\int_0^2 f(x)\, dx = \lim_{h \to 0^+} \left[\int_0^{1-h} f(x)\, dx + \int_{1+h}^2 f(x)\, dx \right]$$

$$= \lim_{h \to 0^+} \left[\int_0^{1-h} dx + \int_{1+h}^2 (x - 1)\, dx \right]$$

$$= \lim_{h \to 0^+} \left[x \Big|_0^{1-h} + \left(\frac{x^2}{2} - x \right) \Big|_{1+h}^2 \right]$$

$$= \lim_{h \to 0^+} \left[1 - h - \frac{(1 + h)^2}{2} + (1 + h) \right] = \frac{3}{2}.$$

Note the important fact that the value of the integral is independent of the value of f at the point of discontinuity, $x = 1$.

We consider one more example.

▶ Example 3. Let f be defined as follows:

$$f(x) = \begin{cases} 2, & 0 \le x < 1, \\ 1, & 1 < x \le 2, \\ \frac{1}{2}, & 2 < x < 3, \\ \frac{1}{4}, & x = 3, \\ x - 3, & 3 < x < 4, \\ 5 - x, & 4 < x \le 5. \end{cases}$$

(See Figure 7.3.4.) Find $\int_0^5 f(x)\, dx$.

Solution. It is clear in this case that the points of discontinuity occur at 1, 2, 3, and 4. In fact, f is not even defined at 1 and 4. Still, on $[0, 5]$, excluding 1 and 4, $|f(x)| \le 2$ and we can evaluate $\int_0^5 f(x)\, dx$. (The value of the integral is completely

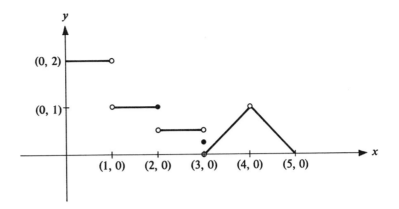

Figure 7.3.4

independent of the value of the function at the points of discontinuity.) We easily extend the method used for one point of discontinuity in the preceding examples.

$$\int_0^5 f(x)\,dx = \lim_{h\to 0^+}\left[\int_0^{1-h} 2\,dx + \int_{1+h}^{2-h} dx + \int_{2+h}^{3-h} \tfrac{1}{2}\,dx \right.$$

$$\left. + \int_{3+h}^{4-h} (x-3)\,dx + \int_{4+h}^5 (5-x)\,dx\right]$$

$$= \lim_{h\to 0^+}\left[2(1-h)+(2-h)-(1+h)+\tfrac{1}{2}(3-h)-\tfrac{1}{2}(2+h)\right.$$

$$\left. + \left(\frac{x^2}{2}-3x\right)\Big|_{3+h}^{4-h} + \left(5x-\frac{x^2}{2}\right)\Big|_{4+h}^5\right]$$

$$= \tfrac{9}{2}.$$

A far more serious problem arises when we consider the integral of a function $f(x)$ that has a finite number of discontinuities on $[a, b]$ and for which $|f(x)|$ assumes arbitrarily large values near these points of discontinuity. That is, f is no longer bounded on $[a, b]$. Suppose we consider the function $f(x) = 1/x$ for $0 < x \le 1$. We note that as $x \to 0$, $f(x)$ becomes very large. We now ask whether we can assign a meaning to $\int_0^1 x^{-1}\,dx$. Let us first observe that

$$\int_b^1 x^{-1}\,dx = \log 1 - \log b = \log 1/b,$$

where $0 < b < 1$. Now, as $b \to 0$, $\log 1/b \to \infty$. In this case, we say that $\int_0^1 dx/x$ has *no* meaning—that is, the function $1/x$ is not integrable on $[0, 1]$. Note, that for any α, $0 < \alpha < 1$, the integral $\int_\alpha^1 x^{-1}\,dx$ exists and has the value $\log 1/\alpha$.

It is remarkable, however, that if we consider the function $f(x) = x^{-1/2}$, $x > 0$, we can assign a meaning to $\int_0^1 x^{-1/2}\,dx$. Both $1/x$ and $x^{-1/2}$ become infinite as $x \to 0$. Thus both have one discontinuity in the interval $[0, 1]$. (See Figure 7.3.5.) Now, we investigate

$$\int_h^1 x^{-1/2}\,dx = 2x^{1/2}\Big|_h^1 = 2 - 2h^{1/2}.$$

As $h \to 0$, $\int_h^1 x^{-1/2}\,dx \to 2$, since $h^{1/2} \to 0$.

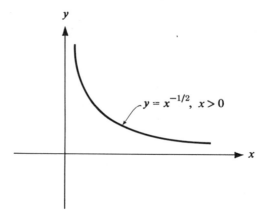

Figure 7.3.5

We say that $\int_0^1 x^{-1/2} \, dx$ exists or converges as an improper integral even though the function is not defined at 0 and is not continuous in the closed interval $[0, 1]$.

In general, consider a function f continuous on $a < x \leq b$. Thus, for every positive number h such that $a + h < b$, f is continuous on $[a + h, b]$. It is also assumed that as $x \to a$, $|f(x)| \to \infty$. We then form the integral, $\int_{a+h}^b f(x) \, dx$. Since f is Newton integrable on $[a + h, b]$,

$$\int_{a+h}^b f(x) \, dx = F(b) - F(a + h),$$

where F is any antiderivative of f on $[a + h, b]$.

DEFINITION. We say that $\int_a^b f(x) \, dx$ exists or converges as an improper integral if and only if $\lim_{h \to 0^+} [F(b) - F(a + h)]$ exists, where F is an antiderivative of f.

In this case, we set

$$\int_a^b f(x) \, dx = F(b) - \lim_{h \to 0^+} F(a + h)$$

by definition. If $\lim_{h \to 0^+} [F(b) - F(a + h)]$ does not exist, we say that $\int_a^b f(x) \, dx$ is *divergent*.

▶ Example 4. Determine whether or not the improper integral $\int_0^1 x^{-1/5} \, dx$ converges. If it does converge, evaluate the integral. (See Figure 7.3.6.)

Solution. First observe that the function $f(x)$ is continuous on $(0, 1]$ but not on $[0, 1]$. Hence, the integral is an improper integral. We then look at

$$\int_h^1 x^{-1/5} \, dx = \tfrac{5}{4}x^{4/5} \Big|_h^1 = \tfrac{5}{4} - \tfrac{5}{4}h^{4/5}.$$

Now, $\lim_{h \to 0^+} (\tfrac{5}{4} - \tfrac{5}{4}h^{4/5}) = \tfrac{5}{4}$. Hence, the improper integral converges and $\int_0^1 x^{-1/5} \, dx = \tfrac{5}{4}$.

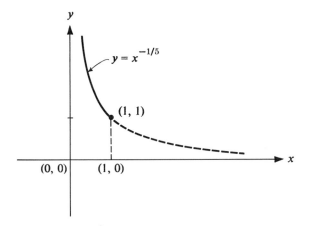

Figure 7.3.6

▶ **Example 5.** Determine whether or not the improper integral $\int_0^1 x^{-2}\,dx$ converges. If so, evaluate it. (See Figure 7.3.7.)

Solution. Again, since $x^{-2} \to \infty$ as $x \to 0$, $\int_0^1 x^{-2}\,dx$ is *improper*. Now, for $h > 0$, x^{-2} is continuous on $[h, 1]$. Hence,

$$\int_h^1 \frac{dx}{x^2} = -x^{-1}\Big|_h^1 = \frac{1}{h} - 1.$$

However, $\lim_{h\to 0^+} (1/h - 1)$ does not exist. Thus, $\int_0^1 x^{-2}\,dx$ is divergent.

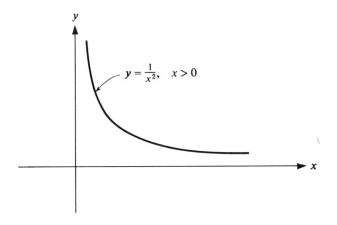

Figure 7.3.7

Suppose we consider a function f continuous on $[a, b)$ but not continuous on $[a, b]$ and assume that as $x \to b^-$, $|f(x)| \to \infty$. Thus, for $h > 0$ and $a < (b - h)$, f is continuous on $[a, b - h]$. We say that the *improper* integral $\int_a^b f(x)\,dx$ is convergent if and only if $\lim_{h\to 0^+} [F(b - h) - F(a)]$ exists, where F is any antiderivative of f on $[a, b - h]$.

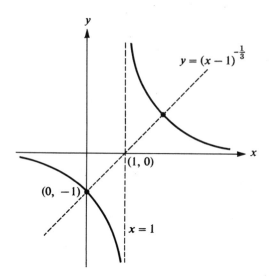

Figure 7.3.8

▶ **Example 6.** Determine whether $\int_{1/2}^{1} (x - 1)^{-1/3}\, dx$ converges. If so, evaluate it. (See Figure 7.3.8.)

Solution. Clearly, the function $f(x) = (x - 1)^{-1/3}$ is continuous on $[\frac{1}{2}, 1)$ but not on $[\frac{1}{2}, 1]$. Hence, the integral is improper. Now,

$$\int_{1/2}^{1-h} \frac{dx}{(x - 1)^{1/3}} = \int_{-1/2}^{-h} u^{-1/3}\, du = \frac{3}{2} u^{2/3} \Big|_{-1/2}^{-h}$$

$$= \frac{3}{2}(-h)^{2/3} - \frac{3}{2^{5/3}}.$$

However,

$$\lim_{h \to 0^+} \left[\frac{3}{2}(-h)^{2/3} - \frac{3}{2^{5/3}} \right] = -\frac{3}{2^{5/3}}.$$

Thus, $\int_{1/2}^{1} (x - 1)^{1/3}\, dx$ converges and has the value $-3/2^{5/3}$.

It is also possible that f is continuous on (a, b) but not on $[a, b]$. Thus, if $h > 0$ with $a + h < b - h$, f is continuous on $[a + h, b - h]$. We say that $\int_a^b f(x)\, dx$ is convergent if and only if $\lim_{h \to 0^+} F(b - h)$ and $\lim_{h \to 0^+} F(a + h)$ exist, where F is any antiderivative of f on $[a + h, b - h]$. In this case,

$$\int_a^b f(x)\, dx = \lim_{h \to 0^+} F(b - h) - \lim_{h \to 0^+} F(a + h).$$

Here we are assuming that as $x \to a^+$ and $x \to b^-$, $|f(x)| \to \infty$.

▶ **Example 7.** Does the integral $\int_0^1 [x(x - 1)]^{-1}\, dx$ converge? If so, evaluate it. (See Figure 7.3.9.)

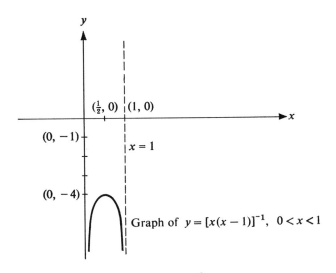

Figure 7.3.9

Solution. It is clear that the function $1/x(x-1)$ is continuous on $(0, 1)$ but not on $[0, 1]$. Thus, the integral is improper. First note that

$$\frac{1}{x(x-1)} = \frac{1}{x-1} - \frac{1}{x}.$$

Thus,

$$\int_0^1 \frac{dx}{x(x-1)} = \int_0^1 \frac{dx}{x-1} - \int_0^1 \frac{dx}{x}.$$

Now,

$$\int_0^1 \frac{dx}{x-1} = \log|x-1| \Big|_0^{1-h} = \log|-h|$$

and

$$\int_h^1 \frac{dx}{x} = \log|x| \Big|_h^1 = -\log|h|.$$

Observe that $\lim_{h \to 0^+} \log|-h|$ and $\lim_{h \to 0^+} \log|h|$ do not exist. Hence, since $\lim_{h \to 0^+} \log|-h|$ and $\lim_{h \to 0^+} \log|h|$ do not exist, $\int_0^1 [x(x-1)]^{-1}\, dx$ is *divergent*.

▶ Example 8. Let

$$f(x) = \begin{cases} x^{-1/2}, & 0 < x < \tfrac{1}{2}, \\ (1-x)^{-1/2}, & \tfrac{1}{2} \le x < 1. \end{cases}$$

(See Figure 7.3.10.) Does $\int_0^1 f(x)\, dx$ exist? If so, evaluate it.

Solution. It is clear from Figure 7.3.10 that as $x \to 0$ and $x \to 1$, $|f(x)| \to \infty$. Hence, the integral is certainly improper. However, f is certainly continuous on $[h, 1-h]$, where $0 < h < 1$. It is clear that

$$\int_h^{1-h} f(x)\, dx = \int_h^{1/2} x^{-1/2}\, dx + \int_{1/2}^{1-h} (1-x)^{-1/2}\, dx.$$

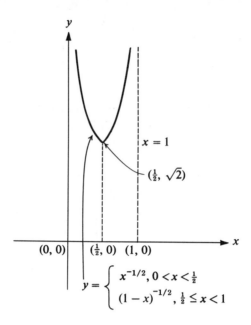

$$y = \begin{cases} x^{-1/2}, 0 < x < \tfrac{1}{2} \\ (1-x)^{-1/2}, \tfrac{1}{2} \le x < 1 \end{cases}$$

Figure 7.3.10

We must check to see if

$$\lim_{h \to 0^+} \int_h^{1/2} x^{-1/2}\, dx + \lim_{h \to 0^+} \int_{1/2}^{1-h} (1-x)^{-1/2}\, dx$$

exists. Now,

$$\int_h^{1/2} x^{-1/2}\, dx = 2x^{1/2}\bigg|_h^{1/2} = \frac{2}{\sqrt{2}} - 2\sqrt{h}$$

and

$$\int_{1/2}^{1-h} (1-x)^{-1/2}\, dx = -2(1-x)^{1/2}\bigg|_{1/2}^{1-h} = -2h^{1/2} + \frac{2}{\sqrt{2}}.$$

Since

$$\lim_{h \to 0^+}\left(\frac{2}{\sqrt{2}} - 2\sqrt{h}\right) = \frac{2}{\sqrt{2}} \quad \text{and} \quad \lim_{h \to 0^+}\left(\frac{2}{\sqrt{2}} - 2\sqrt{h}\right) = \frac{2}{\sqrt{2}},$$

$\int_0^1 f(x)\, dx$ exists and has the value $2/\sqrt{2} + 2/\sqrt{2} = 4/\sqrt{2}$.

Remark. A close inspection of Example 8 may lead the reader to suggest we replace the expression

$$\lim_{h \to 0^+}\left(\int_h^{1/2} x^{-1/2}\, dx\right) + \lim_{h \to 0^+}\left[\int_{1/2}^{1-h} (1-x)^{-1/2}\, dx\right]$$

by

$$\lim_{h \to 0^+}\left[\int_h^{1/2} x^{-1/2}\, dx + \int_{1/2}^{1-h} (1-x)^{-1/2}\, dx\right].$$

In this particular problem, such a change would not make any difference. However, there are cases where it would. We illustrate this point in Exercise 10 at the end of this section. We only comment here that each point of discontinuity should be handled separately.

There are other ways in which $\int_a^b f(x)\,dx$ can be improper. Suppose that f has unbounded discontinuities at a finite set of numbers $c_1, c_2, c_3, \ldots, c_n$, where $a < c_1 < c_2 < c_3 < \cdots < c_n < b$. That is, $|f(x)| \to \infty$ as $x \to c_i$, $i = 1, 2, \ldots, n$. The problem now is to determine whether or not $\int_a^b f(x)\,dx$ exists. We assert that $\int_a^b f(x)\,dx$ is convergent if and only if

$$\int_a^{c_1} f(x)\,dx + \int_{c_1}^{c_2} f(x)\,dx + \cdots + \int_{c_{n-1}}^{c_n} f(x)\,dx + \int_{c_n}^b f(x)\,dx$$

exists. In the event that the sum does *not* exist, we say that $\int_a^b f(x)\,dx$ *diverges*. We consider some examples.

▶ **Example 9.** Does $\int_{-1}^1 x^{-2}\,dx$ converge? If so, evaluate it. (See Figure 7.3.11.)

Solution. Clearly, $1/x^2$ has an unbounded discontinuity at $x = 0$. $\int_{-1}^1 x^{-2}\,dx$ converges if and only if

$$\lim_{h \to 0^+} \left(\int_{-1}^{0-h} \frac{dx}{x^2} \right) + \lim_{h \to 0^+} \left(\int_{0+h}^1 \frac{dx}{x^2} \right)$$

exists. Now,

$$\int_{-1}^{-h} \frac{dx}{x^2} = -x^{-1} \Big|_{-1}^{-h} = \frac{1}{h} - 1$$

and

$$\int_h^1 \frac{dx}{x^2} = -x^{-1} \Big|_h^1 = -1 + \frac{1}{h}.$$

Since $\lim_{h \to 0^+} (1/h - 1)$ and $\lim_{h \to 0^+} (-1 + 1/h)$ do not exist, the integral $\int_{-1}^1 x^{-2}\,dx$ is divergent.

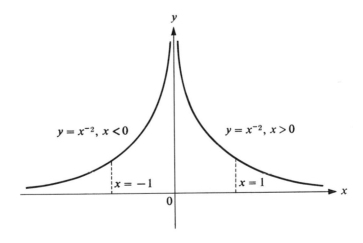

$y = x^{-2},\ x < 0$ $\qquad$ $y = x^{-2},\ x > 0$

$x = -1$ $\qquad$ $x = 1$

Figure 7.3.11

Suppose we were given $\int_{-1}^{1} x^{-2}\, dx$ and neglected to see that f has an unbounded discontinuity at $x = 0$. We would then be tempted to blindly use the formula

$$\int_{-1}^{1} \frac{dx}{x^2} = -x^{-1}\Big|_{-1}^{1} = -1 - 1 = -2$$

and say that $\int_{-1}^{1} x^{-2}\, dx$ certainly exists. However, this is false (as you have seen earlier). It should now be clear that it is essential to check to see whether the integrand has any unbounded discontinuities within the range of integration. Otherwise, one may be led to false conclusions and thus easily misinterpret any data involving such integrals. This is one reason why it is not sufficient to simply memorize formulas without understanding what is happening.

Let us consider one more example.

▶ Example 10. Determine whether the integral $\int_{-1}^{3} |x - 1|^{-1}\, dx$ converges. If so, evaluate it. (See Figure 7.3.12.)

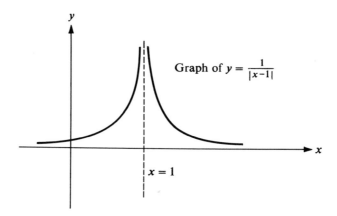

Graph of $y = \frac{1}{|x-1|}$

$x = 1$

Figure 7.3.12

Solution. First, note that $1/|x - 1|$ has an unbounded discontinuity at $x = 1$. Also recall that if $x > 1$,

$$\frac{1}{|x - 1|} = \frac{1}{x - 1},$$

and if $x < 1$,

$$\frac{1}{|x - 1|} = -\frac{1}{x - 1} = \frac{1}{1 - x}.$$

Now, $\int_{-1}^{3} |x - 1|^{-1}\, dx$ exists if and only if

$$\lim_{h \to 0^+} \left(\int_{1}^{1-h} \frac{dx}{1 - x} \right) + \lim_{h \to 0^+} \left(\int_{1+h}^{3} \frac{dx}{x - 1} \right)$$

exists. But,

$$\int_{-1}^{1-h} \frac{dx}{1 - x} = \log |1 - x|\Big|_{-1}^{1-h} = \log 2 - \log |h|$$

while

$$\int_{1+h}^{3} \frac{dx}{x - 1} = \log |x - 1|\Big|_{1+h}^{3} = \log 2 - \log |h|.$$

Neither the limit, $\lim_{h\to 0^+} (\log 2 - \log |h|)$, nor $\lim_{h\to 0^+} (\log 2 - \log |h|)$ exists and thus $\int_{-1}^{3} |x - 1|^{-1} \, dx$ is divergent.

There is one other type of improper integral, dealing with large values. Suppose f is a continuous function on $[a, \infty)$. We say that $\int_a^\infty f(x) \, dx$ exists or is convergent if and only if $\lim_{N\to\infty} [F(N) - F(a)]$ exists, where F is any antiderivative of f on $[a, N]$. We may write

$$\int_a^\infty f(x) \, dx = \lim_{N\to\infty} \int_a^N f(x) \, dx.$$

Note that f is certainly Newton integrable on $[a, N]$.

In like manner, if f is continuous on $(-\infty, b]$, we say that $\int_{-\infty}^b f(x) \, dx$ exists or is convergent if and only if $\lim_{M\to -\infty} [\int_M^b f(x) \, dx]$ exists. If it exists we write

$$\int_{-\infty}^b f(x) \, dx = \lim_{M\to -\infty} \int_M^b f(x) \, dx.$$

▶ **Example 11.** Determine whether the improper integral $\int_0^\infty e^{-x} \, dx$ is convergent. If so, evaluate it.

Solution. We first evaluate

$$\int_0^N e^{-x} \, dx = -e^{-x} \Big|_0^N = 1 - e^{-N}.$$

Now, $\lim_{N\to\infty} (1 - e^{-N}) = 1$. Hence, $\int_0^\infty e^{-x} \, dx$ is convergent and $\int_0^\infty e^{-x} \, dx = 1$.

▶ **Example 12.** Determine whether the improper integral $\int_1^\infty x^{-1} \, dx$ is convergent.

Solution. First, evaluate $\int_1^N x^{-1} \, dx = \log N$. Now, $\lim_{N\to\infty} \log N$ does not exist and hence the integral is divergent.

▶ **Example 13.** Determine whether the improper integral $\int_{-\infty}^0 e^x \, dx$ converges. If so, evaluate it.

Solution. First evaluate

$$\int_M^0 e^x \, dx = e^x \Big|_M^0 = 1 - e^M.$$

Since $\lim_{M\to -\infty} (1 - e^M) = 1$, we see that $\int_{-\infty}^0 e^x \, dx$ is convergent and has the value 1.

▶ **Example 14.** Determine whether the integral $\int_{-\infty}^\infty x e^{-x^2} \, dx$ converges. If so, evaluate it.

Solution. In the case that both limits are infinite, we say that $\int_{-\infty}^\infty x e^{-x^2} \, dx$ converges if and only if

$$\lim_{B\to -\infty} \int_B^0 x e^{-x^2} \, dx \quad \text{and} \quad \lim_{A\to\infty} \int_0^A x e^{-x^2} \, dx$$

exist. Letting $u = x^2$,

$$\int_0^A xe^{-x^2} \, dx = \tfrac{1}{2} \int_0^{A^2} e^{-u} \, du = -\tfrac{1}{2}e^{-A^2} + \tfrac{1}{2}$$

and

$$\int_B^0 xe^{-x^2} \, dx = \tfrac{1}{2} \int_{B^2}^0 e^{-u} \, du = -\tfrac{1}{2} + \tfrac{1}{2}e^{-B^2}.$$

Thus,

$$\lim_{A \to \infty} (-\tfrac{1}{2}e^{-A^2} + \tfrac{1}{2}) = \tfrac{1}{2} \quad \text{and} \quad \lim_{B \to -\infty} (\tfrac{1}{2}e^{-B^2} - \tfrac{1}{2}) = -\tfrac{1}{2}.$$

Hence, $\int_{-\infty}^{\infty} xe^{-x^2} \, dx$ exists and has the value $-\tfrac{1}{2} + \tfrac{1}{2} = 0$.

It is frequently possible to determine the convergence of an improper integral without computing it, by comparing it with another integral that is known to converge. We omit a discussion of this topic, but refer the reader to any of the more complete calculus texts for additional details. In the social and biological sciences, it is often necessary to compute the value of convergent improper integrals. This is seen in the following example from statistics.

▶ **Example 15.** Infinite integrals constantly occur in statistical contexts which arise in many social and biological problems. We say that $f(x)$ is a possible probability density function (or frequency distribution) if it is nonnegative and $\int_{-\infty}^{\infty} f(x) \, dx = 1$. The mean μ is given by

$$\mu = \frac{\int_{-\infty}^{\infty} xf(x) \, dx}{\int_{-\infty}^{\infty} f(x) \, dx} = \int_{-\infty}^{\infty} xf(x) \, dx.$$

In all cases, we must assume convergence of the relevant integrals.

The standard deviation σ (measuring the spread of the distribution about the mean μ) is given by

$$\sigma^2 = \int_{-\infty}^{\infty} (x - \mu)^2 f(x) \, dx.$$

(a) Given $p > 0$, find the value of the constant A such that the function

$$f(x) = \begin{cases} Ax^2 e^{-px}, & x \geq 0, \\ 0 & x < 0, \end{cases}$$

represents a probability density function.

Solution. From the definitions above, we must have

$$\int_{-\infty}^{\infty} f(x) \, dx = 1.$$

Thus,

$$\int_{-\infty}^{\infty} f(x) \, dx = A \int_0^{\infty} x^2 e^{-px} \, dx = 1.$$

Letting $y = -px$, we obtain

$$\int_0^{\infty} x^2 e^{-px} \, dx = -\frac{1}{p^3} \int_0^{-\infty} y^2 e^y \, dy = \frac{1}{p^3} \int_{-\infty}^0 y^2 e^y \, dy.$$

Integrating by parts twice, we see that

$$\int_{-\infty}^{0} y^2 e^y \, dy = \lim_{t \to -\infty} e^y(y^2 - 2y + 2) \Big|_{-t}^{0}.$$

But it is clear that $\lim_{t \to -\infty} e^t(t^2 - 2t + 2) = 0$. Hence, A must satisfy the relation

$$2A/p^3 = 1 \quad \text{or} \quad A = \tfrac{1}{2}p^3.$$

(b) Show that the mean is $3/p$ and the square of the standard deviation $\sigma^2 = 3/p^2$.
The mean μ is given by

$$\mu = \int_{-\infty}^{\infty} x f(x) \, dx = \tfrac{1}{2}p^3 \int_{0}^{\infty} x^3 e^{-px} \, dx.$$

Again, if we set $y = -px$

$$\int_{0}^{\infty} x^3 e^{-px} \, dx = \lim_{t \to -\infty} -\frac{1}{p^4} e^y(y^3 - 3y^2 + 6y - 6) \Big|_{-t}^{0}.$$

Using the fact that $\lim_{t \to -\infty} e^t(t^3 - 3t^2 + 6t - 6) = 0$, we see that

$$\int_{0}^{\infty} x^3 e^{-px} \, dx = \frac{6}{p^4}$$

and hence

$$\mu = \frac{3}{p}.$$

(c) $\sigma^2 = \displaystyle\int_{0}^{\infty} (x - \mu)^2 A x^2 e^{-px} \, dx$

$$= \tfrac{1}{2}p^3 \left(\int_{0}^{\infty} x^4 e^{-px} \, dx - 2\mu \int_{0}^{\infty} x^3 e^{-px} \, dx + \mu^2 \int_{0}^{\infty} x^2 e^{-px} \, dx \right)$$

$$= \tfrac{1}{2}p^3 \left(\int_{0}^{\infty} x^4 e^{-px} \, dx - 2 \cdot \frac{3}{p} \cdot \frac{3 \cdot 2}{p^4} + \frac{9 \cdot 2}{p^2 p^3} \right)$$

[where we have used (a) and (b)]

$$= \tfrac{1}{2}p^3 \left(\frac{24}{p^5} - \frac{36}{p^5} + \frac{18}{p^5} \right)$$

$$= \frac{3}{p^2}.$$

Exercises

1. Let $f(x)$ be defined by

$$f(x) = \begin{cases} x, & 0 \le x < 1, \\ 3, & x = 1, \\ -x + 2, & 1 < x < 2. \end{cases}$$

(a) Sketch the graph of f.

(b) Evaluate $\int_0^2 f(x) \, dx$. Why does this integral exist?

2. Define a function $h(x)$ as follows:

$$h(x) = \begin{cases} 1, & x = 0, \\ x, & 0 < x < 1, \\ 0, & x = 1, \\ -x + 2, & 1 < x < 2, \\ 1, & x = 2. \end{cases}$$

(a) Graph the function.

(b) Evaluate $\int_0^2 h(x)\, dx$. Compare this value with the integral in Exercise 1. Does the result make sense? Why?

3. Define a function $f(x)$ as follows:

$$f(x) = \begin{cases} x^{-1/2}, & 0 < x < 1, \\ x, & 1 \le x < 2, \\ 3, & x = 2. \end{cases}$$

(a) Sketch the graph of f.

(b) Does $\int_0^2 f(x)\, dx$ exist? If so, evaluate it.

4. Determine whether or not the following integrals converge. If so, evaluate them.

(a) $\displaystyle\int_0^1 \frac{1}{x^{1/6}}\, dx.$

(b) $\displaystyle\int_{-1}^1 \frac{dx}{x^3}.$

(c) $\displaystyle\int_0^1 \log x\, dx.$

(d) $\displaystyle\int_0^2 \frac{dx}{x^2 - 2x}.$

[$Hint:$ $\lim_{x \to 0^+} x \log x = 0.$]

(e) $\displaystyle\int_0^5 \frac{dx}{5 - x}.$

(f) $\displaystyle\int_0^2 \frac{dx}{|x - 2|}.$

5. Determine whether the following integrals are convergent. If so, evaluate them.

(a) $\displaystyle\int_0^1 \frac{x\, dx}{\sqrt{1 - x^2}}.$

(b) $\displaystyle\int_{-1}^1 \frac{dx}{\sqrt{|x|}}.$

(c) $\displaystyle\int_0^1 x \log x\, dx.$ [$Hint:$ See Exercise 4(c).]

(d) $\displaystyle\int_0^1 \frac{x\, dx}{1 - x^2}.$

6. Determine whether the following integrals are convergent. If so, evaluate them.

(a) $\displaystyle\int_1^\infty \frac{dx}{x^{2/3}}.$

(b) $\displaystyle\int_0^\infty x^2 e^{-x}\, dx.$

(c) $\displaystyle\int_{-\infty}^0 x e^x\, dx.$

(d) $\displaystyle\int_{-\infty}^\infty x e^{-|x|}\, dx.$

(e) $\displaystyle\int_1^\infty \frac{dx}{x^{0.98}}.$

(f) $\displaystyle\int_1^\infty \frac{1}{x^2}\, dx.$

7. (a) For what numbers n does $\int_0^1 x^{-n}\, dx$ exist? Evaluate the integral for those n.

(b) For what numbers n does $\int_1^\infty x^{-n}\, dx$ exist? Evaluate the integral for those n.

*8. (a) Show that $\int_2^\infty (x \log x)^{-1} \, dx$ is divergent, but $\int_2^\infty x^{-1} (\log x)^{-2} \, dx$ is convergent, and find its value.

(b) For what values of n does $\int_2^\infty x^{-1} (\log x)^{-n} \, dx$ exist?

[*Hint:* Use Exercise 7(b).]

9. Show that the area bounded by the x axis, the curve $y = 1/x$, and the line $x = 1$ is infinite, whereas the volume generated by revolving this area about the x axis is finite. Find this volume.

10. Consider $\int_{-1}^1 [x/(x^2 - 1)^2] \, dx$. Show that the integral diverges. In other words, show that

$$\lim_{h \to 0} \int_{-1+h}^0 \frac{x}{(x^2 - 1)^2} \, dx + \lim_{h \to 0} \int_0^{1-h} \frac{x}{(x^2 - 1)^2} \, dx$$

does not exist. The form of the last expression suggests that we might wish to define

$$\int_{-1}^1 \frac{x}{(x^2 - 1)^2} \, dx = \lim_{h \to 0} \int_{-1+h}^{1-h} \frac{x}{(x^2 - 1)^2} \, dx.$$

Show that this last expression does have a limit. Thus the two definitions are really different. There are several good reasons for our choosing the first definition in this text.

11. The length of life of a radio component (in thousands of hours) has the associated probability density function

$$f(x) = \begin{cases} k(4 - x^2), & 0 \le x \le 2, \\ 0, & \text{otherwise,} \end{cases}$$

where k is unknown. Determine the value of k so that f is indeed a probability density function. Find the mean μ and square of the standard deviation σ associated with this probability density function.

12. Determine the value of A such that f is a probability density function if

$$f(x) = \begin{cases} Ax(1 - x), & 0 < x < 1, \\ 0, & \text{otherwise.} \end{cases}$$

Compute the mean and standard deviation squared associated with f.

13. A certain law of income states that

$$f(x) = \int_x^\infty (p - 1)u^{-p} \, du \qquad (x \ge 1),$$

where x denotes the income level, $p > 1$ is a constant, and $f(x)$ is the proportion of persons whose incomes exceed x. Find the proportion of persons falling in the income bracket between 3 and 4.

*7.4 Numerical Integration

We have now seen that if a function is continuous except at a finite number of points over $[a, b]$, then its Newton integral can be defined. However, not all definite integrals can be evaluated by any of the techniques we have introduced previously.

Examples of such integrals are $\int_0^1 e^{-x^2}\,dx$ and $\int_0^3 \sqrt{x^3 + 1}\,dx$. In such instances methods of approximating integrals have considerable practical value.

The curve $y = e^{-x}$ is the famous bell-shaped curve, or normal distribution, which psychologists encounter a dozen times a day. If you plot IQ scores, height, ability to throw a football, etc. for a population sample, you will obtain a bell-shaped curve as indicated in Figure 7.4.1. (For IQ scores the top of the bell would occur at the point 100 on the x axis.) It is not surprising that psychologists should want to know a numerical value for $\int_a^b e^{-x^2}\,dx$ for various a and b. In fact, $\int_{-a}^a e^{-x^2}\,dx$ is just what they measure in the laboratory in many of their experiments.

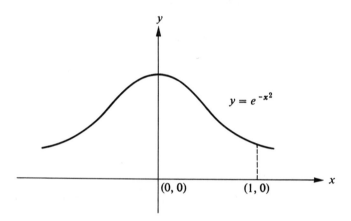

Figure 7.4.1

Let us investigate a numerical scheme to find an approximate value for $\int_0^1 e^{-x^2}\,dx$. First, we graph $y = e^{-x^2}$ and observe that the integral represents the area under the curve $y = e^{-x^2}$ from $x = 0$ to $x = 1$. (See Figure 7.4.1.) We try to find an approximation to this area. First, divide the interval $[0, 1]$ into four parts, say $[0, \frac{1}{4}]$; $[\frac{1}{4}, \frac{1}{2}]$; $[\frac{1}{2}, \frac{3}{4}]$; $[\frac{3}{4}, 1]$ and construct ordinates to the curve at the endpoints of each subdivision. (See Figure 7.4.2.) Next, replace the curve by a broken line by drawing chords as indicated in Figure 7.4.2. Then, we calculate the area of each one of these trapezoids and add the results. Recall that the area of a trapezoid is equal to one-half the sum of the parallel sides multiplied by the distance between them. If we label the four resulting trapezoids T_1, T_2, T_3, and T_4, then the area of each is given by

$$\text{area of } T_1 = A(T_1) = \tfrac{1}{2}(1 + e^{-1/16}) \cdot \tfrac{1}{4},$$
$$A(T_2) = \tfrac{1}{2}(e^{-1/16} + e^{-1/4}) \cdot \tfrac{1}{4},$$
$$A(T_3) = \tfrac{1}{2}(e^{-1/4} + e^{-9/16}) \cdot \tfrac{1}{4},$$
$$A(T_4) = \tfrac{1}{2}(e^{-9/16} + e^{-1}) \cdot \tfrac{1}{4}.$$

From Table A1 in the Appendix, $e^{-1/16} \simeq e^{-0.07} \simeq 0.93$, $e^{-1/4} = e^{-0.25} \simeq 0.78$, $e^{-9/16} \simeq 0.58$, $e^{-1} \simeq 0.37$. Thus,

$$A(T_1) + A(T_2) + A(T_3) + A(T_4) \cong \tfrac{1}{8}(1.93 + 1.71 + 1.36 + 0.95)$$

$$= \frac{5.95}{8} \simeq 0.74.$$

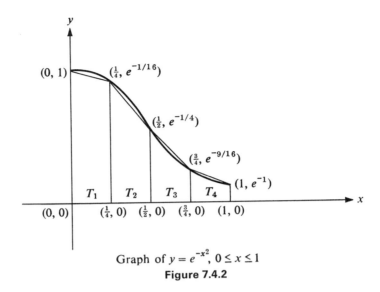

Graph of $y = e^{-x^2}$, $0 \leq x \leq 1$

Figure 7.4.2

We then might say that to two decimal places $\int_0^1 e^{-x^2}\, dx = 0.74$. Obviously we are making some errors in our method of computation. It turns out that the more subdivisions we take for $[0, 1]$, the better our approximate value will be. The technique outlined is called the trapezoid rule for approximating integrals, and can be extended to any continuous function defined on $[a, b]$. The number of subdivisions of $[a, b]$ we choose depends on how accurate we wish our computation. If we are interested in only seeing a very approximate value of integral, then often four subdivisions is enough. We could estimate carefully the error introduced by using the trapezoidal rule, but in the present discussion we choose to omit it. In actual practice, it is important to compute the error. We summarize by stating that to compute the value of $\int_a^b f(x)\, dx$ approximately by the trapezoidal rule we first subdivide $[a, b]$ into n subintervals, say

$$[a, b] = [a, x_1] \cup [x_1, x_2] \cup \cdots \cup [x_{n-2}, x_{n-1}] \cup [x_{n-1}, b]$$

and replace the curve by the chords joining $(x_{i-1}, f(x_{i-1}))$ to $(x_i, f(x_i))$ and then sum up the areas of the resulting trapezoids. Our choice of n is dictated by the accuracy we desire.

The trapezoidal rule is based essentially on the principle of approximating the curve by a series of linear segments. This rule is equivalent to approximating the function to be integrated by a series of linear functions, or polynomials of the first degree. We now consider a method based on approximation by polynomials of the second degree. It turns out in practice that this gives us a better approximation to $\int_a^b f(x)\, dx$ and is very easy to apply. Nowadays, it is convenient to use digital computers to compute definite integrals. The next technique, known as Simpson's rule, is readily adapted for use on computers.

This method is based on the simple formula

$$A_p = \tfrac{1}{3}h(y_0 + 4y_1 + y_2) \qquad (7.4.1)$$

for the area under the arc of the parabola $y = Ax^2 + Bx + C$ between $x = -h$ and $x = h$ (see Figure 7.4.3), where $y_0 = Ah^2 - Bh + C$, $y_1 = C$, and $y_2 = Ah^2 +$

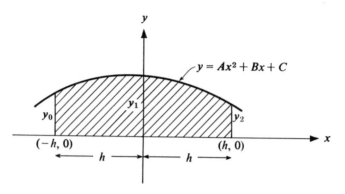

Figure 7.4.3

$Bh + C$. Formula (7.4.1) is easily obtained by computing the area under the parabola and observing that the curve goes through the points $(-h, y_0)$, $(0, y_1)$, and (h, y_2).

Simpson's rule follows by applying (7.4.1) to successive pieces of the curve $y = f(x)$ between $x = a$ and $x = b$. Each separate piece of the curve, covering a subinterval of length $2h$ on the x axis, is approximated by an arc of a parabola through its ends and its midpoint. (See Figure 7.4.4.) The area under each parabolic arc is given by an expression like (7.4.1). If we sum all areas of this type we obtain

$$A_s = \tfrac{1}{3}h[(y_0 + 4y_1 + y_2) + (y_2 + 4y_3 + y_4)$$
$$+ (y_4 + 4y_5 + y_6) + \cdots + (y_{n-2} + 4y_{n-1} + y_n)]$$
$$= \tfrac{1}{3}h[y_0 + 4y_1 + 2y_2 + 4y_3 + 2y_4 + 4y_5 + 2y_6 + \cdots + 2y_{n-2} + 4y_{n-1} + y_n].$$

$$(7.4.2)$$

(Note that the coefficient of the first and last y_i is 1 and the coefficients of the other y_i's are alternately 4's and 2's in the sum, which is taken as an approximate value to $\int_a^b f(x)\,dx$.) In (7.4.2), $y_0, y_1, \ldots, y_n$ are given by

$$y_0 = f(a), \ y_1 = f(a + h), \ y_2 = f(a + 2h), \ldots, \ y_n = f(a + nh) = f(b)$$

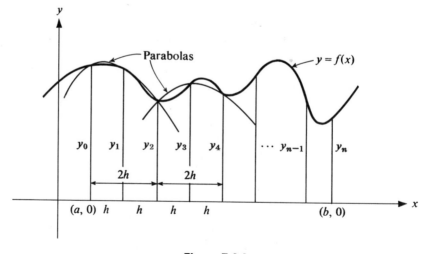

Figure 7.4.4

corresponding to a subdivision of the interval $a \leq x \leq b$ into n, where n is an *even* integer, equal subintervals each of width $h = (b - a)/n$. To apply this rule, we do not have to graph the function f, nor determine the approximating parabolic arcs. The formula (7.4.2) does this for us. It turns out that if f is continuous, the approximations get better as n increases and h gets smaller.

We now give an example of the above.

▶ Example. Approximate $\int_0^1 e^{-x^2} \, dx$ by Simpson's rule using

(a) four subdivisions, that is, $n = 4$;
(b) six subdivisions, that is, $n = 6$.

Solution. (a) We wish to use formula (7.4.2) with $n = 4$, $a = 0$, $b = 1$, and $h = \frac{1}{4}$. Thus,

$$y_0 = 1,$$
$$y_1 = e^{-(0.25)^2},$$
$$y_2 = e^{-(0.50)^2},$$
$$y_3 = e^{-(0.75)^2},$$
$$y_4 = e^{-1}.$$

$$\int_0^1 e^{-x^2} \simeq \tfrac{1}{12}[e^{-0} + 4e^{-(0.25)^2} + 2e^{-(0.50)^2} + 4e^{-(0.75)^2} + e^{-1}].$$

Using Table A1 we obtain

$$\int_0^1 e^{-x^2} \, dx \simeq \tfrac{1}{12}[1 + 4(0.940) + 2(0.779) + 4(0.570) + (0.368)]$$
$$\simeq 0.747.$$

Recall that, using the trapezoidal rule, we obtained 0.74. Thus, by taking $n = 4$ in Simpson's rule, we did gain a little. (Note that we should expect our result to be greater than 0.74.)

(b) In this case we use formula (7.4.2) with $n = 6$, $a = 0$, $b = 1$, and $h = \frac{1}{6}$. Thus,

$$y_0 = 1,$$
$$y_1 = e^{-(1/6)^2},$$
$$y_2 = e^{-(1/3)^2},$$
$$y_3 = e^{-(1/2)^2},$$
$$y_4 = e^{-(2/3)^2},$$
$$y_5 = e^{-(5/6)^2},$$
$$y_6 = e^{-1}.$$

Hence,

$$\int_0^1 e^{-x^2} \simeq \tfrac{1}{18}[1 + 4e^{-(0.0256)} + 2e^{-(0.0999)} + 4e^{-(0.2500)}$$
$$+ 2e^{-(0.4296)} + 4e^{-(0.6739)} + e^{-1}]$$
$$\simeq \tfrac{1}{18}[1 + 4(0.974) + 2(0.906) + 4(0.779) + 2(0.652) + 4(0.510) + (0.368)]$$
$$\simeq 0.758.$$

If we choose more subdivisions, we should expect to get a better result. It is possible to get an estimate for our error by making use of inequalities involving the fourth derivative of f. However, we omit a discussion of this here.

Exercises

1. For a curve which is concave upward, draw the approximating trapezoids using four subdivisions. Is the area under the trapezoids larger or smaller than the area under the curve? What happens if the curve is concave downward?

2. Approximate the following integral by means of the trapezoidal rule using four subdivisions of $[0, 2]$: $\int_0^2 e^{-0.5x^2} \, dx$. (Use Table A1.)

3. Use Simpson's rule instead of the trapezoidal rule in Exercise 2 and compare the results.

4. Approximate $\int_0^1 (1 + x^2)^{-1} \, dx$ by

   (a) The trapezoidal rule with five subdivisions of $[0, 1]$.

   (b) Simpson's rule with four subdivisions.

   Check your results by evaluating the integral with the aid of Table A3.

5. A certain curve is given by the following pairs of rectangular coordinates:

x	1	2	3	4	5	6	7	8	9
y	0	0.6	0.9	1.2	1.4	1.5	1.7	1.8	2

   (a) Approximate the area between the curve, the x axis, and the ordinates $x = 1$ and $x = 9$, using Simpson's rule.

   (b) Approximate the volume of the solid generated by revolving the area in (a) about the x axis, using Simpson's rule.

6. About how much crushed rock will be required to make a roadbed 1 mile long having the cross section shown in Figure 7.4.5 below. [*Hint:* Use Simpson's rule; first find the area, $A(x)$, and then the volume $V = \int A(x) \, dx$.]

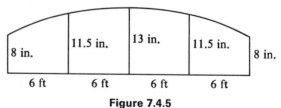

8 in.   11.5 in.   13 in.   11.5 in.   8 in.

6 ft   6 ft   6 ft   6 ft

Figure 7.4.5

7.5 Chapter 7 Summary

1. Familiarize yourself once again with the table of integrals Table A3 in the Appendix.

2. **Improper integrals**

(a) Let f be continuous on (a, b) but *not* on $[a, b]$. Then $\int_a^b f(x)\,dx$ is *convergent* if and only if $\lim_{h\to 0^+} F(b - h)$ and $\lim_{h\to 0^+} F(a + h)$ exist, where F is any antiderivative of f on $[a + h, b - h]$. In this case,

$$\int_a^b f(x)\,dx = \lim_{h\to 0^+} F(b - h) - \lim_{h\to 0^+} F(a + h).$$

(b) If f has unbounded discontinuities at a finite set of numbers $c_1, c_2, \ldots, c_m$, where $a < c_1 < c_2 < \cdots < c_m < b$ (that is, $\lim_{x\to c_i} |f(x)| = \infty$, $i = 1, 2, \ldots, m$), then $\int_a^b f(x)\,dx$ is *convergent* if and only if the sum

$$\int_a^{c_1} f(x)\,dx + \int_{c_1}^{c_2} f(x)\,dx + \cdots + \int_{c_{n-1}}^{c_n} f(x)\,dx + \int_{c_n}^b f(x)\,dx$$

exists [according to the definition given in (a)].

(c) $\int_a^\infty f(x)\,dx$ is *convergent* if and only if

$$\lim_{M\to\infty} \int_a^M f(x)\,dx$$

exists. In that case,

$$\int_a^\infty f(x)\,dx = \lim_{M\to\infty} \int_a^M f(x)\,dx.$$

*3. **Numerical integration**

(a) *Trapezoidal rule.* To compute the value of $\int_a^b f(x)\,dx$, first divide $[a, b]$ into n subintervals, say

$$[a, x_1], [x_1, x_2], \ldots, [x_{n-2}, x_{n-1}], [x_{n-1}, b];$$

replace the curve by the chords joining $(x_{i-1}, f(x_{i-1}))$ to $(x_i, f(x_i))$, $i = 1, 2, \ldots, n$, and then sum up the areas of the resulting trapezoids. The choice of n is dictated by the accuracy we desire.

(b) *Simpson's rule.*

$$\int_a^b f(x)\,dx = \tfrac{1}{3}h[y_0 + 4y_1 + 2y_2 + 4y_3 + 2y_4 + \cdots + 2y_{n-2} + 4y_{n-1} + y_n],$$

where

$$y_0 = f(a),\ y_1 = f(a + h),\ y_2 = f(a + 2h), \ldots, y_n = f(a + nh) = f(b),$$

n is an *even* integer, and $h = (b - a)/n$. Again the choice of n is determined by the accuracy we desire.

Review Exercises

1. Evaluate the following integrals (use Table A3):

(a) $\displaystyle\int \frac{dx}{9 - x^2}$.

(b) $\displaystyle\int \frac{dx}{\sqrt{4x^2 + 9}}$.

(c) $\displaystyle\int \frac{x + 2}{\sqrt{x^2 + 2x - 3}}\,dx$.

(d) $\displaystyle\int \frac{x + 2}{\sqrt{x^2 + 9}}\,dx$.

(e) $\displaystyle\int \sqrt{3x^2 + 5}\, dx.$

(f) $\displaystyle\int \frac{dx}{e^x + 1}.$

(g) $\displaystyle\int (x + 1) \log (2x^2 + 4x + 1)\, dx.$

(h) $\displaystyle\int \frac{dx}{x(3x + 6)^2}.$

(i) $\displaystyle\int x^2\sqrt{25 - x^2}\, dx.$

(j) $\displaystyle\int \frac{e^x\, dx}{\sqrt{e^{2x} + 9}}.$

2. Evaluate the following (if they exist):

(a) $\displaystyle\int_0^4 \frac{dx}{(x - 1)^2}.$

(b) $\displaystyle\int_0^4 \frac{dx}{\sqrt[3]{x - 1}}.$

(c) $\displaystyle\int_{-\infty}^0 e^{3x}\, dx.$

(d) $\displaystyle\int_{-\infty}^{\infty} \frac{dx}{e^x + e^{-x}}.$

(e) $\displaystyle\int_{-1}^1 x^{-2/3}\, dx.$

(f) $\displaystyle\int_2^{\infty} \frac{x^2\, dx}{(x^3 - 1)^{3/2}}.$

3. Find the area bounded by the curve $f(x) = xe^{-x^2}$ and the x axis. (Assume $x \geq 0$.)

4. Show that $\int_0^1 \log x\, dx$ is a convergent integral and find its value. Sketch the integrand for $0 < x \leq 1$.

5. Evaluate

$$\lim_{h \to 0} \frac{1}{h} \int_2^{2+h} e^{-x^2}\, dx.$$

[*Hint:* Define $F(x) = \int_0^x e^{-u^2}\, du$ and observe that $F(2 + h) - F(2) = \int_2^{2+h} e^{-u^2}\, du.$]

6. A plot of land lies between a straight fence and a stream. At distances x yards from one end of the fence, the width of the plot was measured as follows:

x	0	20	40	60	80	100	120
y	0	22	41	53	38	17	0

Use Simpson's rule to approximate the area of the plot.

7. Approximate $\int_0^1 \sqrt{1 + x^3}\, dx$ using

(a) The trapezoidal rule with $n = 5$.

(b) Simpson's rule with $n = 4$.

Compare the results.

8. From our geometric interpretation of the integral, we know that $\int_0^1 \sqrt{1 - x^2}\, dx = \frac{1}{4}\pi$. Find the approximate values of π by using this equation, $n = 4$ and (a) the trapezoidal rule; (b) Simpson's rule.

Elementary Differential Equations

8.1 Introduction

Suppose we denote by $P(t)$ the function that gives the size of a population at time t. The simplest type of population growth is obtained by assuming that the rate of change of the size of a population is proportional to the population size. In terms of the function $P(t)$, this assumption implies that there exists a constant k, called the *coefficient of growth*, such that

$$\frac{dP(t)}{dt} = kP(t). \tag{8.1.1}$$

Thus, in order to find the population at any time t, we must find a solution to Equation (8.1.1). This is an example of a differential equation—that is, an equation involving a derivative. Equation (8.1.1) is a type of equation that arises very often in the discussion of physical and social science phenomena. We show in Section 8.2 that the function

$$P(t) = Ce^{kt} \qquad (C \text{ is an arbitrary constant})$$

is a solution of Equation (8.1.1), since

$$\frac{dP}{dt} = \frac{d}{dx}(Ce^{kt}) = Cke^{kt} = k(Ce^{kt}) = kP(t).$$

Thus, a function P is called a solution of Equation (8.1.1) provided that $P(t)$ satisfies Equation (8.1.1) for every $t \in$ domain P. In general, an equation involving x, y, and the derivatives

$$\frac{dy}{dx}, \frac{d^2y}{dx^2}, \ldots, \frac{d^ny}{dx^n}$$

is called an *ordinary differential equation* of order n. Equation (8.1.1) is an ordinary differential equation of order 1. A function $y = f(x)$ is called a solution of the differential equation if for every x in the *domain of f*, it satisfies the equation. For example, if our equation is of the form

$$x + y + \frac{dy}{dx} + \frac{d^2y}{dx^2} = 0,$$

then $y = f(x)$ is a solution of this second-order equation provided that

$$x + f(x) + \frac{df(x)}{dx} + \frac{d^2f(x)}{dx^2} = 0.$$

In computing antiderivatives of functions in Chapter 5, we were actually finding the solution to certain differential equations. For example, we found previously that $\int (x^2 + 1)\, dx = \frac{1}{3}x^3 + x + C$. Thus, the function $f(x) = \frac{1}{3}x^3 + x + C$ must be a solution of the differential equation $\dfrac{dy}{dx} = x^2 + 1$, since $f'(x) = x^2 + 1$. This is an example of a first-order differential equation. Consider the following differential equation:

$$\frac{dy}{dx} = x^{1/2} + 2.$$

To find a solution to this equation, we must simply find an antiderivative of $x^{1/2} + 2$. Thus a solution is given by

$$y(x) = \int (x^{1/2} + 2)\, dx = \tfrac{2}{3}x^{3/2} + 2x + C,$$

where C is an arbitrary constant. More generally, we can say that a solution to the differential equation

$$\frac{dy}{dx} = f(x), \tag{8.1.2}$$

where f is continuous on some interval, is given by

$$y(x) = \int f(x)\, dx.$$

Notice that in all of our examples above, one arbitrary constant appeared in the solution. In solving problems from physical, biological, and social sciences, we usually have a condition on y that allows us to evaluate that constant. For example, in the first problem considered, suppose we know that at time $t = 0$, the population size is $P(0) = 200$. We see that if $P(t) = Ce^{kt}$, then $P(0) = 200 = C$ and the particular solution we want is $P(t) = 200e^{kt}$.

We have observed that the solution to Equation (8.1.1) involved one arbitrary constant. This is a particular case of the fact that the solution of every first-order differential equation contains one arbitrary constant. It turns out that the solution of every second-order equation contains two arbitrary constants. In fact, we can state that every nth-order equation has a solution containing n arbitrary constants. Such a solution will be called a *general solution* of the differential equation. A solution obtained by giving particular values to the arbitrary constants is called a *particular solution*. Let us now consider some examples to illustrate these statements.

▶ **Example 1.** Find the general solution of the differential equation

$$\frac{d^2y}{dx^2} = x.$$

Solution. We observe that one integration of the equation yields

$$\frac{dy}{dx} = \tfrac{1}{2}x^2 + C_1,$$

and thus integrating once more, we obtain

$$y(x) = \tfrac{1}{6}x^3 + C_1x + C_2.$$

We expected two arbitrary constants since our original equation is a second-order differential equation. We have thus obtained the general solution to the equation. To check, we simply observe that

$$\frac{dy}{dx} = \frac{3x^2}{6} + C_1 \quad \text{and} \quad \frac{d^2y}{dx^2} = x.$$

▶ **Example 2.** Find the particular solution of the equation $\dfrac{d^2y}{dx^2} = 5x^3$ containing the points $(1, 0)$ and $(0, 1)$.

Solution. First, we find the general solution (containing two arbitrary constants) and then evaluate the constants using the facts that $y(1) = 0$ and $y(0) = 1$. Integrating once, we obtain

$$\frac{dy}{dx} = \frac{5x^4}{4} + C_1.$$

Hence, another integration yields

$$y(x) = \tfrac{5}{20}x^5 + C_1x + C_2 = \tfrac{1}{4}x^5 + C_1x + C_2.$$

Since $y(0) = 1$, we see that $y(0) = 0 + 0 + C_2$, and thus $C_2 = 1$. To determine C_1, we use the fact that $y(1) = 0 = \tfrac{1}{4} + C_1 + 1$, and thus $C_1 = -\tfrac{5}{4}$. Hence, the particular solution we seek is given by

$$y(x) = \tfrac{1}{4}x^5 - \tfrac{5}{4}x + 1.$$

▶ **Example 3.** Find the particular solution of the second-order equation

$$\frac{d^2y}{dx^2} = 4x^3 + 3x^2$$

passing through the point $(0, 1)$ and satisfying the condition $y'(0) = 2$.

Solution. First, we find the general solution of the equation. We know that this solution must contain two arbitrary constants. One integration of the equation yields

$$\frac{dy}{dx} = x^4 + x^3 + C_1 \qquad (C_1 \text{ an arbitrary constant}).$$

In order to find $y(x)$, we must integrate once more. Thus,

$$y(x) = \tfrac{1}{5}x^5 + \tfrac{1}{4}x^4 + C_1x + C_2 \qquad (C_2 \text{ an arbitrary constant}).$$

We now proceed to evaluate the C_1 and C_2. The condition that the solution must pass through $(0, 1)$ means that $y(0) = 1$. Hence, $y(0) = C_2 = 1$. Now, $y'(0) = C_1$ and thus $C_1 = 2$. Our particular solution is given by

$$y(x) = \tfrac{1}{5}x^5 + \tfrac{1}{4}x^4 + 2x + 1.$$

Exercises

1. Verify that the function $y(x) = \tfrac{1}{4}x^4 + C_1 x + C_2$ is a solution of the differential equation $y''(x) = 3x^2$.

2. Show that $y(x) = C/x$ is a solution to the differential equation $\dfrac{dy}{dx} = -\dfrac{y}{x}$ for $x \neq 0$.

3. Verify that $y(x) = C_1 e^x + C_2 x e^x$ is a solution of the differential equation $y''(x) - 2y'(x) + y = 0$.

4. Find the general solution to the following differential equations:

   (a) $y'(x) = x^3 - 3x$, (b) $y'(x) = xe^{x^2}$,

   (c) $y''(x) = x^2 + 1$, (d) $y''(x) = 5x^2 + e^x$,

   (e) $y'(x) = x/(x^2 + 1)$.

5. Find the particular solution of the following differential equations satisfying the given conditions:

   (a) $y'(x) = x^4 - 3x^2 + 1$, $y(0) = 1$.

   (b) $y''(x) = 5x^4 + e^x$, $y'(0) = 1$, $y(0) = 3$.

   (c) $y'(x) = xe^{-x^2}$, $y(0) = \tfrac{1}{2}$.

   (d) $y''(x) = \log x$, $y(1) = y'(1) = 0$.

6. Consider the equation $y' + 5y = 2$.

   (a) Show that the function f given by $f(x) = \tfrac{2}{5} + Ce^{-5x}$ is a solution, where C is any constant.

   (b) Assuming every solution has this form find that solution satisfying $f(1) = 2$.

   (c) Find the solution f that satisfies the condition $f(1) = 3f(0)$.

7. Consider the differential equation $y'' = 3x + 1$.

   (a) Find all solutions on the interval $0 \leq x \leq 1$.

   (b) Find the solution f that satisfies $f(0) = 1$, $f'(0) = 2$.

   (c) Find the solution f that satisfies $f(0) = 0$, $f'(1) = 3$.

8.2 Separation of Variables

In this section, we are concerned with a certain type of first-order differential equation, namely, one of the form

$$\frac{dy}{dx} = \frac{g(x)}{h(y)}, \tag{8.2.1}$$

where g and h are functions only of x and y, respectively. There is no known technique that leads to the solution of all first-order equations. Fortunately, a wide class of equations such as Equation (8.2.1), many of which appear in applications, can be tackled by a method known as separation of variables. Any first-order equation that can be written in the form of Equation (8.2.1) is said to have variables separated. In this case our equation may be written in the form

$$h(y)\frac{dy}{dx} = g(x). \tag{8.2.2}$$

For example, consider the differential equation

$$\frac{dy}{dx} = y. \tag{8.2.3}$$

We see that

$$\frac{1}{y}\frac{dy}{dx} = 1 \quad \text{for} \quad y \neq 0$$

and thus the equation is one with variables separated, where $h(y) = 1/y$ and $g(x) = 1$. Observe that

$$\int \frac{1}{y}\frac{dy}{dx}\,dx = \int dx.$$

Thus,

$$\int \frac{1}{y}\frac{dy}{dx}\,dx = x + C_1.$$

Now, in the first integral, let $u = y(x)$; formally, we see that $du = y'(x)\,dx$ and by our change of variable theorem in Chapter 5, the right-hand side becomes $\int (1/u)\,du = x + C_1$. Thus, $\log u = x + C_1$ and hence $\log y(x) = x + C_1$.

We can now find $y(x)$ by changing the last equation to exponential form and observing that

$$y(x) = Ke^x, \tag{8.2.4}$$

where $K = e^{C_1}$, a constant. It is a simple matter to show that this function is indeed a solution to (8.2.3), since $\dfrac{dy}{dx} = Ke^x = y(x)$. This example suggests that to find a solution to Equation (8.2.2) we simply integrate both sides of Equation (8.2.2), use a change of variable, and then obtain a solution. Let us now do this.

If y satisfies Equation (8.2.2), we see that

$$\int h(y)\frac{dy}{dx}\,dx = \int g(x)\,dx. \tag{8.2.5}$$

Letting $u = y(x)$, we obtain via the change of variable theorem

$$\int h(u)\,du = \int g(x)\,dx, \tag{8.2.6}$$

and thus a solution to the differential equation is given by any function $u = y(x)$ satisfying Equation (8.2.6).

Formally, the Leibniz notation for derivatives suggests a way to handle Equation (8.2.6). We begin with Equation (8.2.1), $dy/dx = g(x)/h(y)$, and rewrite this as $h(y)\, dy = g(x)\, dx$, where we emphasize, as in Chapter 5, that this manipulation is strictly a mnemonic device. Integrating formally, we obtain Equation (8.2.6), namely, $\int h(y)\, dy = \int g(x)\, dx$. Thus, we may now state the following theorem, which yields the general solution to Equation (8.2.1).

THEOREM. Let h and g be given continuous functions. Then the general solution of the differential equation

$$y' = g(x)/h(y) \qquad (8.2.7)$$

is given by

$$\int h(y)\, dy = \int g(x)\, dx \qquad (8.2.8)$$

in the sense that any solution $y = f(x)$ must satisfy Equation (8.2.8).

In all cases, we can check that we really have a solution to the given differential equation by seeing if any solution $y = f(x)$ really satisfies the given differential equation. Let us now proceed to some examples.

▶ Example 1. Solve the differential equation $y' = 2x/3y^2$.

Solution. Using the Liebniz notation, we write the differential equation as

$$3y^2 \frac{dy}{dx} = 2x,$$

and hence

$$3y^2\, dy = 2x\, dx.$$

Thus our theorem tells us that the general solution is given by

$$\int 3y^2\, dy = \int 2x\, dx$$

or $y^3 = x^2 + C$, where C is any arbitrary constant. We might check our solution by using implicit differentiation and observing that

$$3y^2 \frac{dy}{dx} = 2x \quad \text{and thus} \quad \frac{dy}{dx} = \frac{2x}{3y^2}.$$

Hence, $y^3 = x^2 + C$ really yields the solution.

We should remark at this point why we obtain only one arbitrary constant C in Example 1. We note that

$$\int 3y^2\, dy = y^3 + C_1 \quad \text{and} \quad \int 2x\, dx = x^2 + C_2,$$

where C_1 and C_2 are arbitrary constants. Hence, we see that $y^3 + C_1 = x^2 + C_2$ or equivalently $y^3 = x^2 + C_2 - C_1$. Now if C_1 and C_2 are arbitrary constants, then so is $C_2 - C_1$. We simply set $C = C_2 - C_1$. In the future, we will always insert only one arbitrary constant into our solution of a separable first-order equation.

▶ **Example 2.** Find all solutions to the differential equation

$$y' = e^{x-y}.$$

Solution. Using the Leibniz notation, we wish to solve $\dfrac{dy}{dx} = e^{x-y}$ or, equivalently,

$e^y \, dy = e^x \, dx$ (since $e^{x-y} = e^x e^{-y}$). Thus, all solutions are given by $\int e^y \, dy = \int e^x \, dx$, from whence we obtain $e^y = e^x + C$. In order to find y, we take the logarithm of both sides and recall that $\log e^y = y$. Thus, $y = \log (e^x + C)$. It is not difficult to check this solution. Differentiating, we obtain

$$\frac{dy}{dx} = \frac{e^x}{(e^x + C)}.$$

But, $e^x + C = e^y$, and thus,

$$\frac{dy}{dx} = \frac{e^x}{e^y} = e^{x-y}.$$

▶ **Example 3.** We now return to the problem first considered in Section 8.1. In that example, it was necessary for us to solve the differential equation

$$\frac{dP(t)}{dt} = kP(t).$$

Hence, we see that a solution must be given by

$$\int \frac{dP}{P} = \int k \, dt \quad \text{or} \quad \log P = kt + C_1,$$

where C_1 is any constant. Thus,

$$P = e^{kt + c_1} = Ce^{kt},$$

where $C = e^{c_1}$ is again an arbitrary constant. If the population at time $t = 0$ is denoted by $P(0)$, we see that $C = P(0)$ and the particular solution we seek is given by

$$P(t) = P(0)e^{kt}.$$

Exercises

1. Find all solutions to the following differential equations:

   (a) $y' = x^2 y.$ (b) $yy' = x.$

   (c) $y' = \dfrac{x + x^2}{y - y^2}.$ (d) $y' = \dfrac{e^{x-y}}{1 + e^x}.$

   (e) $y' = x^3 y - 4x^3.$ (f) $y' = \dfrac{1}{xy}.$

2. (a) Find the solution of the differential equation $y' = 2y^{1/2}$ passing through the point $(1, 4)$.

   (b) Find all solutions of this equation passing through $(a, 0)$.

3. A firm observes that the rate of decrease of sales with respect to price is directly proportional to the sales volume and inversely proportional to the sales price plus a constant; that is,

$$\frac{ds}{dp} = -\frac{Bs}{A + p},$$

where s is the sales volume, p is the unit sales price, and A and B are constants. Find the relationship between sales, volume, and price.

4. Find the particular solutions to the following differential equations:

(a) $\dfrac{dy}{dx} = \dfrac{x^2}{y^2}$, where $y(0) = 1$.

(b) $\sqrt{1 + x^2}\,\dfrac{dy}{dx} + 3xy^2 = 0$, where $y(1) = 1$.

(c) $\dfrac{dy}{dx} = \dfrac{xy + y}{x + xy}$, where $y(1) = 1$.

(d) $\dfrac{dy}{dx} = \dfrac{e^x}{y}$, where $y(0) = 4$.

(e) $\dfrac{dy}{dx} = y \log x$ for $x > 0$, where $y(1) = 1$.

5. The supply curve for the product Ordinarium satisfies the relation $dp/ds = 10e^{-s}$, where s is the supply and p is the price. (Thus here p is a function of s.) If $p(0) = 0$, find the supply curve $p(s)$. Note that the supply grows arbitrarily large as the price approaches 10. Such behavior is not unreasonable for simple products such as lead pencils or bobby pins.

8.3 First-Order Equations Solved with Integrating Factors

Another type of differential equation that often arises in applications is

$$\frac{dy}{dx} + p(x)y = q(x), \tag{8.3.1}$$

where $p(x)$ and $q(x)$ are known functions. An equation of this form is a first-order linear differential equation. It is important to be able to recognize an equation of this form.

If $q(x) \neq 0$, then it is not possible to separate variables in Equation (8.3.1). Let us try a different approach. We will take Equation (8.3.1) which we can't solve (as it stands) and multiply it by a function f (called an integrating factor) to obtain an equation we *can* solve. More precisely, if we could find a function f with the property that

$$\frac{d}{dx}\left[f(x)y(x)\right] = f(x)\frac{dy}{dx} + f(x)p(x)y,$$

then it would be an easy matter to solve (8.3.1). Indeed, we would just multiply both sides of (8.3.1) by $f(x)$ to obtain

$$f(x)\frac{dy}{dx} + f(x)p(x)y = f(x)q(x)$$

and then rewrite this as

$$\frac{d}{dx}[f(x)y(x)] = f(x)q(x). \tag{8.3.2}$$

Now we can solve Equation (8.3.2) by integrating both sides to obtain

$$f(x)y(x) = \int f(x)q(x)\,dx$$

or equivalently

$$y(x) = \frac{1}{f(x)}\int f(x)q(x)\,dx. \tag{8.3.3}$$

The big question is: Can we find a function $f(x)$ with these magical properties? The answer is yes. Indeed the function

$$f(x) = e^{\int p(x)\,dx}$$

does all that we require. For those who do not regard this rabbit-out-of-a-hat technique kindly, we now show how we obtained this function. Let us begin with a specific example. Consider

$$\frac{dy}{dx} + xy = x. \tag{8.3.4}$$

In this case $p(x) = q(x) = x$. First, we multiply both sides of Equation (8.3.4) by a function $f(x)$ to obtain

$$f(x)\frac{dy}{dx} + xyf(x) = xf(x). \tag{8.3.5}$$

Now we wish to write the left-hand side of Equation (8.3.5) as follows:

$$f(x)\frac{dy}{dx} + xyf(x) = \frac{d}{dx}[f(x)y].$$

Thus, we want

$$f(x)\frac{dy}{dx} + xyf(x) = \frac{df(x)}{dx}y + f(x)\frac{dy}{dx}.$$

In order to obtain this equality it is clear that we must set

$$xf(x) = \frac{df(x)}{dx}. \tag{8.3.6}$$

Therefore, if we can find any function f satisfying the first-order equation (8.3.6), we can immediately solve our given differential equation. Observe that Equation (8.3.6) is really an equation whose variables are separated, since we may write Equation (8.3.6) as

$$\frac{1}{f(x)}\frac{df}{dx} = x.$$

Hence, we see that the solution to the above is obtained by integration. Namely,

$$\int\frac{df}{f} = \int x\,dx, \quad \text{and thus} \quad \log f = \tfrac{1}{2}x^2 + C.$$

However, since *any* solution of Equation (8.3.6) will do, we can take $C = 0$. The function f we seek is just given by

$$\log f = \tfrac{1}{2}x^2 \quad \text{or equivalently,} \quad f(x) = e^{x^2/2}.$$

Returning to Equation (8.3.4), we can obtain y by observing that

$$\left(\frac{dy}{dx} + xy\right) e^{x^2/2} = \frac{d}{dx}(e^{x^2/2}y) = xe^{x^2/2}.$$

Integrating once again, we see that

$$e^{x^2/2}y = \int xe^{x^2/2}\, dx = e^{x^2/2} + C.$$

Thus,

$$y(x) = 1 + Ce^{-x^2/2}$$

is the general solution of Equation (8.3.4). To check that our results are correct, we just show that this form of the solution satisfies Equation (8.3.4). Note that $\frac{dy}{dx} = -Cxe^{-x^2/2}$ and thus

$$\frac{dy}{dx} + xy = -Cxe^{-x^2/2} + x(1 + Ce^{-x^2/2}) = x.$$

We can easily generalize the scheme we have used in this example. We must first find a function f such that

$$f(x)y'(x) + p(x)f(x)y = f(x)y'(x) + yf'(x).$$

It is clear that the above equation is satisfied if

$$p(x)f(x) = \frac{df}{dx}. \tag{8.3.7}$$

As before, a solution to Equation (8.3.7) is obtained by integration:

$$\int \frac{1}{f}\, df = \int p(x)\, dx.$$

Thus,

$$\log f(x) = \int p(x)\, dx \quad \text{and} \quad f(x) = e^{\int p(x)\, dx}. \tag{8.3.8}$$

The function $f(x) = e^{\int p(x)\, dx}$ is called an *integrating factor* of the differential equation

$$\frac{dy}{dx} + p(x)y = q(x).$$

The general solution is found by first multiplying both sides of the equation by $e^{\int p(x)\, dx}$ and then integrating the result. After we multiply by $e^{\int p(x)\, dx}$, our equation becomes

$$e^{\int p(x)\, dx}y'(x) + e^{\int p(x)\, dx}p(x)y = e^{\int p(x)\, dx}q(x).$$

We now rewrite the left-hand side to obtain

$$\frac{d}{dx}[e^{\int p(x)\, dx}y] = e^{\int p(x)\, dx}q(x).$$

Thus,

$$e^{\int p(x)\,dx}y = \int [e^{\int p(x)\,dx}q(x)]\,dx + C.$$

Finally we multiply by $e^{-\int p(x)\,dx}$ to obtain

$$y = e^{-\int p(x)\,dx}\int [e^{\int p(x)\,dx}q(x)]\,dx + Ce^{-\int p(x)\,dx},$$

where C is an arbitrary constant.

Let us now consider several examples.

▶ **Example 1.** Find the general solution to the differential equation $y' = ky$, where k is a fixed number.

Solution. This differential equation has separable variables and could be solved by methods of Section 8.2. However, we choose to use the techniques in this section and show that the result is, of course, equivalent. Rewrite the equation in the form $y' - ky = 0$. In this case, $p(x) = -k$ and $q(x) = 0$. Hence, an integrating factor is given by

$$f(x) = e^{-\int k\,dx} = e^{-kx}.$$

Multiplying both sides of the equation by $f(x)$, we have

$$\left(\frac{dy}{dx} - ky\right)e^{-kx} = \frac{d}{dx}ye^{-kx} = 0.$$

Hence

$$ye^{-kx} = C \quad \text{and} \quad y = Ce^{kx},$$

where C is an arbitrary constant. This is the result obtained in Section 8.2.

As you can see, a given differential equation may be solved by more than one method. When this occurs, equivalent solutions are always obtained.

▶ **Example 2.** Find the general solution of the differential equation

$$\frac{dy}{dx} - 2y = 1.$$

Solution. In this example, $p(x) = -2$ and $q(x) = 1$. Hence, an integrating factor is given by

$$f(x) = e^{-\int 2\,dx} = e^{-2x}.$$

Multiplying both sides of the equation by $f(x)$, we have

$$\left(\frac{dy}{dx} - 2y\right)e^{-2x} = \frac{d}{dx}(ye^{-2x}) = e^{-2x}$$

and hence,

$$ye^{-2x} = \int e^{-2x}\,dx + C.$$

However,

$$\int e^{-2x}\,dx = -\tfrac{1}{2}e^{-2x}.$$

Therefore,

$$ye^{-2x} = -\tfrac{1}{2}e^{-2x} + C \quad \text{and thus} \quad y = -\tfrac{1}{2} + Ce^{2x},$$

where C is an arbitrary constant. It is a simple matter to check that this is a solution simply by substituting this expression for y into the left-hand side of the differential equation and observing that the result is 1.

▶ **Example 3.** A differential equation relating net profits P and advertising effort x is given by

$$\frac{dP}{dx} + aP = b - ax,$$

where a and b are constants. Suppose that at $x = 0$, $P = 2 \cdot (b + 1)/a$. Find the net profit in terms of the advertising effort x.

Solution. In this example, $p(x) = a$ and $q(x) = b - ax$. Hence an integrating factor is

$$f(x) = e^{\int a\,dx} = e^{ax}.$$

Multiplying both sides of the equation by $f(x)$, we have

$$\left(\frac{dP}{dx} + aP\right)e^{ax} = \frac{d}{dx}\,(Pe^{ax}) = (b - ax)e^{ax},$$

and hence

$$Pe^{ax} = \int (b - ax)e^{ax}\,dx + C.$$

Now,

$$\int (b - ax)e^{ax}\,dx = \frac{b}{a}\cdot e^{ax} - a\left[\frac{x}{a}\cdot e^{ax} - \frac{1}{a^2}\cdot e^{ax}\right]$$

$$= \frac{1 + b}{a}\cdot e^{ax} - xe^{ax} + C = \left(\frac{1 + b}{a} - x\right)e^{ax} + C.$$

Therefore,

$$P = \left(\frac{1 + b}{a} - x\right) + Ce^{-ax},$$

where C is an arbitrary constant. Using the condition $P(0) = 2(b + 1)/a$ we can evaluate the constant C.

$$P(0) = \frac{2(1 + b)}{a} = \frac{1 + b}{a} + C.$$

Thus, $C = (1 + b)/a$ and the particular solution is given by

$$P(x) = \frac{1 + b}{a} - x + \frac{1 + b}{a}\,e^{-ax}.$$

▶ **Example 4.** Find the particular solution of the differential equation

$$\frac{dy}{dx} + xy = (2 + x)e^{2x},$$

satisfying the condition $y(0) = 2$.

Solution. For this example, $p(x) = x$ and $q(x) = (2 + x)e^{2x}$. Thus, an integrating factor is given by

$$f(x) = e^{\int x \, dx} = e^{x^2/2}.$$

Therefore,

$$\left(\frac{dy}{dx} + xy\right) e^{x^2/2} = \frac{d}{dx}(ye^{x^2/2}) = (2 + x)e^{2x}e^{x^2/2},$$

and hence

$$ye^{x^2/2} = \int (2 + x)e^{2x+(x^2/2)} \, dx = e^{2x+(x^2/2)} + C$$

and

$$y(x) = e^{2x} + Ce^{-x^2/2},$$

where C is an arbitrary constant. The condition $y(0) = 2$ yields $2 = 1 + C$, implying that $C = 1$. The particular solution is then given by

$$y = e^{2x} + e^{-x^2/2}.$$

The methods presented in Sections 8.2 and 8.3 apply only to differential equations of the appropriate form [the form of Equation (8.3.1) or Equation (8.2.1)]. Differential equations of these two forms occur frequently in applications to the physical, social, and biological sciences. We consider several examples in Section 8.4. However, differential equations in many other forms arise in various applications and have been thoroughly studied by mathematicians. The reader interested in learning more about these differential equations is referred to the many elementary textbooks on the subject.

Exercises

1. Find the general solution to the following differential equations:

   (a) $\dfrac{dy}{dx} - y = -2e^{-x}$.

   (b) $\dfrac{dy}{dx} + 2xy = 4x$.

   (c) $\dfrac{dy}{dx} - \dfrac{y}{x} = x^2$.

   (d) $(x - 2)\dfrac{dy}{dx} = y + 2(x - 2)^3$.

   [*Hint:* Assuming $x \neq 2$, first divide through by $(x - 2)$ and then write the equation in the form

   $$\frac{dy}{dx} + p(x)y = q(x).]$$

   (e) $\dfrac{dy}{dx} + \dfrac{y}{x^2} = e^{1/x}$.

   (f) $\dfrac{dy}{dx} + \dfrac{1}{x \log x} y = \dfrac{1}{x}$.

   (g) $\dfrac{dy}{dx} + 3x^2 y = x^2$.

2. Find the particular solution to the following differential equations:

(a) $\dfrac{dy}{dx} + 2xy = 2x,\quad y(0) = 2.$

(b) $\dfrac{dy}{dx} = ay + b\quad (a, b\ \text{constants}),\quad y(0) = -\dfrac{b}{a}.$

(c) $\dfrac{dy}{dx} + \dfrac{y}{x} = x,\quad y(1) = 1.$

(d) $x\dfrac{dy}{dx} + y = 3x^3 - 1\quad (\text{for } x > 0),\quad y(1) = 1.$

(e) $\dfrac{dy}{dx} + 2xy = xe^{-x^2},\quad y(0) = 1.$

(f) $\dfrac{dy}{dx} - 2y = x^2 + x,\quad y(0) = \tfrac{1}{2}.$

3. Manufacturing costs K are related to the number of items produced x by the equation

$$\frac{dK}{dx} + K = b + cx,$$

where b and c are constants. If $K = 0$ when $x = 0$, find K as a function of x.

4. Consider the differential equation

$$x^2\frac{dy}{dx} + 2xy = 1,\quad \text{where}\quad 0 < x < \infty.$$

(a) Show that every solution tends to zero as $x \to \infty$.

(b) Find the solution f that satisfies the condition $f(2) = 2f(1)$.

*5. Various equations can be made linear by suitable changes of variables. For example, the equation

$$2xy\frac{dy}{dx} + y^2 = x \tag{5a}$$

becomes, if we let $Y = y^2$, the linear equation

$$x\frac{dY}{dx} + Y = x. \tag{5b}$$

(a) Check that the change of variable $Y = y^2$ transforms (5a) into (5b).

(b) Find the general solution to (5a) by first solving (5b).

*6. Consider the differential equation

$$\frac{dy}{dx} + p(x) = q(x)e^{my}. \tag{6a}$$

(a) Show that the change of variable $Y = e^{-my}$ transforms Equation (6a) into

$$-\frac{1}{m}\frac{dY}{dx} + p(x)Y = q(x). \tag{6b}$$

(b) Now find a form for the general solution of (6a) by first solving (6b).

(c) Using the technique indicated in (a), find the general solution to the differential equation

$$x \frac{dy}{dx} + 1 = x^2 e^{2y}.$$

8.4 Applications of Differential Equations

In this section we are concerned with various applications of first-order differential equations to problems in business and the social and biological sciences.

▶ Application 1. (*Growth*) Any important bit of news spreads quickly. The number of people who know it grows with time. In fact, the number $N = N(t)$ of people who know the news at time t should be proportional to the original number N_0 who were told the news at time t_0 and who could not help but spread it. Thus, the older the news, the more people know it. If $N(t)$ people know the news at time t, a natural question to ask is how many know it at a slightly later time, say $t = t_1$. It is not unreasonable to expect that the number $N(t_1) - N(t)$ of people who learn the news in the time interval $[t, t_1]$ is approximately proportional to both $N(t)$ and the lapse of time $t_1 - t$. If we accept this idea as our initial assumption, its mathematical reformulation becomes

$$N(t_1) - N(t) = kN(t)(t_1 - t), \qquad N(t_0) = N_0, \tag{8.4.1}$$

where k is a positive constant called the *growth coefficient*. We now wish to discover how N is related to the initial number N_0. With (8.4.1) as our model, we let the time lapse $t_1 - t$ tend to zero and obtain a differential equation for N.

Rewriting (8.4.1) as

$$\frac{N(t_1) - N(t)}{t - t} = kN(t),$$

we see that

$$\lim_{t_1 \to t} \frac{N(t_1) - N(t)}{t_1 - t} = kN(t).$$

Hence, we obtain the differential equation

$$\frac{dN(t)}{dt} = kN(t), \qquad \text{with} \quad N(t_0) = N_0. \tag{8.4.2}$$

Equation (8.4.2) tells us simply that the rate of change of N is proportional to N. This is the basic mathematical model for growth. By solving Equation (8.4.2), we can determine N in terms of the original number of people N_0 who were first told the news. If k should be negative, Equation (8.4.2) becomes the basic equation for decay. If $k = 0$, then N must be constant and there is no spread of news. For convenience, we always take $t = 0$ so that the condition $N(t_0) = N_0$ becomes $N(0) = N_0$.

From our results in Section 8.2, we see that

$$N(t) = N_0 e^{kt}, \tag{8.4.3}$$

where t is time elapsed since the beginning of the process.

The differential equation (8.4.2) and its solution (8.4.3) obviously have many other applications than the spreading of news. They represent a model for the growth of timber and vegetation, population growth (both people and bacteria), the growth of money in bonds (those that credit the interest to the capital continuously), the growth of a substance in the course of a chemical reaction, radioactive decay, and other processes.

▶ **Example 1.** Assume that the rate of change of the world population is proportional to the population size (which unfortunately seems to be true). Suppose that in 1950 the world population was 2 billion (2×10^9) and in 1960 the world population was 3 billion (3×10^9). What will be the world population in the year 2000?

Solution. Our basic differential equation is again Equation (8.4.2). Now, let $t = 0$ correspond to the year 1950 and $t = 1$ correspond to the year 1960. We are taking our unit of time as 10 years. $N(0) = 2 \times 10^9$. Thus, from Equation (8.4.3),

$$N(t) = (2 \times 10^9)e^{kt}.$$

However, the fact that the population was 3 billion in 1960 implies that $N(1) = 3 \times 10^9$. Hence,

$$N(1) = 3 \times 10^9 = (2 \times 10^9)e^k \quad \text{and} \quad e^k = \tfrac{3}{2} \text{ or } k = \log \tfrac{3}{2}.$$

It is essential for us to correctly determine k in order to solve the problem. The population at any time t (where t is measured in units of 10 years) is given by

$$N(t) = (2 \times 10^9)e^{t \log 3/2}.$$

We wish to know what the population will be in the year 2000. This corresponds to $t = 5$.

$$N(5) = (2 \times 10^9)e^{5 \log 3/2}$$
$$= (2 \times 10^9)e^{\log (3/2)^5}$$
$$= (2 \times 10^9)(\tfrac{3}{2})^5 = \tfrac{243}{16} \times 10^9.$$

Thus, the population in the year 2000 should be approximately 15,200,000,000.

Note that in this example, as well as in Example 3, it is best to leave k as $\log \tfrac{3}{2}$ rather than evaluating it by means of the table.

▶ **Example 2.** If we deposit $10 in a bank that offers interest and this interest is compounded instantaneously at 5%, how many years will it take for the total to reach $20? (Various banks compute interest yearly, semiannually, monthly, or daily. Compounding interest instantaneously is a natural extension of this.)

Solution. We have not defined what we mean by compounding interest instantaneously. One reasonable definition states that $\dfrac{dN(t)}{dt} = kN(t)$, where $N(t)$ is the amount of money at time t, and k is the (instantaneous) interest rate. This equation is just our old friend, Equation (8.4.2). It is also possible to approach the notion of instantaneous compound interest in a different way. Suppose we compute interest at k percent compounded n times per year. If we start with the amount P_0, then our principal P at the end of a year is given by the expression $P = P_0(1 + k/n)^n$. If we let n tend to infinity, which corresponds to our intuitive notion of compounding interest instantaneously; we discover that both definitions lead to the same answer. (Since they agree, we will not pursue the discussion of the definitions further but instead return to our original problem.) Using the first definition, we see from Equation (8.4.3) that $N(t) = N_0 e^{kt}$. Since we are given $k = 0.05$ and $N_0 = 10$, our equation becomes $N(t) = 10e^{0.05t}$. We now want to find when $N(t)$ will equal 20. We thus set $20 = 10e^{0.05t}$ or $e^{0.05t} = 2$. Taking the logarithm of both sides, we find that $t = 20 \log 2$. Since, by Table A2, $\log 2 = 0.69$, this means $t = 13.9$, the number of years it takes for the total to reach \$20.

In Chapter 5, we represented the solution to this problem in terms of an integral. That integral is just another form of the solution to the differential equation studied here.

▶ **Example 3.** In the chemical reaction that produces prudentium, the amount produced at a given moment is proportional to the amount present. If there is one gram of prudentium at time zero and 10 grams 2 min later, find the equation for the amount at time t.

Solution. The basic differential equation is again Equation (8.4.2), where N_0 represents the initial amount of prudentium and $N(t)$ represents the amount of prudentium at time t. First we must find the growth coefficient. Since $N_0 = 1$, our equation is $N(t) = e^{kt}$. Since $N(2) = 10$, we find that $10 = e^{2k}$. Now we take the logarithm of both sides of this equation and obtain $2k = \log 10$ or $k = \frac{1}{2} \log 10$. The answer to the problem can thus be written as $N(t) = e^{[(1/2) \log 10]t}$.

▶ **Application 2.** (*Decay*) Various radioactive elements show marked differences in their rate of decay, but they all share the property that the rate at which a given substance decomposes at any instant is proportional to the amount present at that instant. If we denote by $N(t)$ the amount present at time t, then the law of decay states that

$$\frac{dN}{dt} = -kN(t), \tag{8.4.4}$$

where k is a positive constant (called the decay constant) whose actual value depends upon the particular element that is decomposing. Note that the minus sign appears since N decreases as t increases and hence $\dfrac{dN}{dt}$ must always be negative.

If $N(0) = N_0$, then every solution of Equation (8.4.4) must be of the form

$$N(t) = N_0 e^{-kt}. \tag{8.4.5}$$

Observe that there can be *no* finite time at which $N(t) = 0$, since e^{-kt} is never zero for finite t. Thus, there is no use in studying the "total lifetime" of a radioactive substance. It is of course possible to determine the time required for any particular *fraction* of a sample to decay. For convenience the fraction $\frac{1}{2}$ is chosen, and the time T at which $N(T)/N_0 = \frac{1}{2}$ is called the *half-life* of the substance. This is determined by solving the equation

$$e^{-kT} = \tfrac{1}{2}$$

for T. We see immediately by using logarithms that

$$-kT = \log \tfrac{1}{2} = -\log 2,$$

and therefore the equation

$$T = (1/k) \log 2$$

relates the half-life to the decay constant k.

▶ Example 4. The half-life for radium is approximately 1,600 years. Find what percentage of a given quantity of radium disintegrates in 100 years.

Solution. From our previous discussion, we see that we can determine k from the equation

$$T = 1,600 = \frac{1}{k} \log 2 = \frac{1}{k} (0.6931).$$

Thus,

$$k = \frac{0.6931}{1,600} = 0.000433.$$

Now, at the end of 100 years,

$$N(100) = N_0 e^{-0.000433(100)} \quad \text{or} \quad N(100) \simeq 0.958 N_0$$

where $\simeq$ is read "is approximately." We have resorted to Table A1 to determine these numbers. Thus, at the end of 100 years we have approximately $0.958 N_0$ amount of radium left. This tells us that only 4.2% disappears in 100 years.

The simple decay model that we have considered describes the essential features of many other phenomena as well.

▶ Application 3. (*Bounded Growth*) If for some reason the population cannot exceed a certain maximum number, say M (for example, after M people the food supply is exhausted), we may reasonably assume that the rate of growth is jointly proportional to both $N(t)$ and $M - N(t)$. Thus, we have a second type of growth law,

$$\frac{dN(t)}{dt} = kN(t)[M - N(t)], \qquad N(0) = N_0, \qquad (8.4.6)$$

where k is the "growth" constant and, say, M is a constant.

We solve Equation (8.4.6) using the separation of variables technique discussed in Section 8.2. After the first step we obtain

$$\int \frac{dN}{N(M - N)} = k \int dt.$$

We note that

$$\frac{1}{N(M - N)} = \frac{1}{M} \left[\frac{1}{N} + \frac{1}{M - N} \right]$$

and, hence,

$$\int \frac{dN}{N(M - N)} = \frac{1}{M} [\log N - \log (M - N)].$$

Thus,

$$\frac{1}{M} [\log N - \log (M - N)] = kt + C, \qquad (8.4.7)$$

where C is an arbitrary constant. From Equation (8.4.7), we see that

$$\log \frac{N}{M - N} = Mkt + C_1,$$

where C_1 is the constant CM. The above result yields

$$\frac{N(t)}{M - N(t)} = C_2 e^{Mkt},$$

where C_2 is the constant e^{C_1}. Using the fact that $N(0) = N_0$ and $C_2 = N_0/(M - N_0)$, we see that

$$N(t) = [M - N(t)] \frac{N_0}{M - N_0} e^{Mkt},$$

or

$$\frac{MN_0}{M - N_0} e^{Mkt} = N(t) \left[1 + \frac{N_0}{M - N_0} \right] e^{Mkt}.$$

A final simplification yields

$$N(t) = \frac{MN_0 e^{Mkt}}{M + N_0(e^{Mkt} - 1)}. \tag{8.4.8}$$

We notice the following facts about Equation (8.4.8). First, it may be rewritten in the form

$$N(t) = \frac{MN_0}{N_0 + (M - N_0)e^{-Mkt}}.$$

This is particularly useful if we wish to determine the behavior of the population as t becomes very large. In fact, the solution tells us that as $t \to \infty$, $N(t) \to M$. This is certainly in agreement with our original statement of the problem, which said the population can never be larger than M.

▶ Application 4. (*Production Capacity*)* The Red, White, and Blue Company manufactures a variety of products, one of which is an American flag. The price is competitive with respect to flags of the same size and quality, and owing to increased emphasis on patriotism, there has been a large demand for these flags. Since the company has reason to expect this demand to continue, they wish to expand their production capacity. At present, 500 machines are capable of supplying current demand. However, since the demand is expected to double over the next 10 years, the management wants to institute plans for gradual plant expansion. Toward that end, Red, White, and Blue has decided to purchase 20 machines in the first year of the expansion program, 40 in the second year, 60 in the third year and so on. At first glance, this rate of expansion seems too great since more than 500 new machines will be purchased by the end of the eighth year. However, the management is well aware that machines tend to wear out, and that productivity decreases with time. In fact, according to the manufacturer's own data, it can be expected that productivity will decrease by 5% per year. We will equate a loss in productivity with a loss in

---

* Example taken from Mark E. Stern, *Mathematics for Management* (Prentice-Hall, Englewood Cliffs, N.J., 1963), Chapter 6.

machines. A moment's thought shows that it does not make any difference whether a 5% loss in productivity means a 5% decrease for *each* machine or a 5% breakdown rate for the machines themselves (the remaining 95% running perfectly). As you will see, the production decay factor, in this case 5%, will turn out to be *identical* with the decay constant k in the differential equations we have been studying. In order to compensate for this loss, more machines must be purchased each year. Moreover, by the tenth year, Red, White, and Blue must double its plant capacity to keep up with increased demand. A main source of concern is the validity of the machine manufacturer's estimate of the annual productivity decrease factor. Obviously, if this figure were 4% or 6% Red, White, and Blue might find itself overexpanded or under-expanded at the end of the 10-year period. Thus, management wishes to obtain a relationship for plant capacity in any given year of the expansion program in terms of the decay factor in order to analyze the effect of possible errors in the estimate of the annual decrease in machine productivity.

The simplest method of determining the number of years required to double plant capacity assumes that all *new* equipment is purchased at the end of the year. Thus, if year 1 is begun with 500 machines, and 5% of them will become unproductive during the year, then only 475 machines will be producing at full capacity at the end of year 1. With the purchase of 20 new machines the year end total will be 495. (See Table 8.4.1.)

TABLE 8.4.1	Year	Begin	Loss	Gain	End
	1	500	25	20	495
	2	495	25	40	510
	3	510	26	60	544
	4	544	27	80	597
	5	597	30	100	667
	6	667	33	120	754
	7	754	38	140	856
	8	856	43	160	973
	9	973	49	180	1104
	10	1104	55	200	1249

Of the 495 machines producing at the beginning of year 2, 95% of them, or 470, will be producing at the end of the year. After 40 new machines are purchased, the year-end total will be 510. In this manner, the year-end production capacity may be computed for as many years as desired. From the table, it appears as if the objective of doubling machine capacity is very nearly achieved at the end of year 8. There are several difficulties that arise in this tabular method of determining plant capacity. The most important deficiencies are: (1) productivity decrease is not considered as a continuous phenomenon and (2) it does not lend itself to the derivation of a formula by which the plant capacity can be represented as a function of the decay factor and the year of the expansion program. We now show how to derive such a formula.

Let C represent the plant capacity that depends on time t [that is, the plant capacity $C = C(t)$, a function of time t]. The time rate of change of capacity resulting from the productivity decay factor λ is $-\lambda C(t)$. This is equivalent to the statement that the capacity is continuously decaying at a rate $\lambda C(t)$. $\lambda = 0.05$ does, indeed, correspond to a 5% per year rate of decrease in productivity. In order to find the continuous rate at which additions to capacity are being made, note that, in the discrete case, the added capacity after the year t is

$$20(1 + 2 + 3 + \cdots + t) = 20[\tfrac{1}{2}t(t + 1)] = 10t^2 + 10t.$$

Now, it can be shown that if t is a positive integer,

$$1 + 2 + 3 + \cdots + t = \tfrac{1}{2}t(t + 1).$$

(You can check this for various values of t.) Hence,

$$20(1 + 2 + 3 + \cdots + t) = 20[\tfrac{1}{2}t(t + 1)] = 10t^2 + 10t.$$

The instantaneous rate of plant expansion is then given by the derivative with respect to t of the above, namely, $20t + 10$. Since the rate of change of plant capacity is equal to the loss through decrease of productivity plus the instantaneous rate of expansion, we have

$$\frac{dC}{dt} = -\lambda C + 20t + 10,$$

or

$$\frac{dC}{dt} + \lambda C = 20t + 10, \tag{8.4.9}$$

where C is the plant capacity and λ the productivity decay factor.

Equation (8.4.9) is a first-order differential equation that can be solved via the methods given in Section 8.3. An integrating factor $f(t)$ is

$$f(t) = e^{\int \lambda \, dt} = e^{\lambda t}.$$

Hence, from Section 8.3, we have

$$C(t)e^{\lambda t} = \int (20t + 10)e^{\lambda t} \, dt = 20\left(\frac{t}{\lambda} - \frac{1}{\lambda^2}\right)e^{\lambda t} + \frac{10}{\lambda} e^{\lambda t} + k,$$

where k is an arbitrary constant. Therefore,

$$C(t) = ke^{-\lambda t} + \frac{10}{\lambda}(2t + 1) - \frac{20}{\lambda^2} = ke^{-\lambda t} + \frac{20t}{\lambda} + \frac{10\lambda - 20}{\lambda^2}. \tag{8.4.10}$$

Using the condition that $C(0) = 500$ (our assumption was that we began with 500 machines), we can evaluate k.

$$500 = k + \frac{10\lambda - 20}{\lambda^2}, \quad \text{and hence} \quad k = 500 - \frac{10\lambda - 20}{\lambda^2}.$$

Thus, the complete solution is given by

$$C(t) = \left(500 - \frac{10\lambda - 20}{\lambda^2}\right)e^{-\lambda t} + \frac{20t}{\lambda} + \frac{10\lambda - 20}{\lambda^2}. \tag{8.4.11}$$

This allows us to compute the plant capacity at any time $t \geq 0$ for a given value of λ.

If the manufacturer's statement on decay is correct, that is, $\lambda = 0.05$, then (8.4.11) becomes

$$C(t) = 8,300e^{-0.05t} + 400t - 7,800.$$

For $t = 8$, we find that

$$C(8) = 8,300e^{-0.4} + 3,200 - 7,800 = 8,300(0.67032) - 4,600 = 963.$$

A closer approximation to the time when $C = 1,000$ (double the initial capacity) can be found by trying $t = 8.3$. $C(8.3) = 1,000$. Thus it is reasonable to expect that the objective of doubling plant capacity will be achieved at the beginning of the second quarter of year 9 of the expansion program. Note that $C(8) = 963$ is 12 units less than the estimate made in the discrete table. The main difference in the results can be attributed to the fact that we consider a continuous rather than discrete analysis of the table.

After reading through the previous four applications of first-order differential equations, you may suspect that as we apply mathematics to the various sciences we reach a point where there appear to be only a few different kinds of processes occurring at the present mathematical level. The equations remain the same, only the names of the functions and variables change from science to science. You would be right.

Exercises

1. The rate of growth of a colony of bacteria is proportional to the population. Initially, there are 1,000 bacteria and one day later, 3,000. What will the population of the colony be after four days if it grows without restrictions?

2. Write the differential equation for the growth of a bacterial population which increases 2.5% every hour. If there are N_0 bacteria at the start, how many are there at the end of 10 hours?

3. A man invests $300 at 6% per year compounded instantaneously. How much will the investment be worth in 10 years?

4. A firm conducts pricing experiments and finds that an increase of $1 in the unit sales price results in a constant decrease of b units in sales volume. What is the relationship between sales and price? Find the relationship if we are told that the sales volume is V when the price is $5. [*Hint:* Suppose there exists a relationship between the sales volume s and price p. Then, the rate of change of sales with respect to price is ds/dp, and we are told that $ds/dp = -b$.]

5. A resort hotel hires 100 waiters and a constant number quit each month. If 82 waiters are still employed at the end of one year, how many will be employed after two years? After x months? What is the maximum length of service of the waiters? [*Hint:* We approximate the discrete case with a continuous variable. Let y be the number of waiters and x the number of months since first employed; then $dy/dx = -a$, where a is the number who quit each month.]

6. If waiters quit at rates proportional to the number of months of service and if 82 out of 100 waiters are still employed at the end of one year, what is the

expected number remaining after two years? What is the maximum length of service of the waiters? [*Hint:* Again approximate the discrete case with a continuous variable. In this case $dy/dx = -ax$ (using the same notation as in Exercise 5).]

7. If $2,000 is deposited at 7% interest per year and the interest is compounded instantaneously, what is the amount in 25 years? How many years will it take for the total to reach $20,000?

8. In a certain first-order chemical reaction, half of the original substance has decomposed in 10 sec. What length of time is required for 90% of the substance to decompose?

9. Carbon-14, C^{14}, is a radioactive isotope of carbon that has a half-life of approximately 5,600 years. What percent of the original substance is left at the end of 56,000 years? Why is carbon dating only effective back to about 50,000 B.C.?

10. In a certain stimulus-sampling theory, it is assumed that there is a continuous function N such that if T is the number of trials, then $N(T)$ is the expected number of elements conditioned. Let S be the number of stimulus elements effective at any given time (which is assumed to be constant). Moreover, it is assumed that

$$\frac{dN}{dx} = \frac{1}{S}[S - N(x)],$$

for x in the domain of N. Find N for any number of trials x.

11. Let N be the number of persons each with an income of M or more dollars. Clearly N decreases as M increases. More specifically, it has been estimated that

$$\frac{dN}{dM} = kN,$$

where k is a constant. Solve the differential equation. Write N as a function of M. Let N_0 be the total population, i.e., all those people with incomes of zero or more dollars. If precisely half the population has income of $10,000 or more, show that $k = \frac{1}{10,000} \log \frac{1}{2}$. What fraction of the population has an income of $20,000 or more?

12. The rate of increase in profit per dollar increase in sales price dP/dp is given by

$$dP/dp = (a - 1) + c/p + \log p,$$

where P is the profit, p the sales price, and a and c are constants. Show that profit is given by

$$P = (a - 1)p + c \log p + p \log p - p + \text{constant}.$$

13. Show that the solution to the differential equation for bounded growth Equation (8.4.6) can be expressed in the form

$$N(t) = \frac{M}{1 + e^{-a(t - t_0)}},$$

where a is a constant and t_0 is the time at which $N(t) = \frac{1}{2}M$.

14. Suppose in the fourth application to production capacity, the Red, White, and Blue Company decides to increase its capacity by purchasing n machines at the end of each year, where n is a number to be determined. (The management has decided it is not feasible to increase its purchase order by 20 machines each year.) Write a differential equation relating C to n, λ, and t. Solve this differential equation assuming that initially the Red, White, and Blue Company has 500 machines. If in 10 years the Red, White, and Blue Company wants to double its capacity, what number n of machines should it add at the end of each year? (Always assume that manufacturer's statement on decay is correct; that is, $\lambda = 0.05$.)

8.5 Chapter 8 Summary

1. The solution to the differential equation

$$\frac{dy}{dx} = f(x),$$

where f is continuous on some interval, is given by

$$y = \int f(x)\, dx.$$

A differential equation of the form

$$\frac{d^n y}{dx^n} = f(x)$$

(n a positive integer) requires n integrations to determine $f(x)$. In this case the solution will contain n arbitrary constants.

2. The general solution to the differential equation

$$h(y)\frac{dy}{dx} = g(x), \tag{2a}$$

where h and g are continuous functions, is given by

$$\int h(y)\, dy = \int g(x)\, dx \tag{2b}$$

in the sense that any solution $y = f(x)$ must satisfy Equation (2b). One arbitrary constant is contained in the solution.

3. Given the differential equation

$$\frac{dy}{dx} + p(x)y = q(x),$$

we call the function $f(x) = e^{\int p(x)\, dx}$ an *integrating factor* for the equation. The general solution is given by

$$y(x) = e^{-\int p(x)\, dx}\left[\int q(x)e^{\int p(x)\, dx}\, dx\right] + Ce^{-\int p(x)\, dx},$$

where C is an arbitrary constant. (We incorporate all of the constants involved in the integrations into C.)

4. The basic differential equations for various applications are given as follows:

(a) *Growth*

$$\frac{dN}{dt} = kN,$$

where k is called the growth coefficient.

(b) *Decay*

$$\frac{dN}{dt} = -aN,$$

where a is called the decay factor.

In both (a) and (b), the constants k and a are determined by the initial condition $N(0) = N_0$.

(c) *Bounded growth*

If the function $N(t)$ cannot exceed a certain number M, then the equation for bounded growth is given by

$$\frac{dN}{dt} = kN(t)[M - N(t)].$$

k is again a growth constant determined from the condition $N(0) = N_0$.

5. Remember that in applying differential equations to various fields of social and natural sciences, the basic mathematical equations are the same; only the names of the functions and variables change from one field to the other.

Review Exercises

1. Find the general solution to the following differential equations:

(a) $y''(x) = x^2 + x^3 + 3e^x$.

(b) $y' + xy = 1$.

(c) $\dfrac{dy}{dx} - \dfrac{y}{x} = 1 + \sqrt{x}$.

(d) $y' - \dfrac{2y}{x} = 3x^3$.

(e) $\log y \dfrac{dy}{dx} = \dfrac{y}{x}$.

(f) $(x - 1)\dfrac{dy}{dx} + y = x + 1$.

(g) $\dfrac{dy}{dx} = \dfrac{x^2 + x}{y^2 - y - 2}$.

(h) $2xy' + y = \dfrac{2ax^2 + 1}{x}$, where a is a constant.

2. Find the particular solution of the following:

(a) $xy' - 2y = x^5$, with $y(1) = 1$.

(b) $y' + xy = x$, with $y(0) = 0$.

(c) $\dfrac{dp}{dt} + p = e^{2t}$, with $p(0) = 1$.

(d) $xy' + (1 - x)y = e^{2x}$, with $y(1) = b$.

(e) $yy' = 5x$, with $y(0) = 1$.

(f) $y' + 2xe^y = 0$, with $y(0) = 0$.

3. Find all solutions of $xy' + y = xe^{-x^2}$ on the interval $(0, \infty)$. Prove that all solutions approach 0 as $x \to \infty$.

4. Solve the differential equation $(1 + y^2e^{2x})y' + y = 0$ by introducing a change of variable of the form $y = u(x)e^{mx}$, where m is a constant and u is a new unknown function.

5. Scientists at an atomic works station isolated one gram of a new radioactive element called deterium. It was found to decay at a rate proportional to the square of the amount present. After one year, one-half gram remained. Set up and solve the differential equation for the amount of deterium remaining at time t.

6. (a) Discuss the growth of a population in which the birth rate and death rate per year remain constant.

   (b) Discuss the growth if the birth rate per thousand, which is now higher than the death rate, decreases uniformly with time.

7. (a) An amount P is invested at 5% compounded instantaneously. In 10 years t has grown to $1,000. Find P.

   (b) At the end of what year will the principal have doubled if it is invested at 5% compounded instantaneously?

8. Equipment maintenance and operating costs C are related to the overhaul interval x by the equation

$$x^2 \frac{dC}{dx} - (n - 1)xC = -na,$$

where a and n are constants and $C = C_0$ when $x = x_0$. Find C as a function of x.

9. Consider the following generalization of the bounded growth equation. Suppose that the population increases at a rate proportional to the product of two factors,

$$\frac{dN}{dt} = aN\left(1 - \frac{N}{b}\right),$$

where a, b are constants. This equation is often referred to as the "logistic" equation. Solve the differential equation.

10. Find all differentiable functions f that satisfy the equation

$$f(x) = \int_0^x f(t) \, dt.$$

[*Hint*: Differentiate both sides to obtain a differential equation.]

Functions of Several Variables

9.1 Introduction

Although many situations can be described with the use of functions of one variable (such as those applications considered in Chapters 4, 5, and 8), there are also problems in which we must consider more than one variable. For instance, the cost of a product may depend both on the price of raw materials and the cost of labor; the profits of a resort hotel may depend on the state of the economy in general, prices charged by competing hotels, the weather, and numerous other factors; supply may depend on the size of its potential market, its retail price, and also the price of competing markets. A need for several variables also occurs in biology. Epidemics such as influenza appear to travel like a wave across the earth. One needs to describe changes in the proportion of people having influenza as a function of time and distance.

To be more specific, we consider the equation $z = x + 2y - 1$, which relates the cost of a product to x, the cost of raw materials and y, the cost of labor. Typically, one must determine the minimum price the manufacturer can charge and still make a reasonable profit. In order to solve such problems, it is necessary for us to first carefully discuss the concept of a function of two variables (x, y in the above examples) and the rate of change of the function with respect to both of these variables. In order to graph or represent pictorially functions of two variables, we need a spatial, three-dimensional coordinate system.

We now describe a method for attaching to every point in three-dimensional space an ordered triple (x, y, z). This has already been discussed briefly in an appendix to Chapter 5. First, we introduce the x, y, and z axes as in Figure 9.1.1. The three axes are mutually perpendicular and form a right-handed system. (This means that if we

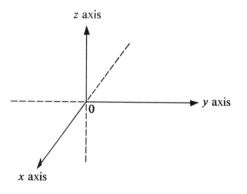

Figure 9.1.1

turn the x axis toward the y axis, the positive z axis points in the direction the z axis would go if it were a right-hand screw.) The point of intersection of the x, y, and z axes is called the origin, 0. The plane containing the x axis and the y axis is called the xy plane. (Two distinct lines determine a plane.) The plane containing the x axis and the z axis is called the xz plane. The plane containing the y axis and the z axis is called the yz plane. (See Figure 9.1.2.)

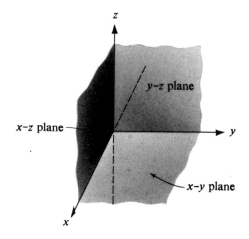

Figure 9.1.2

To every ordered triple (a, b, c) we will associate a point P in three-space. Consider the triple $(3, 1, 2)$. To find the associated point we first count off 3 units on the positive x axis; then move 1 unit in the direction of the positive y axis (1 unit in the direction parallel to the y axis and perpendicular to the xz plane); and finally we count up 2 units in the direction of the positive z axis (2 units up in the direction perpendicular to the xy plane).

See Figure 9.1.3 for a pictorial version of this counting system. Also examine Figure 9.1.1 again.

Thus, if the triple (a, b, c) represents the point P, then P is a units from the yz plane, b units from the xz plane, and c units from the xy plane.

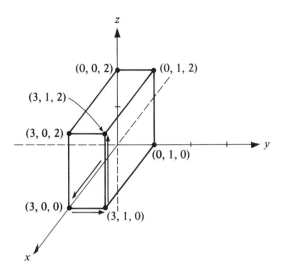

Figure 9.1.3

Not only does every triple (a, b, c) represent a point, but every point can be represented as a triple. Given P, to find its coordinates (a, b, c), one simply measures the perpendicular distances to the yz, xz, and xy planes, respectively. This correspondence between points and ordered triples is unique. Thus, there is exactly one ordered triple for each point and exactly one point for each ordered triple. From now on we identify points and ordered triples by writing $P = (a, b, c)$.

The planes in three-space which are perpendicular to one of the coordinate axes are of particular interest. By the plane $x = 1$, one means the set of points $\{(x, y, z) \mid x = 1\}$, that is, all points of the form $(1, y, z)$. This set of points is a plane perpendicular to the x axis and passing through $(1, 0, 0)$. (See Figure 9.1.4.)

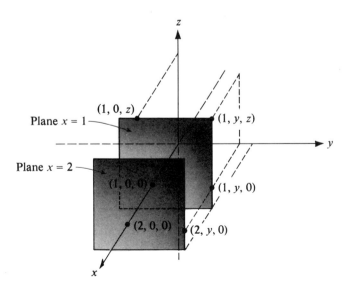

Figure 9.1.4

Let us now consider some examples illustrating the remarks made in this section.

▶ Example 1. Draw a three-dimensional coordinate system and plot the points (3, 2, 4) and (−4, −3, 5).

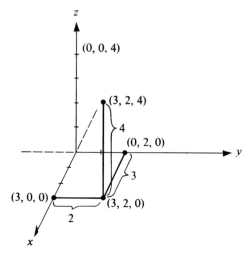

Figure 9.1.5

Solution. To plot the point (3, 2, 4), first measure three units along the positive *x* axis, two units parallel to the positive *y* axis, and then measure four units up in the direction of the positive *z* axis. (See Figure 9.1.5.) The point (−4, −3, 5) is plotted in Figure 9.1.6. In this case, first measure four units along the negative *x* axis, then three units in the direction of the negative *y* axis, and then five units up in the direction of the positive *z* axis.

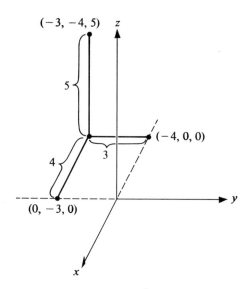

Figure 9.1.6

◀ Example 2.   Sketch the plane $y = 2$.

Solution.   We are really asked to plot the set of all triples of the form $(x, 2, z)$. First draw the line $y = 2$ in the xy plane and then construct a plane containing the line $y = 2$ parallel to the xz plane. (See Figure 9.1.7.)

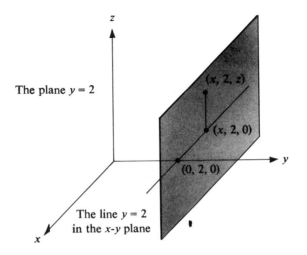

Figure 9.1.7

Exercises

1. Graph the following points:

   (a) $(1, -1, 3)$.

   (b) $(0, 1, -4)$.

   (c) $(-1, -1, 1)$.

   (d) $(3, -1, 0)$.

   (e) $(5, 4, 6)$.

   (f) $(-2, 1, 2)$.

   (g) $(4, -1, 1)$.

   (h) $(1, 1, 1)$.

2. Sketch the following:

   (a) the plane $x = -1$.

   (b) the plane $x = 0$.

   (c) the plane $y = 1$.

   (d) the plane $y = 0$.

   (e) the plane $y = -2$.

   (f) the plane $z = 2$.

   (g) the plane $z = 0$.

   (h) the plane $z = -3$.

3. Using a three-dimensional system of coordinate axes, plot the points $(3, 0, 0)$, $(0, 4, 0)$, and $(0, 0, 5)$ and connect them with straight lines to indicate the plane on which they all lie. Judging from this diagram, does the plane also contain the point $(0, 0, 1)$? Does it seem to contain the point $(1\frac{1}{2}, 0, 2\frac{1}{2})$?

*4. Use the Pythagorean theorem to show that the distance (geometric) between the points $P = (x_1, y_1, z_1)$ and $Q = (x_2, y_2, z_2)$ is given by

$$\text{distance } [P, Q] = [(x_1 - x_2)^2 + (y_1 - y_2)^2 + (z_1 - z_2)^2]^{1/2}.$$

5. Use Exercise 4 to show that the point

$$\left(\frac{x_1 + x_2}{2}, \frac{y_1 + y_2}{2}, \frac{z_1 + z_2}{2}\right)$$

is the midpoint of the line segment joining the points (x_1, y_1, z_1) and (x_2, y_2, z_2).

9.2   Definition of a Function of Two Variables and Some Graphing

By a function of two variables we mean

1. a region D in the xy plane called the domain and
2. a rule which associates with every point in D exactly one real number.

Customarily, such functions are written as $z = f(x, y)$. For example, if we write $z = 1 + x + y^2$, then the domain is understood to be the xy plane. We associate the real number $1 + (2) + (1)^2 = 4$ to the point $(2, 1)$ in the xy plane. With the point $(-5, 9)$ in the xy plane we associate the real number $1 + (-5) + (9)^2 = 77$.

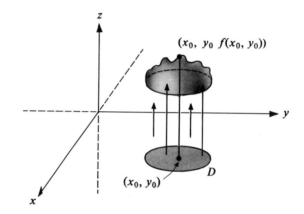

Figure 9.2.1

Geometrically we can think of the graph of a function $z = f(x, y)$ as a surface. If the domain of the function is D, then to obtain the graph of f think of D lifted up as in Figure 9.2.1 and stretched into an irregular shape. In Figure 9.2.1 we have represented $z = f(x, y)$ by its graph. The graph of $f(x, y)$ is simply $\{(x, y, z) \mid z = f(x, y)\}$. One can think of the function f as sending the point (a, b) in the xy plane into the point $(a, b, f(a, b))$ in three-space. If $z = f(x, y)$ is a function, then f can lift, push down, stretch, wrinkle, or tear D. It can *not* fold D. (If it did, then f would associate more than one real number with some point in the domain.)

It is natural to ask how one goes about graphing a function $z = f(x, y)$. There is no easy answer to this question. We shall first give several examples and then briefly discuss one technique. If no domain is specified, then it is understood to be the xy plane.

▶ Example 1. $z = \sqrt{1 - (x^2 + y^2)}$. The graph of this function is the top half of a sphere. Domain $= \{(x, y) \mid x^2 + y^2 \leq 1\}$. (See Figure 9.2.2.)

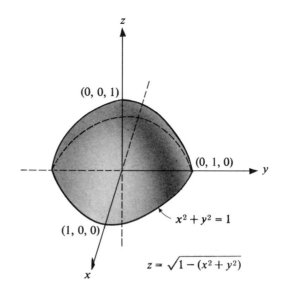

Figure 9.2.2

▶ Example 2. $z = 1 - x - y$. The graph of this function z is a plane. (See Figure 9.2.3.)

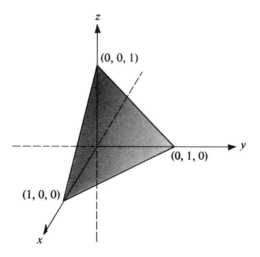

Figure 9.2.3

We now discuss a technique of graphing, known as the profile method, in some detail.

▶ Example 3. Consider the function $z = x^2 + y^2$, where the domain is the xy plane. Since x^2 and y^2 are both positive, clearly $z = x^2 + y^2 \geq 0$, so the graph of the surface will lie above the xy plane. What we do now is fix one of the variables and see what we get under these circumstances. For example, let $z = 5$. Then

$5 = x^2 + y^2$ is the equation of a circle. Setting $z = 5$ is the same as intersecting the surface with the plane $z = 5$. (See Figure 9.2.4.) We now know that the circle in Figure 9.2.4 is part of the surface $z = x^2 + y^2$. This may not seem like much

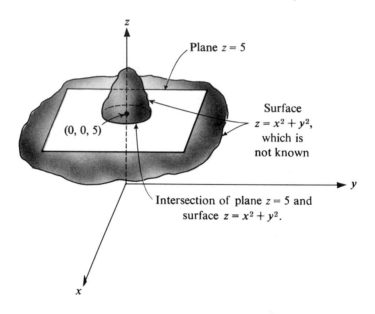

Figure 9.2.4

information, but you can consider it to be one clue in tracking down the actual shape of the surface. Note that for any $c > 0$, the intersection of the plane $z = c$ with the surface $z = x^2 + y^2$ is a circle; in fact, it is the circle $c = x^2 + y^2$ ($z = c$). Now, we know that the surface is formed by a bunch of circles piled on top of one another. However, there are many different ways they could be arranged. Several

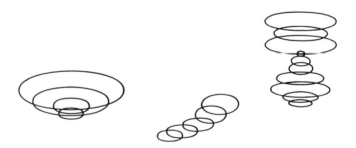

Figure 9.2.5

such arrangements appear in Figure 9.2.5. To determine the correct arrangement, we examine the intersection of the plane $x = 0$ (the yz plane) with the surface. To do this we simply set $x = 0$ in the equation $z = x^2 + y^2$ and obtain the parabola $z = y^2$ in the yz plane. The profile of the surface in the plane $x = 0$ is given in Figure 9.2.6. If we combine our two sets of profiles, we see that the surface has the

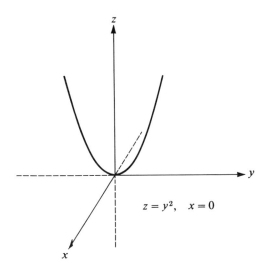

Figure 9.2.6

shape given in Figure 9.2.7. The surface $z = x^2 + y^2$ thus bears some resemblance to the end of a football. This surface is fairly simple, and we were able to determine it with only a few profiles. For a complicated surface, many profiles may be needed.

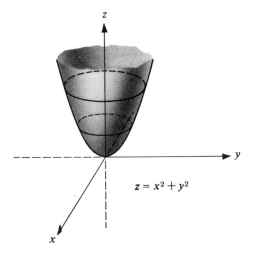

Figure 9.2.7

▶ Example 4. Consider the surface $z = y^2 - x^2$, where the domain is the xy plane. Sketch a graph of this surface.

Solution. Let us first consider the profile obtained by intersecting the surface $z = y^2 - x^2$ with the planes $y = b$. To do this, we simply substitute $y = b$ to obtain $z = -x^2 + b^2$. The curve $z = -x^2 + b^2$ is a parabola in the $y = b$ plane with its vertex pointing up. The vertices of the parabola lie in the plane $x = 0$. For $x = 0$ the profile we obtain is $z = y^2$, another parabola. Putting this information into a picture, we obtain Figure 9.2.8. Figure 9.2.8 already gives us a pretty good

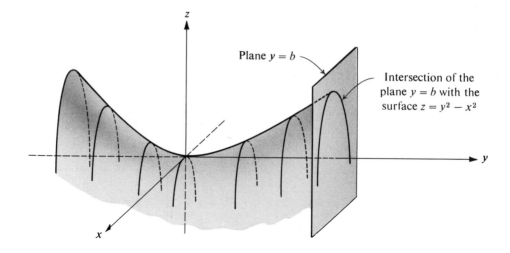

Plane $y = b$

Intersection of the plane $y = b$ with the surface $z = y^2 - x^2$

Figure 9.2.8

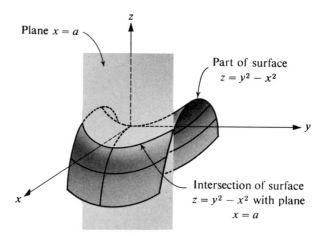

Plane $x = a$

Part of surface $z = y^2 - x^2$

Intersection of surface $z = y^2 - x^2$ with plane $x = a$

Figure 9.2.9

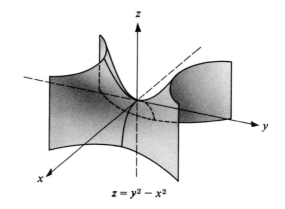

$z = y^2 - x^2$

Figure 9.2.10

idea of what the graph will look like. Let us now consider the profile obtained by intersecting the surface $z = y^2 - x^2$ with the plane $x = a$. We thus obtain $z = y^2 - a^2$. This is just the equation of a parabola in the $x = a$ plane with vertex pointing down. (See Figure 9.2.9.) The vertices of these parabolas (for different a's) lie along the curve $z = -x^2$ in the plane $y = 0$. We now can draw a reasonable picture of this surface. (See Figure 9.2.10.) It is saddle-shaped and thus is sometimes called a saddle surface. A person standing at the origin on this surface would see himself as crossing a mountain pass between two higher peaks.

Exercises

1. Can a sphere be represented as the graph of a function $z = f(x, y)$? Why?

2. Can a torus (doughnut or inner tube) be represented as the graph of a function $z = f(x, y)$? Why?

3. Graph the function $z = -x + 2y + 5$, given that it is a plane.

4. If a company spends x dollars on research and development and y dollars on advertising, its profit (in dollars) is given by

$$P(x, y) = 40{,}000 + 50x + 30y + \frac{xy}{100}$$

   for all positive integers x and y less than 25,000.

   (a) What is the company's profit if it spends $2,000 on research and development and $5,000 on advertising?

   (b) What will be the company's profit if it spends $8,000 on research and development and $6,000 on advertising?

   (c) If the company is planning to spend $4,000 on research and development and hopes to make a profit of $590,000, how much should it spend on advertising?

5. If $g(x, y) = x^2 + y^2 + 2$, for all real values of x and y, find

   (a) $g(1, 0)$.   (b) $g(-1, -1)$.

   (c) $g(3, 4)$.   (d) $g(1, 1)$.

   (e) $g(-3, 4)$.   (f) $g(-2, 0)$.

6. Graph the following functions by the profile method:

   (a) $z = \sqrt{x^2 + y^2}$, domain xy plane.

   (b) $z = -\sqrt{1 - x^2 - y^2}$, domain $\{(x, y) \mid x^2 + y^2 \leq 1\}$.

   (c) $z = (x^2 + y^2)^{1/4}$, domain xy plane (bugle surface).

   (d) $z = \sqrt{1 - y^2}$, domain $-\infty < x < \infty, |y| \leq 1$.

   (e) $z = x + y^2$, domain xy plane.

   (f) $z = |x| + |y|$, domain xy plane (inverted pyramid).

(g) $z = 2x - y + 1$. [*Hint:* This surface is a plane. Three points determine a plane.]

(h) $z = 1 + 2x^2 + y^2$. [*Hint:* $a^2x^2 + b^2y^2 = c^2$ is the equation of an ellipse.]

(i) $z = -x - 4y + 2$, domain xy plane.

(j) $z = x^2 - y^2$, domain xy plane.

(k) $z = -2 + x^2 + y^2$, domain xy plane.

9.3 Limits and Continuity: A Brief Discussion

We will now discuss the notion of limit for a function of two variables. When we write

$$\lim_{(x, y) \to (a, b)} f(x, y) = L,$$

in a very crude sense we mean that for points (x, y) very close to but not equal to (a, b), the value $f(x, y)$ of the function is very close to L [or as (x, y) approaches (a, b), $f(x, y)$ approaches L.]

An equivalent interpretation of the expression

$$\lim_{(x, y) \to (a, b)} f(x, y) = L$$

which sounds more precise, but is still not completely satisfactory, goes as follows. For $d > 0$, let C_d be the disk $\{(x, y) \mid (x - a)^2 + (y - b)^2 < d^2\}$. (See Figure 9.3.1.) By moving C_d up and down, we sweep out a cylindrical region as in the figure. Think of the walls of the cylinder as if they were the walls of a tin can of infinite height. We now "can" the part of the graph of f over C_d. For convenience we assume that (a, b) is not an element of domain f. Choose a bottom B and top T for the can so that the graph of f for points in C_d lies above B and below T; that is, the graph of f for points in C_d is contained in the tin can we have just constructed.

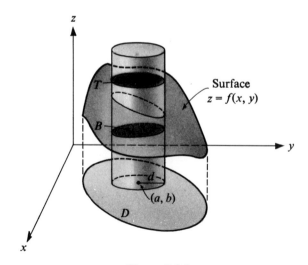

Figure 9.3.1

(See Figure 9.3.1.) If this process can be done for every disk C_d in such a way that as d tends to zero the height of the corresponding can tends to zero, we say $\lim_{(x,y)\to(a,b)} f(x, y)$ exists. The value L of the limit will be the z value where the top and bottom of the cans meet.

We will assume that the reader now has a working definition of limits for a function of two variables. To obtain a precise definition of $\lim_{(x,y)\to(a,b)} f(x, y) = L$, one must reverse the procedure above. Thus, for any $D > 0$, the top of the can is specified to be at a height $L + D$ and the bottom of the can at a height $L - D$. Then the disk C_d with center at (a, b) must be chosen so that the graph of the function over C_d is "inside" the tin can. This definition is given in Exercise 4.

It is now possible to talk about continuity for a function of two variables. Thus, $f(x, y)$ is continuous at (a, b) if: (1) $(a, b) \in$ domain of f and (2) $\lim_{(x,y)\to(a,b)} f(x, y) = f(a, b)$. The function f is continuous in the domain D if it is continuous at every point of D. Geometrically speaking, f is continuous on D if the graph of f has no tears. This is analogous to the situation where a function of one variable is continuous if there are no jumps in its graph. All the functions graphed in Section 9.2 are continuous. Let us give an example of a function that is not continuous.

▶ **Example 1.** Consider the function $z = f(x, y)$ given by

$$z = \begin{cases} 1, & 0 \le x \le 1; 0 \le y < \tfrac{1}{2}, \\ 2, & 0 \le x \le 1; \tfrac{1}{2} \le y \le 1. \end{cases}$$

Domain $= \{(x, y) \mid 0 \le x \le 1; 0 \le y \le 1\}$.

The graph of z is shown in Figure 9.3.2. This function is not continuous at any point of the form $(x, \tfrac{1}{2})$ for $0 \le x \le 1$.

▶ **Example 2.** We might point out that a function of two variables being continuous is different from a function being continuous in each variable holding the other fixed, To illustrate, consider

$$f(x, y) = \begin{cases} \dfrac{xy}{x^2 + y^2}, & (x, y) \ne (0, 0), \\ 0, & (x, y) = (0, 0). \end{cases}$$

Observe that if y is held fixed, $\lim_{x\to 0} f(x, y) = 0 = f(0, y)$ and if x is fixed, $\lim_{y\to 0} f(x, y) = 0 = f(x, 0)$. Thus this function is continuous in the separate variables at $(0, 0)$. However, $\lim_{(x,y)\to(0,0)} f(x, y)$ does not exist. To see this let $y = x$. Then

$$f(x, x) = \begin{cases} \tfrac{1}{2}, & x \ne 0, \\ 0, & x = 0. \end{cases}$$

Now

$$\lim_{\substack{(x,y)\to(0,0)\\ y=x}} f(x, y) = \frac{1}{2} \ne f(0, 0).$$

If we let $y = 2x$ we see that

$$\lim_{\substack{(x,y)\to(0,0)\\ y=2x}} f(x, y) = \frac{2}{3}.$$

Hence, $\lim_{(x,y)\to(0,0)} f(x, y)$ could not exist.

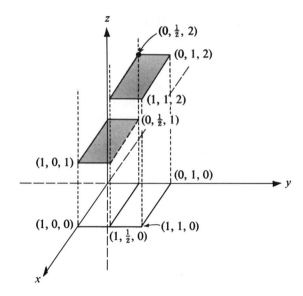

Figure 9.3.2

Thus a limit cannot exist if $\lim_{(x,y)\to(a,b)} f(x, y)$ has different values as we let y be a function of x or x a function of y. We see that the value of $\lim_{(x,y)\to(a,b)} f(x, y)$ must be the same no matter how we approach the point (a, b).

Exercises

1. Draw a picture of a function $z = f(x, y)$ which is discontinuous at $(1, 2)$.

2. Is the following function continuous at $(0, 0)$?

$$z = \begin{cases} \sqrt{1 - (x^2 + y^2)}, & 0 \leq x^2 + y^2 \leq 1, \\ 2, & x = y = 0. \end{cases}$$

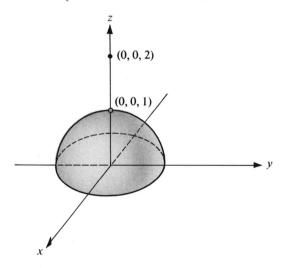

Figure 9.3.3

3. Consider the function defined by

$$f(x, y) = \frac{x^3 + y^3}{x^2 + y^2}.$$

Since f is *not* defined at $(0, 0)$, f is not continuous there. How do you think f should be defined at $(0, 0)$ in order to make f continuous there? (Do not attempt to provide a rigorous explanation—follow the explanation in Example 2.)

*4. Let $f(x, y)$ be a function with domain R. Then

$$\lim_{(x, y) \to (a, b)} f(x, y) = L$$

if for every $D > 0$ there exists $d > 0$ such that

$$|f(x, y) - L| < D \qquad \text{for} \quad 0 < (x - a)^2 + (y - b)^2 < d^2.$$

Use this definition to show that the function in Exercise 2 has a limit at the point $(0, 0)$ but is not continuous there.

*5. Let

$$f(x, y) = \begin{cases} 0, & x = y = 0, \\ \dfrac{2xy}{x^2 + y^2}, & \text{for} \quad x^2 + y^2 > 0. \end{cases}$$

Is $f(x, y)$ continuous at $(0, 0)$? [*Hint:* See Example 2 in this section.]

*6. Graph the function $f(x) = \sqrt{x^2 + y^2 - 1}$ for $x^2 + y^2 \geq 1$. [*Hint:* Write $z = \sqrt{x^2 + y^2 - 1}$ as $z^2 = x^2 + y^2 - 1$. Now look at what f does to the points $x^2 + y^2 = R^2$, R fixed. Finally consider what happens when $x = 0$. This surface is sometimes called the bugle surface.]

9.4 Partial Derivatives

As we have already seen, it is very informative when studying curves of the form $y = f(x)$ to examine the slope of the curve at every point. This was done by looking at the derivative. Let us try to generalize this notion to surfaces.

A mountain climber pauses at point P as in Figure 9.4.1. There are three paths

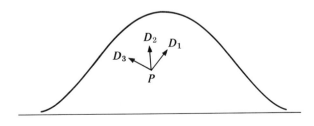

Figure 9.4.1

to the summit of varying steepness. As before, we could measure the steepness. The important thing is to notice that the slope of the mountain at point P depends on the direction taken (that is, whether we go in the direction D_1, D_2, or D_3 in Figure 9.4.1). Let us give another example which illustrates this behavior more clearly.

In the Figure 9.4.2, the path in direction D_2 is quite steep, whereas the path in direction D_1 is level, i.e., has zero slope. (Anyone who skis is acutely aware of this phenomenon.) By means of a mathematical version of this simple notion (called the partial derivative), we investigate surfaces.

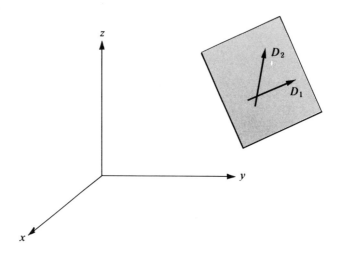

Figure 9.4.2

Let $z = f(x, y)$ be a function with domain D, and let (a, b) be a point in D. Consider the intersection of the surface $z = f(x, y)$ with the plane $y = b$. (See Figure 9.4.3.) The intersection of the plane and the surface is a curve. Indeed, it is a curve in the plane $y = b$, and its equation is $z = f(x, b)$. Let us pull the plane $y = b$ out of Figure 9.4.4.

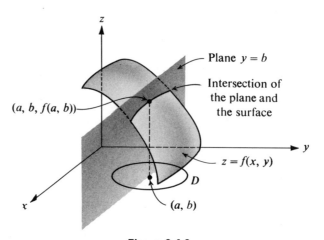

Figure 9.4.3

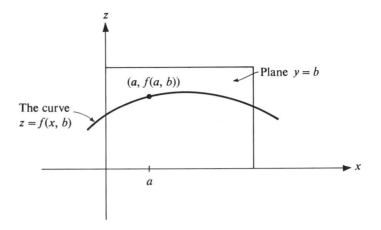

Figure 9.4.4

Now we can easily find the slope of the curve $z = f(x, b) = g(x)$ at the point a. It is just $g'(a)$. By definition,

$$g'(a) = \lim_{x \to a} \frac{g(x) - g(a)}{x - a} = \lim_{x \to a} \frac{f(x, b) - f(a, b)}{x - a}.$$

We have arrived at the definition of the partial derivative.

DEFINITION. Let $z = f(x, y)$. Then

$$\left.\frac{\partial f}{\partial x}\right|_{(a,b)} = \left.\frac{\partial z}{\partial x}\right|_{(a,b)} = \frac{\partial f}{\partial x}(a, b)$$

$$= \lim_{x \to a} \frac{f(x, b) - f(a, b)}{x - a}$$

[*read the partial of f with respect to x at the point* (a, b)]. *Similarly,*

$$\left.\frac{\partial f}{\partial y}\right|_{(a,b)} = \left.\frac{\partial z}{\partial y}\right|_{(a,b)} = \lim_{y \to b} \frac{f(a, y) - f(a, b)}{y - b}.$$

To summarize; $\left.\dfrac{\partial f}{\partial x}\right|_{(a,b)}$ is the slope of the surface $z = f(x, y)$ in the x direction at

(a, b) or the slope of the intersection of the surface and the plane $y = b$, and $\left.\dfrac{\partial f}{\partial y}\right|_{(a,b)}$ is the slope of the surface in the y direction at (a, b). (See Figure 9.4.5.)

At this point, you might think it is going to be difficult to calculate $\dfrac{\partial f}{\partial x}$ or $\dfrac{\partial f}{\partial y}$ given $z = f(x, y)$. The next pleasant surprise is that it is very easy.

RULE. Let $z = f(x, y)$.

(1) To find $\left.\dfrac{\partial f}{\partial x}\right|_{(a,b)}$

(1) Differentiate f with respect to x, treating all the y's as though they were constants.

(2) Substitute a for x and b for y.

(2) To find $\dfrac{\partial f}{\partial y}\Big|_{(a,b)}$

(1) Differentiate f with respect to y treating all the x's as though they were constants.

(2) Substitute a for x and b for y.

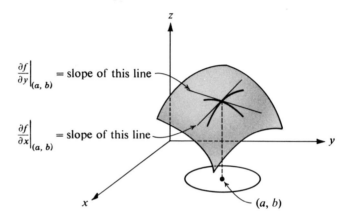

Figure 9.4.5

▶ **Example 1.** If $z = f(x, y) = x^2y + e^y$, find $\dfrac{\partial f}{\partial x}\Big|_{(1,0)}$ and $\dfrac{\partial f}{\partial y}\Big|_{(2,0)}$.

Solution. Following the rule step by step, we obtain

Step (1) $\dfrac{\partial f}{\partial x} = \dfrac{\partial}{\partial x}(x^2y) + \dfrac{\partial}{\partial x}(e^y)$ so $\dfrac{\partial f}{\partial x} = 2xy + 0$.

Step (2) $\dfrac{\partial f}{\partial x}\Big|_{(1,0)} = 2(1)(0) = 0$.

Similarly, for the second half of the problem,

Step (1) $\dfrac{\partial f}{\partial y} = x^2 + e^y$.

Step (2) $\dfrac{\partial f}{\partial y}\Big|_{(2,0)} = (2)^2 + e^0 = 5$.

▶ **Example 2.** If $z = f(x, y) = x^3y^2 + \log(xy)$, find $\dfrac{\partial f}{\partial x}\Big|_{(1,2)}$ and $\dfrac{\partial f}{\partial y}\Big|_{(1,2)}$.

Solution. Applying our rule for partial differentiation, we obtain

Step (1) $\dfrac{\partial f}{\partial x} = \dfrac{\partial}{\partial x}(x^3y^2) + \dfrac{\partial}{\partial x}\log(xy)$

$$= 3x^2y^2 + \dfrac{1}{xy}\dfrac{\partial}{\partial x}(xy)$$

$$= 3x^2y^2 + \dfrac{1}{xy}\cdot y = 3x^2y^2 + \dfrac{1}{x}.$$

Step (2) $\left.\dfrac{\partial f}{\partial x}\right|_{(1,2)} = 3(1)^2(2)^2 + \dfrac{1}{1} = 13.$

Repeating this process for the second half of the problem, we find

Step (1) $\dfrac{\partial f}{\partial y} = \dfrac{\partial}{\partial y}(x^3y^2) + \dfrac{\partial}{\partial y}\log(xy),$ so

$$\dfrac{\partial f}{\partial y} = 2x^3y + \dfrac{1}{xy}\dfrac{\partial}{\partial y}(xy)$$

$$= 2x^3y + \dfrac{1}{xy}\cdot x = 2x^3y + \dfrac{1}{y}.$$

Step (2) $\left.\dfrac{\partial f}{\partial y}\right|_{(1,2)} = 2(1)^3(2) + \tfrac{1}{2} = 4\tfrac{1}{2}.$

Of course, we can find $\dfrac{\partial f}{\partial x}$ or $\dfrac{\partial f}{\partial y}$ at a general point (x, y).

▶ **Example 3.** If $z = f(x, y) = \sqrt{x^2 + y^2} + xe^y$, find $\dfrac{\partial f}{\partial x}$ and $\dfrac{\partial f}{\partial y}$.

Solution. We simply evaluate $\dfrac{\partial f}{\partial x}$ by our foregoing rule.

$$\dfrac{\partial f}{\partial x} = \dfrac{\partial}{\partial x}\sqrt{x^2 + y^2} + \dfrac{\partial}{\partial x}(xe^y) = \tfrac{1}{2}(x^2 + y^2)^{-1/2}\dfrac{\partial}{\partial x}(x^2 + y^2) + e^y$$

$$= \dfrac{1}{2}\dfrac{1}{\sqrt{x^2 + y^2}}\cdot 2x + e^y = \dfrac{x}{\sqrt{x^2 + y^2}} + e^y.$$

Similarly, we see that

$$\dfrac{\partial f}{\partial y} = \dfrac{\partial}{\partial y}(\sqrt{x^2 + y^2}) + \dfrac{\partial}{\partial y}(xe^y) = \tfrac{1}{2}(x^2 + y^2)^{-1/2}\dfrac{\partial}{\partial y}(x^2 + y^2) + xe^y$$

$$= \dfrac{y}{\sqrt{x^2 + y^2}} + xe^y.$$

Having defined partial derivatives, we can extend our definitions to higher-order derivatives. Thus, if $z = f(x, y)$,

(1) $\dfrac{\partial^2 f}{\partial x^2} = \dfrac{\partial^2 z}{\partial x^2} = \dfrac{\partial}{\partial x}\left(\dfrac{\partial f}{\partial x}\right).$

(2) $\dfrac{\partial^2 f}{\partial x\, \partial y} = \dfrac{\partial}{\partial x}\left(\dfrac{\partial f}{\partial y}\right).$

(3) $\dfrac{\partial^2 f}{\partial y\, \partial x} = \dfrac{\partial}{\partial y}\left(\dfrac{\partial f}{\partial x}\right).$

(4) $\dfrac{\partial^3 f}{\partial y^3} = \dfrac{\partial}{\partial y}\dfrac{\partial}{\partial y}\left(\dfrac{\partial f}{\partial y}\right).$

We remark that for all functions we are likely to encounter, (2) and (3) are equal. Let us consider the following example.

▶ **Example 4.** If $f(x, y) = x^2 y + y e^x$, find $\dfrac{\partial^2 f}{\partial x^2}, \dfrac{\partial^2 f}{\partial y^2}, \dfrac{\partial^2 f}{\partial x\, \partial y}$, and $\dfrac{\partial^2 f}{\partial y\, \partial x}$.

Solution. First we compute $\dfrac{\partial f}{\partial x}$ and $\dfrac{\partial f}{\partial y}$. We have

$$\frac{\partial f}{\partial x} = 2xy + ye^x \quad \text{and} \quad \frac{\partial f}{\partial y} = x^2 + e^x.$$

Now,

$$\frac{\partial^2 f}{\partial x^2} = \frac{\partial}{\partial x}\frac{\partial f}{\partial x} = 2y + ye^x \quad \text{and} \quad \frac{\partial^2 f}{\partial y^2} = \frac{\partial}{\partial y}\frac{\partial f}{\partial y} = 0,$$

since no y term appears in $\dfrac{\partial f}{\partial y}$. Finally,

$$\frac{\partial^2 f}{\partial y\, \partial x} = \frac{\partial}{\partial y}\frac{\partial f}{\partial x} = 2x + e^x$$

and

$$\frac{\partial^2 f}{\partial x\, \partial y} = \frac{\partial}{\partial x}\frac{\partial f}{\partial y} = 2x + e^x.$$

Observe that $\dfrac{\partial^2 f}{\partial x\, \partial y} = \dfrac{\partial^2 f}{\partial y\, \partial x}$. This is not just a coincidence. It turns out that if we consider a function $f(x, y)$ which is continuous and if $\dfrac{\partial^2 f}{\partial x\, \partial y}$ and $\dfrac{\partial^2 f}{\partial y\, \partial x}$ are also continuous, then $\dfrac{\partial^2 f}{\partial x\, \partial y} = \dfrac{\partial^2 f}{\partial y\, \partial x}$.

We next give several examples which show how one interprets partial derivatives in economic applications.

▶ **Example 5.** Consider the equation

$$z = 100 + 6x + 10y,$$

which relates z, the cost of a certain product in dollars, to x, the cost of raw materials in dollars per pound, and y, the cost of labor in dollars per hour. Note that

$$\frac{\partial z}{\partial x} = 6 \quad \text{and} \quad \frac{\partial z}{\partial y} = 10$$

for all x and y. This means that when the cost of labor is held fixed, an increase of $1.00 per pound in the cost of raw materials causes an increase of $6.00 in the cost of the product, and that when the cost of raw materials is held fixed, an increase of $1.00 in the hourly cost of labor brings about an increase of $10.00 in the cost of the product.

▶ Example 6. (*Marginal Demands*)   Let the demands for two different commodities be d_1 and d_2 and let the respective prices be x and y. If the demand functions for two related commodities are given in the form

$$d_1 = f(x, y) \quad \text{and} \quad d_2 = g(x, y), \tag{1}$$

then

$\dfrac{\partial f}{\partial x}$ is the (partial) marginal demand of d_1 with respect to x,

$\dfrac{\partial f}{\partial y}$ is the (partial) marginal demand of d_1 with respect to y,

$\dfrac{\partial g}{\partial x}$ is the (partial) marginal demand of d_2 with respect to x,

$\dfrac{\partial g}{\partial y}$ is the (partial) marginal demand of d_2 with respect to y.

In order to correspond to normal economic situations, we put the following restrictions on d_1, d_2.

Restriction 1.   All the variables are assumed to be positive.
Restriction 2.   If $y = b$, where b is a fixed constant, then d_1 *must* be a decreasing function of x. (See Figure 9.4.6.)
Restriction 3.   If $x = a$, where a is a fixed constant, then d_2 *must* be a decreasing function of y.
Restriction 4.   The functions f and g and their domains in the xy plane must be such that it is possible to solve equations (1) for x, y in terms of d_1, d_2. That is, there exist functions F and G such that $x = F(d_1, d_2)$ and $y = G(d_1, d_2)$.

The two commodities are said to be *competitive* for all prices (a, b) for which $\dfrac{\partial f}{\partial y}\bigg|_{(a,b)}$ and $\dfrac{\partial g}{\partial x}\bigg|_{(a,b)}$ are both positive. This means that an increase (decrease) in demand for one will cause decrease (increase) in demand for the other. The two commodities are said to be *complementary* for all prices (a, b) for which $\dfrac{\partial f}{\partial y}\bigg|_{(a,b)}$ and $\dfrac{\partial g}{\partial x}\bigg|_{(a,b)}$ are both negative. This means that an increase (decrease) in demand for one will cause an increase (decrease) in the demand for the other.

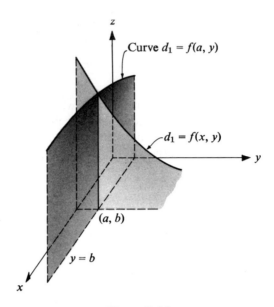

Figure 9.4.6

As a numerical example, consider the following. Determine the four partial marginal demands of the following pair of linear functions and discuss the nature of the relation between the two commodities.

Solution. $f(x, y) = 17 - 2x - y$; $g(x, y) = 14 - x - 2y$. Now,

$$\frac{\partial f}{\partial x} = -2; \quad \frac{\partial f}{\partial y} = -1; \quad \frac{\partial g}{\partial x} = -1; \quad \frac{\partial g}{\partial y} = -2.$$

Since $\frac{\partial f}{\partial y}$ and $\frac{\partial g}{\partial x}$ are *both* negative for *all* values of x and y, the two commodities must be complementary.

▶ **Example 7.** (*Marginal Cost*)   If the joint cost function of producing the quantities x and y of two commodities is given by $C = C(x, y)$, then the partial derivatives of C are called the *marginal cost functions*. Thus, if $C(x, y) = x \log (2 + y)$, then

$$\frac{\partial C}{\partial x} = \log (2 + y) \quad \text{and} \quad \frac{\partial C}{\partial y} = \frac{x}{2 + y}$$

are the marginal cost functions for the commodity.

Exercises

1. If $z = x + x^2 y + 3x^3 y^2$, find $\dfrac{\partial z}{\partial x}\bigg|_{(-1,2)}$ and $\dfrac{\partial z}{\partial y}\bigg|_{(8,-1)}$.

2. Find $\dfrac{\partial z}{\partial x}$ and $\dfrac{\partial z}{\partial y}$ for each of the following functions:

    (a) $z = e^{2xy}$.   (b) $z = x^2 + y^3$.

    (c) $z = xy^2 + y$.   (d) $z = x/y$.

(e) $z = y^5 x^6$.

(f) $z = x \log y$.

(g) $z = \dfrac{4x}{y} + y^2$.

(h) $z = \sqrt{\dfrac{x + y}{x - y}}$.

(i) $z = y \log (x^2 + y)$.

3. If $f(x, y) = \dfrac{x + 1}{\sqrt{y^2 + 1}}$, find $\dfrac{\partial f}{\partial x}\Big|_{(1,2)}$ and $\dfrac{\partial f}{\partial y}\Big|_{(2,1)}$.

4. Any function of the form $z = Ax + By + C$ (A, B, C constants) has a plane as its graph. Find $\dfrac{\partial z}{\partial x}\Big|_{(a,b)}$ and $\dfrac{\partial z}{\partial y}\Big|_{(a,b)}$. Is your answer independent of the point (a, b)? Give a geometric interpretation of this fact. (Look at Example 5 and compare.)

5. If $z = x^2 y^5 + e^{(x+y)}$, find $\dfrac{\partial^2 z}{\partial x^2}\Big|_{(1,0)}$, $\dfrac{\partial^2 z}{\partial x \, \partial y}\Big|_{(2,3)}$, and $\dfrac{\partial^2 z}{\partial y^2}\Big|_{(-1,1)}$.

6. If $z = \sqrt{x^2 + y}$, find $\dfrac{\partial^2 z}{\partial x \, \partial y}$ and $\dfrac{\partial^2 z}{\partial y \, \partial x}$.

7. If $f(x, y) = e^x \cos y$, show that

$$\frac{\partial^2 f}{\partial x^2} + \frac{\partial^2 f}{\partial y^2} = 0.$$

8. If $f(x, y) = e^{(x^2 - y^2)} \cos 2xy$, show that

$$\frac{\partial^2 f}{\partial x^2} + \frac{\partial^2 f}{\partial y^2} = 0.$$

9. If $f(x, y) = x^2 y$, show that

$$x \cdot \frac{\partial^2 f}{\partial x^2} - y \cdot \frac{\partial^2 f}{\partial x \, \partial y} = 0.$$

10. A banker determines the amount of money he is willing to loan on a one-family house by means of the formula

$$z = 2x + 0.02y^2,$$

where x is the size of the down payment, and y is the applicant's monthly salary, and all figures are in dollars. Calculate the two partial derivatives $\dfrac{\partial z}{\partial x}$ and $\dfrac{\partial z}{\partial y}$ and explain quantitatively what they mean.

11. In Whiplash, North Dakota, the demand for ice cream cones in the month of July is given by the equation

$$z = 1{,}000 + 10x^3 - 25y \log x - 8y^2,$$

where x represents the temperature in degrees, and y represents the cost (in cents) of the cone. Calculate the two partial derivatives $\dfrac{\partial z}{\partial x}$ and $\dfrac{\partial z}{\partial y}$ and explain quantitatively what they mean.

12. For each of the following pairs of linear functions, determine the four partial marginal demands and discuss the nature of the relation between the two commodities.

(a) $f(x, y) = 5 - 2x + y, \quad g(x, y) = 6 + x - y.$

(b) $f(x, y) = 2 - 2x + y, \quad g(x, y) = 6 - 2x - 3y.$

(c) $f(x, y) = x^{-1.7}y^{0.8}, \quad g(x, y) = x^5y^{-0.2}.$

13. If a production function is given in the form $z = f(u, v)$, then $\dfrac{\partial f}{\partial u}$ and $\dfrac{\partial g}{\partial v}$ are called the marginal productivities. Find the marginal productivities of the following production functions:

(a) $z = 5 - (1/u) - (1/v)$ at $(u = 1, v = 1).$

(b) $z = 4(4uv - u^2 - 3v^2)$ at $(u = 1, v = \frac{1}{2}).$

(c) $z = 5uv - 2u^2 - 2v^2$ at $(u = 1, v = 1).$

9.5 Maxima and Minima. An Introduction

In this section and the next, we develop techniques for finding the maximum and minimum values of a function $z = f(x, y)$. The increase from one to two variables is accompanied by several complications.

Given a domain D in the xy plane, we refer to its "boundary." This term has the ordinary meaning one associates with it. (See Figure 9.5.1.) The domain D is said to be *bounded* if $D \subset \{(x, y) \mid x^2 + y^2 \leq R\}$ for some $R > 0$. Thus D is bounded if it is contained in some "big" disk in the plane.

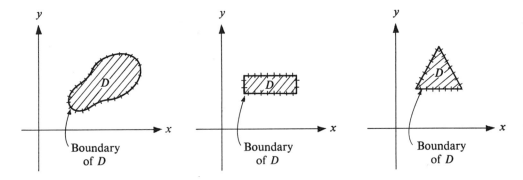

Figure 9.5.1

When we refer to the points near (a, b) or in a neighborhood of (a, b), we mean the points in a "small" disk about (a, b) (Figure 9.5.2), that is, all those points (x, y) whose distance from (a, b) is less than r or, algebraically,

$$[(x - a)^2 + (y - b)^2]^{1/2} < r.$$

A point (a, b) in D is an *interior point* of D if there exists a neighborhood of (a, b) contained entirely in D.

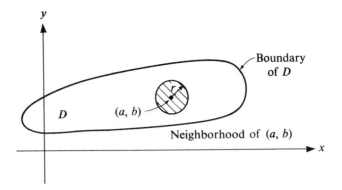

Figure 9.5.2

While this discussion is not very complete, it is sufficient for our needs. We begin with a definition of relative maximum and minimum.

DEFINITION 1. Let $z = f(x, y)$ be a function with domain D and let (a, b) be a point of D. Then f has a relative maximum [minimum] at (a, b) if $f(x, y) \leq f(a, b)$ $[f(x, y) \geq f(a, b)]$ for all (x, y) in some neighborhood of (a, b).

Roughly speaking, at a relative maximum the graph of the function looks like a mountain peak and at a relative minimum the graph looks like a bowl. Examples of a relative maximum and a relative minimum are shown in Figures 9.5.3 and 9.5.4.

Just as in the case of one variable, we can discuss the concept of absolute maxima and minima.

DEFINITION 2. Let $z = f(x, y)$ be a function with domain D. Then f has an absolute maximum [minimum] at (a, b) if $f(x, y) \leq f(a, b)$ $[f(x, y) \geq f(a, b)]$ for all (x, y) in D.

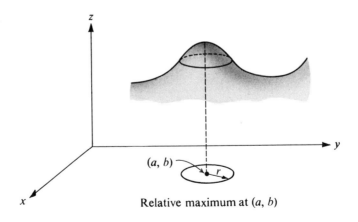

Relative maximum at (a, b)

Figure 9.5.3

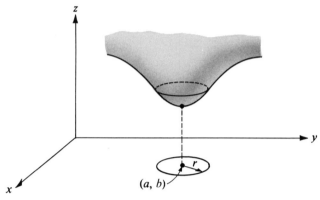

Relative minimum at (a, b)

Figure 9.5.4

Note that a function f can have more than one absolute maximum or minimum. Indeed, if f is a constant function $[f(x, y) = k]$, then every point in its domain is both an absolute maximum and an absolute minimum. The absolute maximum and minimum is also referred to as simply maximum and minimum in accordance with our definitions in Chapter 4. See Figure 9.5.5.

In Chapter 4, we observed that a necessary condition for a differentiable function f of one variable to have a relative maximum or minimum at an interior point a is that $f'(a) = 0$. The following theorem yields an analogous situation for a function of two variables.

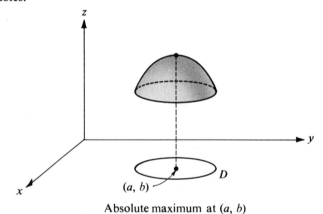

Absolute maximum at (a, b)

Figure 9.5.5

THEOREM 1. *Let $z = f(x, y)$ be a function with domain D, and (a, b) be an interior point of D. If f has a relative maximum or a relative minimum at (a, b) and if $\dfrac{\partial f}{\partial x}$ and $\dfrac{\partial f}{\partial y}$ exist at (a, b), then*

$$\frac{\partial f}{\partial x}\bigg|_{(a,b)} = \frac{\partial f}{\partial y}\bigg|_{(a,b)} = 0.$$

The proof is not hard. It is completely analogous to the proof for a function of one variable. (See Exercises 4 and 5.) What this theorem says is the following: if you are trying to find relative (or absolute) maxima and minima, then you should look at the common zeros of $\dfrac{\partial f}{\partial x}$ and $\dfrac{\partial f}{\partial y}$, provided, of course, these partial derivatives exist. We call the points (a, b) such that

$$\left.\frac{\partial f}{\partial x}\right|_{(a,b)} = \left.\frac{\partial f}{\partial y}\right|_{(a,b)} = 0,$$

critical points of the function f.

Consider the following two examples.

▶ **Example 1.** Find the critical points of the function

$$f(x, y) = y^2 + 3x^4 - 4x^3 - 12x^2 + 24.$$

Solution. First we find

$$\frac{\partial f}{\partial x} = 12x^3 - 12x^2 - 24x \quad \text{and} \quad \frac{\partial f}{\partial y} = 2y.$$

Solving the first equation after setting it equal to zero, we see that

$$12x(x^2 - x - 2) = 0 \quad \text{or} \quad x = 0, -1, 2.$$

Clearly $\dfrac{\partial f}{\partial y} = 0$ precisely when $y = 0$. Thus, the relative maxima and minima must occur at $(0, 0)$, $(-1, 0)$, and $(2, 0)$ if there are any.

▶ **Example 2.** Find the critical points of the function

$$f(x, y) = x^2 - xy + \tfrac{1}{2}y^2 + 5y.$$

Solution. Again we find

$$\frac{\partial f}{\partial x} = 2x - y \quad \text{and} \quad \frac{\partial f}{\partial y} = -x + y + 5.$$

Thus, $2x - y = 0$, or $y = 2x$ and $-x + y + 5 = 0$. Substituting $y = 2x$ in this last equation, we get $-x + 2x + 5 = 0$ or $x = -5$. Since $y = 2x$, the only point where f can have a relative maximum or minimum is $(-5, -10)$.

A word of caution is in order at this point. Just because

$$\left.\frac{\partial f}{\partial x}\right|_{(a,b)} = \left.\frac{\partial f}{\partial y}\right|_{(a,b)} = 0$$

does not mean f has a relative maximum or minimum at (a, b). We present an example to illustrate this phenomenon.

Consider the function $z = f(x, y) = xy$. Its graph is known as a hyperbolic paraboloid. (See Figure 9.5.6.) (The graph of this function is very similar to the one in Example 4 of Section 9.2.) Near the origin, this surface is saddle-shaped as shown in Figure 9.5.6. Observe that both $\dfrac{\partial f}{\partial x}$ and $\dfrac{\partial f}{\partial y}$ are zero at $(0, 0)$, but, as we shall show, there is neither a relative maximum nor a relative minimum there. If (x, y) lies in either the first or third quadrant but not on the axis, then $xy > 0$ and hence $f(x, y) > 0$; while if (x, y) lies in the second or fourth quadrant but not on an axis, then $xy < 0$ and thus $f(x, y) < 0$. Thus, any neighborhood of the origin $(0, 0)$ will contain points for which $f(x, y) < f(0, 0)$ and points such that $f(x, y) > f(0, 0)$. Hence, the origin can neither be a relative maximum nor a relative minimum. The origin, in this example, is called a *saddle-point* of the surface.

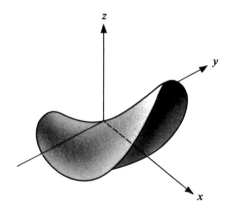

Figure 9.5.6

Exercises

1. Find the critical points of the following functions:

   (a) $f(x, y) = x^3 - y^2 + x + 4y$.

   (b) $f(x, y) = x\sqrt{y^3 - y^2}$.

   (c) $f(x, y) = x^2 + y^2 + 4xy + x$.

   (d) $f(x, y) = 3x^2 + 2xy^3 - 2y$.

   (e) $f(x, y) = x^2 + 8y^2 + 24xy$.

   (f) $f(x, y) = 2x^2 - 3y^2 - xy - 3x + 7y$.

   (g) $f(x, y) = x^3 - 3xy^2 + y^3$.

2. Consider the function given by $z = f(x, y) = x^3 - 3xy^2$. Can this function have a relative maximum or minimum at the origin? Why? (Its graph is given in Figure 9.5.7.)

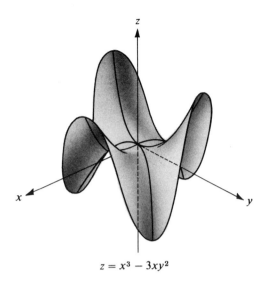

$$z = x^3 - 3xy^2$$

Figure 9.5.7

3. What are the possible candidates for relative maxima or minima for the following functions?

   (a) $f(x, y) = e^{-(x^2+y^2)}$. (The domain of f is taken to be the entire xy plane.)

   (b) $f(x, y) = x^2y^2$.

   (c) $f(x, y) = (x + y)e^{-xy}$.

   (d) $f(x, y) = (x - y)^4$.

   Can you give intuitive arguments which tell you which points are the maxima or minima in the above examples? (You need not construct the surfaces.)

*4. Show that if f has a relative maximum at (a, b) and $\dfrac{\partial f}{\partial x}$ and $\dfrac{\partial f}{\partial y}$ exist at (a, b), then they are 0. [*Hint:* Note that $f(a, b) \geq f(x, y)$ for (x, y) in a neighborhood of (a, b). Hence

$$\frac{f(x, y) - f(a, b)}{x - a} \geq 0 \qquad \text{for} \quad x < a.$$

Why? Similarly,

$$\frac{f(x, y) - f(a, b)}{x - a} \leq 0 \qquad \text{for} \quad x > a.$$

Why? Thus,

$$\lim_{x \to a^-} \frac{f(x, y) - f(a, b)}{x - a} \geq 0$$

and

$$\lim_{x \to a^+} \frac{f(x, y) - f(a, b)}{x - a} \leq 0.$$

Why? Since

$$\lim_{x \to a} \frac{f(x, y) - f(a, b)}{x - a} = \frac{\partial f}{\partial x}\bigg|_{(a, b)}$$

exists, it must be 0. Why? (It may be convenient to refer to left- and right-hand limits here.) The argument that $\dfrac{\partial f}{\partial y}\bigg|_{(a, b)} = 0$ is similar.]

5. Show that Exercise 4 is true if the word maximum is replaced by the word minimum.

*6. Consider the point P on the inner tube. Does the surface have a relative maximum, relative minimum, or a saddle point at P? (The inner tube is not the graph of a function $z = f(x, y)$ since it is not single-valued; however, this is not relevant to the problem.) (See Figure 9.5.8.)

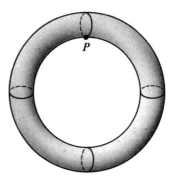

Figure 9.5.8

7. A general fact about surfaces: Imagine an island in the middle of an ocean. We shall assume for the moment that everyone can recognize a mountain peak, a mountain pass, and a pit. Count the number of each on the island. Upon tabulating you will discover the following: number of peaks + number of pits − number of passes = 1. It is surprising that one can make any statement about a general surface (island) and the above relation is certainly remarkable. (It is due to an American mathematician Marston Morse.) By placing advantageous wagers with gullible tourists one might finance an around the world trip via this fact.

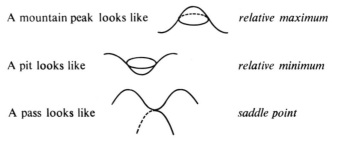

Draw a variety of island landscapes and check the relation on them.

9.6   More on Maxima and Minima

In this section we will study techniques for determining the absolute maximum and minimum for functions of several variables. We shall also discuss some applications to economics and the social sciences. Recall that in our discussion of functions of one variable, we had to distinguish between closed and open intervals in order to describe where the maxima and minima occurred. We must make an analogous distinction when considering functions of two variables. This amounts to considering domain D, with and without boundary.

Recall the following theorem from the single-variable calculus: If f has a derivative on the closed interval $[a, b]$, then f has an absolute maximum and an absolute minimum on $[a, b]$. Moreover, the absolute maximum and absolute minimum occur either at points when f' is zero or at an endpoint of $[a, b]$. We now present the analogous result for two variables.

THEOREM 1.   Let $z = f(x, y)$ be a function on the bounded domain D, where $D contains its boundary. Suppose that f is continuous on D and that $\dfrac{\partial f}{\partial x}$ and $\dfrac{\partial f}{\partial y}$ both exist at all points of D. Then f has an absolute maximum and an absolute minimum on D. Moreover, the absolute maximum [minimum] occurs either at a critical point (a, b), that is, a point where

$$\frac{\partial f}{\partial x}\bigg|_{(a,b)} = \frac{\partial f}{\partial y}\bigg|_{(a,b)} = 0$$

or at a point on the boundary of D.

Remark.   A function may have more than one absolute maximum or minimum. Indeed if $f(x, y) = M$, where M is a constant, then every point in the domain of f is both an absolute maximum and an absolute minimum.

Although this theorem has many hypotheses, it is easy to apply in practice, as the following example shows.

▶ Example 1.   Find the absolute maximum of the function

$$f(x, y) = 2x + 4y - 2x^2 - 3y^2$$

when the domain of f is the disk $x^2 + y^2 \leq 100$. Note that f is certainly continuous for $(x, y) \in D$, where

$$D = \{(x, y) \mid x^2 + y^2 \leq 100\}.$$

The boundary of the domain is the circle $x^2 + y^2 = 100$.

Solution.   First we find the common zeros of the partial derivative

$$\frac{\partial f}{\partial x} = 2 - 4x \quad \text{and} \quad \frac{\partial f}{\partial y} = 4 - 6y.$$

Setting these equal to zero, we obtain

$$2 - 4x = 0 \quad \text{or} \quad x = \tfrac{1}{2}$$

and

$$4 - 6y = 0 \quad \text{or} \quad y = \tfrac{2}{3}.$$

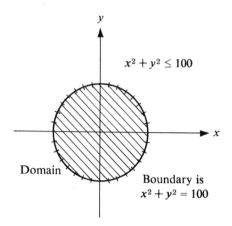

x² + y² ≤ 100

Domain

Boundary is
x² + y² = 100

Figure 9.6.1

Thus the only point where $\dfrac{\partial f}{\partial x} = \dfrac{\partial f}{\partial y} = 0$ is $(\tfrac{1}{2}, \tfrac{2}{3})$. Moreover,

$$f(\tfrac{1}{2}, \tfrac{2}{3}) = 2(\tfrac{1}{2}) + 4(\tfrac{2}{3}) - 2(\tfrac{1}{2})^2 - 3(\tfrac{2}{3})^2 = \tfrac{11}{6}.$$

(See Figure 9.6.1.) We now have to check f on the boundary. Note that

$$f(x, y) = 2x + 4y - (2x^2 + 3y^2), \quad \text{and} \quad 2x^2 + 3y^2 \geq x^2 + y^2 = 100$$

on the boundary; thus $-(2x^2 + 3y^2) \leq -100$ on the boundary. (Why?) For (x, y) in the boundary of D, $|x| \leq 10$ and $|y| \leq 10$. (Why?) Hence,

$$f(x, y) = 2x + 4y - (2x^2 + 3y^2) \leq 20 + 40 - 100 = -40$$

for any point (x, y) on the boundary. Since $f(x, y) \leq -40$ on the boundary, the absolute maximum must occur at $(\tfrac{1}{2}, \tfrac{2}{3})$.

The following theorem is analogous to the result for open intervals in our discussion of functions of one variable in Chapter 4.

THEOREM 2. Let $z = f(x, y)$ be defined in a domain D without boundary. Then f may or may not have an absolute maximum [minimum] on D. If f does have an absolute maximum [minimum], it occurs at a critical point (a, b)—that is, a point (a, b) such that

$$\left.\frac{\partial f}{\partial x}\right|_{(a,b)} = \left.\frac{\partial f}{\partial y}\right|_{(a,b)} = 0.$$

We assume, of course, that $\dfrac{\partial f}{\partial x}$ and $\dfrac{\partial f}{\partial y}$ exist for points in D.

We omit the proof of Theorem 2 and continue with some further examples.

▶ Example 2. A rectangular box is to be made out of 150 square inches of paper. What is the maximum volume of the box?

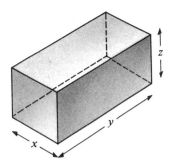

Figure 9.6.2

Solution. First we draw a picture of the box and label it (Figure 9.6.2). The surface area of the box is easily seen to be $2xy + 2yz + 2xz$. Thus $2xy + 2yz + 2xz = 150$. We solve this for z to obtain

$$z = \frac{75 - xy}{x + y}.$$

The volume V of the box is equal to $x \cdot y \cdot z$. Thus,

$$V = x \cdot y \cdot z = \frac{xy(75 - xy)}{x + y} \quad \text{or} \quad V(x, y) = \frac{75xy - x^2y^2}{x + y}.$$

The domain of V is $0 < x < \infty$ and $0 < y < \infty$. (If $x = 0$, then we would not really have a box; moreover, that choice of x clearly does not maximize the volume.) Computing partial derivatives, we find

$$\frac{\partial V}{\partial x} = \frac{(x + y)(75y - 2xy^2) - (75xy - x^2y^2)}{(x + y)^2}$$

and

$$\frac{\partial V}{\partial y} = \frac{(x + y)(75x - 2x^2y) - (75xy - x^2y^2)}{(x + y)^2}.$$

Setting $\dfrac{\partial V}{\partial x} = 0$, we obtain

$$0 = \frac{75y^2 - 2xy^3 - x^2y^2}{(x + y)^2}$$

and hence

$$y^2(75 - 2xy - x^2) = 0.$$

We can ignore the solution $y = 0$ (it's not in D), and we are left with $75 - 2xy - x^2 = 0$. If we set $\dfrac{\partial V}{\partial y} = 0$ and solve, we find that $75 - 2xy - y^2 = 0$. Rather than solving one of these equations, let us subtract the first from the second. This yields $x^2 - y^2 = 0$. Thus $x = \pm y$. But both x and y must be positive so we can reject $x = -y$. Substituting $x = y$ in the first equation, we have $75 - 2x^2 - x^2 = 0$ or $3x^2 = 75$, and hence $x = 5$. Thus, the only point in the domain where $\dfrac{\partial V}{\partial x} = \dfrac{\partial V}{\partial y} =$

0 is (5, 5). Thus if there is a largest box, its dimensions are $x = y = 5$. But there is clearly a largest box on intuitive grounds. Solving the equation

$$2xy + 2xz + 2yz = 150$$

when $x = y = 5$, we find that $z = 5$. Thus the most efficient shape for a box under these circumstances is a cube.

The following remark is extremely useful. In the preceding example, we made use of the fact there had to be a maximum from physical considerations. In many of the maximum-minimum problems in two variables, great simplifications can be achieved by such reasoning. In the case of a function without boundary, one should be on the "look out" for such an argument. The problem may be extremely difficult otherwise.

▶ Example 3. A company which manufactures blazers of both wool and Dacron knows from past experience that if it produces x dozen blazers out of wool and y dozen blazers out of Dacron, the blazers will sell for $80 - 3x$ and $60 - 2y$ dollars per dozen, respectively. How many dozen of each kind should it schedule for production to maximize profit, knowing that the cost of manufacturing x dozen blazers out of wool and y dozen blazers out of Dacron is $12x + 8y + 4xy$ dollars?

Solution. Clearly, the amount of money the company receives for the blazers made out of wool is given by $x(80 - 3x)$ and for those made out of Dacron is $y(60 - 2y)$. The total sales revenue is thus $x(80 - 3x) + y(60 - 2y)$. The profit $P(x, y)$ is then given by

$$P(x, y) = x(80 - 3x) + y(60 - 2y) - (12x + 8y + 4xy)$$

$$= 80x - 3x^2 + 60y - 2y^2 - 12x - 8y - 4xy.$$

We see that

$$\frac{\partial P}{\partial x} = 80 - 6x - 12 - 4y \quad \text{and} \quad \frac{\partial P}{\partial y} = 60 - 4y - 8 - 4x.$$

Their common zeros are given by the solution to the equations

$$6x + 4y = 68 \quad \text{and} \quad 4x + 4y = 52.$$

Solving for x and y, we obtain $x = 8$ and $y = 5$. It is clear from our problem that $0 \le x \le \frac{80}{3}$ and $0 \le y \le 30$. $P(x, y)$ is certainly continuous for $0 \le x \le \frac{80}{3}$, $0 \le y \le 30$. An easy argument shows that $x = 8$ and $y = 5$ must yield the absolute maximum. Let us see why. Since the sales revenue is given by the function $x(80 - 3x) + y(60 - 2y)$, the company is losing money if it sells more than $\frac{80}{3}$ dozen wool blazers or 30 dozen Dacron blazers. Therefore, we may assume that the domain of f is the rectangle $\{(x, y) \mid 0 \le x \le \frac{80}{3} \text{ and } 0 \le y \le 30\}$. (See Figure 9.6.3.) The function we are trying to maximize is

$$P(x, y) = [x(80 - 3x) + y(60 - 2y)] - (12x + 8y + 4xy).$$

For x, y small, we can ignore the term $4xy$ since it would be small. Thus

$$P(x, y) \simeq [x(80 - 3x) - 12x] + [y(60 - 2y) - 8y].$$

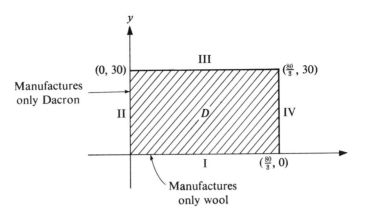

Figure 9.6.3

Since each term is positive for x and y small, it is advantageous to manufacture some of each, not just wool blazers or Dacron blazers exclusively. Thus the maximum cannot occur on sides I or II. Compare the profit on side I with side III. The profit reduces to

$$P(x, 0) = [x(80 - 3x) - 12x]$$

and

$$P(x, 30) = [x(80 - 3x) - 12x] - 240,$$

since $y = 0$ on side I and $y = 30$ on side III, clearly side I is more profitable for any given x. The same reasoning holds for side II versus side IV. Thus the maximum cannot occur on the boundary (side I is better than side III and side II is better than side IV, but the maximum does not occur on sides I or II).

We might remark that there do exist tests for maxima and minima for functions of several variables involving the second partial derivatives (as in the case of one variable). A discussion of these tests brings us beyond the scope of this text. The interested reader is referred to Apostol.*

Exercises

1. Determine whether the following functions have maxima and minima. If they do, find them.

   (a) $f(x, y) = x(1 - x) + y(1 - y)$ when the domain of f is the square $\{(x, y) \mid 0 \le x \le 1 \text{ and } 0 \le y \le 1\}$.

   (b) $f(x, y) = 2x^2 - 2xy + y^2 - 16x + 10y + 12$, where the domain of f is the entire xy plane.

   (c) $f(x, y) = x^2 - 3xy - 2y^2 - 5x + y$, where the domain of f is given by $\{(x, y) \mid |x| \le 1, |y| \le 1\}$.

---

* T. Apostol, *Calculus* (Xerox, Lexington, Mass., 1969), 2nd ed., Vol. II, p. 205.

(d) $f(x, y) = e^{-(x^2+y^2)}$, where the domain of f is given by $\{(x, y) \mid x^2 + y^2 \le 100\}$.

(e) $f(x, y) = 2x - y + 5$, where the domain of f is given by the rectangle $\{(x, y) \mid 0 \le x \le 3 \text{ and } 2 \le y \le 5\}$.

(f) $f(x, y) = xy(12 - 3x - 4y)$, where the domain of f is the entire xy plane.

2. The function $f(x, y) = xe^{-(x^2+y^2)}$ has domain $\{(x, y) \mid x^2 + y^2 \le 100\}$. Determine the maximum of the function.

3. Does the function $f(x, y) = e^{(x+y)}$ have a maximum or a minimum? The domain is the xy plane.

4. A box is to have a volume of 60 cu in. The sides cost 2¢ per square inch, the ends cost 3¢ per square inch, and the top and bottom cost 1¢ per square inch. What are the most economical dimensions for the box?

5. A rectangular box with an open top is to have a volume of 600 cu in. What dimensions require the least amount of material for its construction?

6. An automobile dealer uses two kinds of advertising, television commercials and newspaper advertisements. Let x be the amount of money spent on TV and y the amount spent on newspaper ads. The number of sales per month S is then given by

$$S(x, y) = 40[1 - e^{-[x+(y^2/4)]}].$$

(Note that if he spends nothing, he sells no cars.) His profit per car is $100. We ignore overhead. Thus the total profit is simply 100 times (number of cars sold) minus advertising costs or

$$P(x, y) = 4{,}000[1 - e^{-[x+(y^2/4)]}] - x - y.$$

How much should he spend on the different types of advertising to make his profit greatest? [*Hint:* Solve the simultaneous equations by eliminating the most complicated term.] Test for a maximum on the boundary. The boundary consists of the positive x axis and the positive y axis.

7. Otter Tail Power, a power company, uses coal, wood, and gas to run its generators. Because of the amount of fuel used, the company exerts a definite influence on the price of these items. Assume the quantities have been so normalized that Otter Tail Power needs a total of 100 units of fuel each day; i.e., $x + y + u = 100$, where x is the amount of coal, y is the amount of wood, and u is the amount of gas. The cost of each unit of fuel is given by the functions

$$C_c(x) = 2x - 20, \qquad \text{coal}$$
$$C_w(y) = y - 10, \qquad \text{wood}$$
$$C_g(u) = u, \qquad \text{gas}$$

where x, y, u are the amounts purchased. Thus the cost of x units of coal is just $x \cdot C_c(x)$. How many units of each fuel should the company purchase to minimize its fuel costs? [*Hint:* Express u in terms of x and y.]

*8. A function $z = f(x, y)$ is continuous at (a, b) if, given $r > 0$, there exists a $d > 0$ such that $|f(x, y) - f(a, b)| < r$ whenever $|x - a| + |y - b| < d$ [when (x, y) is in the domain D of f]. Let $\dfrac{\partial f}{\partial x}, \dfrac{\partial f}{\partial y}$ be defined on all of D, and let $\left|\dfrac{\partial f}{\partial x}\right| \leq M$ and $\left|\dfrac{\partial f}{\partial y}\right| \leq M$ at all points of D when M is a constant. Show that f is continuous at all points of D.

[*Hint:* Let (a, b) be a fixed point of D. Given $r > 0$, choose $d = r/2M$. Note that $|x - a| + |y - b| < d$ implies $|x - a| < d$ and $|y - b| < d$. Now

$$|f(x, y) - f(a, b)| = |f(x, y) - f(x, b) + f(x, b) - f(a, b)|$$
$$\leq |f(x, y) - f(x, b)| + |f(x, b) - f(a, b)|.$$

Apply the mean value theorem to each of the preceding terms to obtain

$$\left|\left(\frac{\partial f}{\partial x}\right)_{(x, y_1)}(y - b)\right| + \left|\left(\frac{\partial f}{\partial x}\right)_{(x_1, y)}(x - a)\right|$$
$$\leq M|y - b| + M|x - a| < M\frac{r}{2M} = r$$

(where y_1 is a point between y and b and x_1 is a point between x and a) for $|x - a| + |y - b| < d.$]

9.7 Lagrange Multipliers

In this section we will describe another method for finding maxima and minima. The method, that of Lagrange multipliers, is designed to handle the situation when we wish to maximize a function $z = f(x, y)$ subject to a side condition (constraint) $G(x, y) = 0$. For example, we may wish to maximize production of an item when we place a constraint on total costs. Usually, we consider a function $w = F(x, y, z)$. Since the graph of such a function is a surface in four-space, the geometric point of view is not very useful here. As an example, $t = F(a, b, c)$ could be the temperature at the point (a, b, c) in three-space. We will need to compute partial derivatives of such functions. Thus, if

$$w = 3xyz + z^2 + x^2 e^y,$$

then

$$\frac{\partial w}{\partial x} = 3yz + 2xe^y,$$

$$\frac{\partial w}{\partial y} = 3xz + x^2 e^y,$$

$$\frac{\partial w}{\partial z} = 3xy + 2z.$$

Note that to compute $\dfrac{\partial w}{\partial x}$, one simply differentiates w with respect to x and considers all other variables to be constants. The same comment is true for $\dfrac{\partial w}{\partial y}$ and $\dfrac{\partial w}{\partial z}$. Rather than quote theorems, let us plunge right in and illustrate the method of Lagrange multipliers by an example. After doing a few examples, we discuss why the method works.

▶ Example 1.   We will do a problem we have already seen. What is the largest box that can be produced from 150 sq in. of material?

Solution.   The volume V that we wish to maximize is equal to xyz when x, y, and z are the dimensions; more succinctly, $V(x, y, z) = xyz$. (See Figure 9.7.1.)

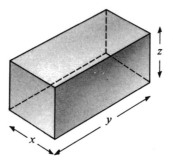

Figure 9.7.1

Since the surface area of the box is 150 square inches, we have the side condition

$$2xy + 2yz + 2xz = 150$$

or

$$2xy + 2yz + 2xz - 150 = 0.$$

Set

$$G(x, y, z) = 2xy + 2yz + 2xz - 150.$$

We now make up a new function of *four* variables as follows. Set

$$F(x, y, z, \lambda) = V(x, y, z) - \lambda G(x, y, z);$$

that is,

$$F(x, y, z, \lambda) = xyz - \lambda(2xy + 2yz + 2xz - 150).$$

Now we find $\dfrac{\partial F}{\partial x}, \dfrac{\partial F}{\partial y}, \dfrac{\partial F}{\partial z}, \dfrac{\partial F}{\partial \lambda}$; set all these terms equal to zero and solve. Thus,

(1) $\dfrac{\partial F}{\partial x} = yz - \lambda(2y + 2z) = 0,$

(2) $\dfrac{\partial F}{\partial y} = xz - \lambda(2x + 2z) = 0,$

(3) $\dfrac{\partial F}{\partial z} = xy - \lambda(2x + 2y) = 0,$

(4) $\dfrac{\partial F}{\partial \lambda} = 2xy + 2yz + 2xz - 150 = 0.$

If we set (1) = (2) and solve, we get $x = y$ or $z = 2\lambda$. If we assume $z = 2\lambda$ and substitute this in (1), we obtain $4\lambda^2 = 0$ or $\lambda = 0$. But $\lambda = 0$ implies $xy = xz = yz = 0$, and this does not satisfy (4). Thus, we conclude that $x = y$. If we set

(2) = (3), we get $y = z$ or $x = 2\lambda$. By the same argument, we conclude that $y = z$ so $x = y = z$. It is now easy to solve (4), $2xy + 2yz + 2xz - 150 = 0$, to obtain $x = y = z = 5$.

Before doing another example, we exhibit a general scheme for the Lagrange multiplier method.

▶ **Problem.** Find the maximum [minimum] of the function $w = F(x, y, z)$ subject to the side condition $G(x, y, z) = 0$.

Suggested Approach.

(1) Set $H(x, y, z, \lambda) = F(x, y, z) - \lambda G(x, y, z)$.

(2) Find $\dfrac{\partial H}{\partial x}, \dfrac{\partial H}{\partial y}, \dfrac{\partial H}{\partial z}, \dfrac{\partial H}{\partial \lambda}$.

(3) Find all points (x, y, z) that satisfy the equations

$$\frac{\partial H}{\partial x} = 0; \qquad \frac{\partial H}{\partial y} = 0; \qquad \frac{\partial H}{\partial z} = 0; \qquad \frac{\partial H}{\partial \lambda} = 0. \tag{9.7.1}$$

(4) Although the maximum [minimum] need not occur at a solution to the equations in (3), in most cases of interest it does. Moreover, one can usually decide from the problem itself at which of the solutions to (3) the maximum occurs.

▶ **Example 2.** Find the maximum of the expression x^2yz subject to the constraint $x^2 + y^2 + z^2 - 16 = 0$. (In other words, how large can the product x^2yz be if x, y, z lie on the surface of the sphere of radius 4?)

Solution. Set $F(x, y, z) = x^2yz$ and $G(x, y, z) = x^2 + y^2 + z^2 - 16$. We set

$$H(x, y, z, \lambda) = x^2yz - \lambda(x^2 + y^2 + z^2 - 16).$$

Then,

(1) $\dfrac{\partial H}{\partial x} = 2xyz - 2\lambda x = 0,$

(2) $\dfrac{\partial H}{\partial y} = x^2z - 2\lambda y = 0,$

(3) $\dfrac{\partial H}{\partial z} = x^2y - 2\lambda z = 0,$

(4) $\dfrac{\partial H}{\partial \lambda} = x^2 + y^2 + z^2 - 16 = 0.$

If we solve (1), we get $x = 0$ or $\lambda = yz$. If we multiply (2) by y and (3) by z, we get

$$x^2yz - 2\lambda y^2 = 0; \qquad x^2yz - 2\lambda z^2 = 0.$$

Subtracting, we get $2\lambda(2y^2 - x^2) = 0$. Thus $\lambda = 0$ or $y = \pm z$. If we multiply (1) by x and (2) by $2y$, we get

$$2x^2yz - 2\lambda x^2 = 0; \qquad 2x^2yz - 4\lambda y^2 = 0.$$

Subtracting, we get $2\lambda(2y^2 - x^2) = 0$. Thus $\lambda = 0$ or $x^2 = 2y^2$. The points where $x = 0$, $\lambda = xy = 0$ are clearly not maxima. Thus, we need only consider the points where $x^2 = 2y^2 = 2z^2$. In this case, by (4),

$$16 = x^2 + y^2 + z^2 = 2y^2 + y^2 + y^2 = 4y^2, \quad \text{or} \quad y = \pm 2.$$

The possible points for maxima and minima are $-(\pm 2\sqrt{2}, \pm 2, \pm 2)$. (Actually there can be no others, but this depends on a nontrivial fact.) If we evaluate $F(x, y, z) = x^2 yz$ at the eight points $(\pm 2\sqrt{2}, \pm 2, \pm 2)$, we obtain 32 at four of them and -32 at the other four. Thus the points $(+2\sqrt{2}, \pm 2, \pm 2)$, $(-2\sqrt{2}, \pm 2, \mp 2)$ correspond to maxima, and the points $(-2\sqrt{2}, \pm 2, \pm 2)$, $(+2\sqrt{2}, \pm 2, \mp 2)$ correspond to minima. [The point $(0, 0, 0)$ is a *saddle point*.]

Remark. Let us now return to the general problem stated in this section. Namely, find the minimum of the function $w = F(x, y, z)$ subject to the constraint $G(x, y, z) = 0$. Assume that $G(x, y, z) = 0$ can be solved for z as a function of (x, y). That is,

$$z = h(x, y) \tag{9.7.2}$$

throughout some neighborhood of the point $P = (a, b)$. Further we assume that this point maximizes the function

$$w = F(x, y, h(x, y)). \tag{9.7.3}$$

In addition we use the notation

$$\begin{aligned}
F_1(x, y, z) &= F_x(x, y, z), \\
F_2(x, y, z) &= F_y(x, y, z), \\
F_3(x, y, z) &= F_z(x, y, z)
\end{aligned} \tag{9.7.4}$$

and assume that $G(x, y, z) \neq 0$ in a neighborhood of $P = (a, b)$.

Recall that a necessary condition for (a, b) to maximize the function $F(x, y, h(x, y))$ is that at (a, b),

$$\frac{\partial}{\partial x} F(a, b, h(a, b)) = 0, \quad \frac{\partial}{\partial y} F(a, b, h(a, b)) = 0.$$

Now,

$$\frac{\partial}{\partial x} F(x, y, h(x, y)) = F_1 + F_3 h_1$$

and

$$\frac{\partial}{\partial y} F(x, h, h(x, y)) = F_2 + F_3 h_2.$$

We also note that the equation $G(x, y, z) = 0$ implies that

$$G_1 + G_3 h_1 = 0, \quad G_2 + G_3 h_2 = 0 \tag{9.7.5}$$

and hence

$$h_1 = -(G_1/G_3), \quad h_2 = -(G_2/G_3).$$

Thus at P,

$$F_1 - F_3(G_1/G_3) = 0 \quad \text{and} \quad F_2 - F_3(G_2/G_3) = 0$$

or equivalently

$$F_1 = G_1(F_3/G_3), \quad F_2 = G_2(F_3/G_3)$$

and
$$F_3 = G_3(F_3/G_3).$$

Let us denote the ratio F_3/G_3 by λ. In this case we see that at the point P (which maximizes F),

$$F_1 - \lambda G_1 = 0, \qquad F_2 - \lambda G_2 = 0, \quad \text{and} \quad F_3 - \lambda G_3 = 0$$

(9.7.6)

and
$$G(x, y, z) = 0.$$

Observe that this is precisely (9.7.1), where we have called $H(x, y, z, \lambda) = F(x, y, z) - \lambda G(x, y, z)$.

The foregoing provides some explanation for the validity of the method of Lagrange multipliers. The results (9.7.6) could also be made plausible by resorting to a geometrical argument. An excellent discussion can be found in Apostol.*

Exercises

1. Find the maximum value of the expression $H(x, y, z) = xyz$ if $x^2 + y^2 + z^2 = 1$. [*Hint*: The maximum must occur at a point where all the derivatives vanish. To solve the four equations, multiply $\dfrac{\partial H}{\partial x}, \dfrac{\partial H}{\partial y}$, and $\dfrac{\partial H}{\partial z}$ by x, y, and z, respectively, and add. The rest should be clear.]

2. A box is to be constructed out of three different kinds of material. The top and bottom cost 5¢ per square inch, the sides cost 2¢ per square inch, and the ends cost 1¢ per square inch. What is the largest box (volume) that can be constructed for $4.00?

3. Do Exercise 5 from Section 9.6 by the method of Lagrange multipliers.

4. Do Exercise 7 from Section 9.6 by the method of Lagrange multipliers.

5. Vacutronics, a small West Coast electronics firm, manufactures three kinds of vacuum tubes: diodes, triodes and just plain odes. Let x, y, and z be the number of odes, diodes, and triodes manufactured each day. Because of a shortage of both skilled workers and equipment, the total number of tubes produced each day is limited by the equation $x + y^2 + 2z^2 = 80$. The company sells all it produces, and its profit margin is $5 per ode, $40 per diode, and $100 per triode. How many of each kind of tube should they produce to make their profit greatest?

6. A rectangular box, open at the top, is to hold 256 cu in. Find the dimensions of the box for which the surface area is a minimum.

7. The temperature T at any point (x, y, z) in space is $T = 400xyz^2$. Find the highest temperature on the surface of the unit sphere $x^2 + y^2 + z^2 = 1$.

---

* T. Apostol, *Calculus* (Xerox, Lexington, Mass., 1969), 2nd ed., Vol. II, p. 207.

9.8 Fitting a Line to a Set of Points by the Method of Least Squares

The Bubble Company, a soap manufacturer, keeps monthly records on soap sales versus total advertising expenditures. They have long been aware that when these figures are graphed they fall roughly along a straight line. Thus if x_i represents advertising expenditures in the ith month and y_i equals sales in the ith month, then we can graph the points (x_i, y_i) thus obtained. (See Figure 9.8.1.) For planning purposes and cost analysis, the company would like to fit a straight line to the data. To do this, one first needs a definition of what it means for a line to "fit" the data. In Figure 9.8.2 we can all agree that, on intuitive grounds, line A is a bad fit and line B is a reasonably good fit.

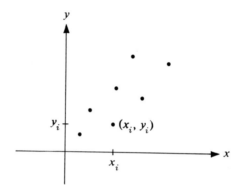

Figure 9.8.1

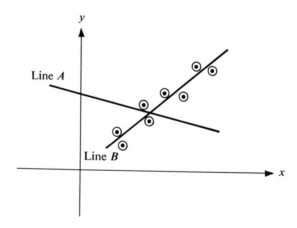

Figure 9.8.2

We now present two different criteria for judging how well a line fits a set of points and settle on one of them.

Let $(x_1, y_1), (x_2, y_2), \ldots, (x_n, y_n)$ be n points in the plane. Let $y = mx + b$ be the equation of a straight line. Set $\hat{y}_i = mx_i + b$. (Figure 9.8.3.) For the line to fit the data well, it is not unreasonable to ask that $\hat{y}_i$ be close to y_i. This suggests that we consider the sum

$$|y_1 - \hat{y}_1| + |y_2 - \hat{y}_2| + \cdots + |y_n - \hat{y}_n|$$

and try to make it small. This criterion is used in some cases, but there is a slightly different criterion that is preferable. Namely, we shall consider the sum

$$(y_1 - \hat{y}_1)^2 + (y_2 - \hat{y}_2)^2 + \cdots + (y_n - \hat{y}_n)^2$$

and try to make it small. We do not wish to argue the virtues of one criterion versus another here. There are several good reasons (among them, ease of computation) to support our choice. (Moreover, the first method can lead to bizarre results; see Exercise 5.) Finally, the line that the latter method selects to fit the data is seen to be geometrically plausible in all examples.

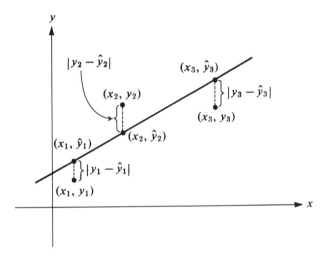

Figure 9.8.3

To restate the problem: We are given n points in the plane, $(x_1, y_1), (x_2, y_2), \ldots, (x_n, y_n)$. We wish to find a line $y = mx + b$ such that

$$(y_1 - \hat{y}_1)^2 + (y_2 - \hat{y}_2)^2 + \cdots + (y_n - \hat{y}_n)^2$$

is smallest, where $\hat{y}_i = mx_i + b$.

The line thus obtained, $y = mx + b$ is called the *regression line*, or *least-squares regression line*, and the technique is called the *method of least squares*.

Note that to find the line we simply have to determine the constants m and b. Set

$$F(b, m) = (y_1 - \hat{y}_1)^2 + \cdots + (y_n - \hat{y}_n)^2.$$

Since $y_i = mx_i + b$, we can rewrite the equation as

$$F(b, m) = (mx_1 + b - y_1)^2 + \cdots + (mx_n + b - y_n)^2.$$

(Think of b and m as being variables for the moment.) We wish to minimize $F(b, m)$ as a function of b and m. To this end we find the partial derivatives of $F(b, m)$ and set them equal to zero.

(1) $\quad \dfrac{\partial F}{\partial b} = 2(mx_1 + b - y_1) + \cdots + 2(mx_n + b - y_n) = 0,$

(2) $\quad \dfrac{\partial F}{\partial m} = 2(mx_1 + b - y_1)x_1 + \cdots + 2(mx_n + b - y_n)x_n = 0.$

These equations may be rewritten as

(1') $(x_1 + \cdots + x_n)m + nb = (y_1 + \cdots + y_n)$

(2') $(x_1^2 + \cdots + x_n^2)m + (x_1 + \cdots + x_n)b = (x_1 y_1 + \cdots + x_n y_n)$

by simple algebra.

Since we know all the x_i's and y_i's, i.e., everything except m and b, (1') and (2') have the form

$$a_1 m + a_2 b = a_3; \qquad c_1 m + c_2 b = c_3.$$

Thus we have two equations in two unknowns and it is easy to solve for m and b. What is more, Equations (1') and (2') will *always* have a unique solution for m and b unless all the x_i's are equal. (In that case we should obviously choose the vertical line through the x_i's as our regression line.)

Since the domain of $F(b, m)$ is the entire xy plane and $F(b, m)$ is large for b or m large, the minimum of $F(b, m)$ must occur at the point where $\dfrac{\partial F}{\partial b} = \dfrac{\partial F}{\partial m} = 0$.

The foregoing may appear complicated, but note that we have arrived at a general formula.

RULE. Let $(x_1, y_1), \ldots, (x_n, y_n)$ be given. Then the least squares regression line has the equation $y = mx + b$. The constants m and b are the unique solutions to the equations

(1) $(x_1 + \cdots + x_n)m + nb = (y_1 + \cdots + y_n)$,

(2) $(x_1^2 + \cdots + x_n^2)m + (x_1 + \cdots + x_n)b = (x_1 y_1 + \cdots + x_n y_n)$.

▶ Example. Find the regression line to the points $(0, 1)$, $(1, 1)$, $(2, 2)$, $(3, 4)$.

Solution. It is convenient to do the arithmetic first, and for this it is handy to have the following table. (Note that $n = 4$.)

x	0	1	2	3
y	1	1	2	4

Then,

$$x_1 + \cdots + x_4 = 6; \qquad y_1 + \cdots + y_4 = 8.$$
$$x_1^2 + \cdots + x_4^2 = 14; \qquad x_1 y_1 + \cdots + x_4 y_4 = 17.$$

Our equations for m and b are

$$6m + 4b = 8 \quad \text{and} \quad 14m + 6b = 17.$$

Solving for m and b, we find that $m = 1$ and $b = \frac{1}{2}$. Thus the regression line has the equation

$$y = x + \tfrac{1}{2}.$$

As a check on our computation and to satisfy our curiosity we graph the points and the line in Figure 9.8.4.

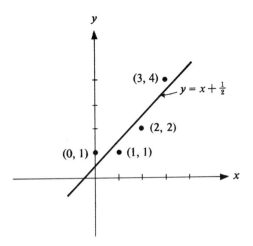

Figure 9.8.4

Exercises

1. Find the regression line for the following points:

x	1	3	5	7	9
y	1	4	4	6	10

    Sketch your results.

2. Find the regression line for the following points:

x	−3	−2	−1	0	1	2
y	4	2	1	−1	−2	−3

3. Find the regression line for the following points:

x	2	−1	2	−2	3
y	−1	3	0	5	−1

4. A manufacturer keeps a record of sales per month and advertising costs. For the first five months of 1975, he finds they are running as follows:

Advertising expenditures	1	2	3	4	5
Sales	2	4	7	9	11

    Find the regression line for these data. On the basis of the regression line, predict his sales if advertising expenditures are set at 6 and 10 units per month.

*5. Assume we are given the following four points: (1, 3), (1, 1), (5, 3), and (5, 1). Let $y = mx + b$, and set $y_1 = m \cdot 1 + b$ and $y_2 = m \cdot 5 + b$. Draw a picture of the four points. Convince yourself that

$$|y - 1| + |y_1 - 3| + |y_2 - 1| + |y_2 - 3|$$

is constant, as long as the line $y = mx + b$ lies below (1, 3) and (5, 3) and above (1, 1) and (5, 1). Convince yourself that the minimum value of

$$|y_1 - 1| + |y_1 - 3| + |y_2 - 1| + |y_2 - 3|$$

is 4 over all possible choices of m and b. (This is most easily done by considering a picture.) What does all this imply about fitting a line to the four points given above when the test of a good fit is the smallness of

$$[|y_1 - \hat{y}_1| + |y_2 - \hat{y}_2| + |y_3 - \hat{y}_3| + |y_4 - \hat{y}_4|]?$$

9.9 Chapter 9 Summary

In this chapter we introduced functions of the form $z = f(x, y)$, that is, functions of two variables, and developed techniques for finding the absolute maxima and minima of such functions. As a first step we defined the partial derivatives

$$\frac{\partial f}{\partial x}\bigg|_{(a, b)} = \lim_{x \to a} \frac{f(x, b) - f(a, b)}{x - a}$$

and

$$\frac{\partial f}{\partial y}\bigg|_{(a, b)} = \lim_{x \to a} \frac{f(a, y) - f(a, b)}{y - b}.$$

Suppose we let $z = f(x, y)$ be a function with domain D and let (a, b) be an interior point of D. If f has a *relative* maximum or *relative* minimum at (a, b) and if $\frac{\partial f}{\partial x}\bigg|_{(a, b)}$ and $\frac{\partial f}{\partial y}\bigg|_{(a, b)}$ exist, then

$$\frac{\partial f}{\partial x}\bigg|_{(a, b)} = \frac{\partial f}{\partial y}\bigg|_{(a, b)} = 0.$$

This means that as long as the partial derivatives exist, the relative maxima and minima are found among those points such that $\frac{\partial f}{\partial x} = \frac{\partial f}{\partial y} = 0$. These points are called *critical points*.

If $z = f(x, y)$ is defined on a bounded domain D with a boundary and either f is continuous on D or $\left|\frac{\partial f}{\partial x}\right| \leq M$ and $\left|\frac{\partial f}{\partial y}\right| \leq M$ at all points of D for some constant M, then f has both an absolute maximum and an absolute minimum on D. If $\frac{\partial f}{\partial x}$ and $\frac{\partial f}{\partial y}$ exist at every point of D, then the absolute maximum [minimum] occurs either on the boundary of D or at a point where both $\frac{\partial f}{\partial x}$ and $\frac{\partial f}{\partial y}$ equal zero. It should be clear that every absolute maximum or minimum point is also a relative maximum or minimum point.

If $z = f(x, y)$ is defined on a domain without a boundary, then f may not have an absolute maximum or minimum. If $\frac{\partial f}{\partial x}$ and $\frac{\partial f}{\partial y}$ exist at all points of D, then an absolute maximum [minimum] must occur at a point where both $\frac{\partial f}{\partial x}$ and $\frac{\partial f}{\partial y}$ equal zero.

The method of Lagrange multipliers is particularly helpful if we wish to maximize or minimize a quantity w when $w = F(x, y, z)$ and the variables x, y, and z must satisfy a side condition $G(x, y, z) = 0$. Thus we wish to find the point (a, b, c) in the domain of F where $w = F(x, y, z)$ is the largest [smallest] and $G(a, b, c) = 0$ as well. To do this, we set up the auxiliary equation

$$F(x, y, z) - \lambda G(x, y, z) = 0$$

and find $\dfrac{\partial}{\partial x}, \dfrac{\partial}{\partial y}, \dfrac{\partial}{\partial z}$, and $\dfrac{\partial}{\partial \lambda}$. This yields four equations

(1) $\quad \dfrac{\partial F}{\partial x} - \lambda \dfrac{\partial G}{\partial x} = 0,$

(2) $\quad \dfrac{\partial F}{\partial y} - \lambda \dfrac{\partial G}{\partial y} = 0,$

(3) $\quad \dfrac{\partial F}{\partial z} - \lambda \dfrac{\partial G}{\partial y} = 0,$

(4) $\quad G(x, y, z) = 0,$

in the four unknowns x, y, z, and λ. We now solve (1), (2), (3), and (4) simultaneously. The point (a, b, c) we are looking for will usually be found among the set of solutions to (1), (2), (3), (4). The decision as to which solution is the right one must be made on the basis of information particular to each problem. Usually this is not difficult.

Review Exercises

1. Find $\dfrac{\partial f}{\partial x}$ and $\dfrac{\partial f}{\partial y}$ for the following functions. The domain of the function is the xy plane unless otherwise stated.

(a) $\quad z = 2x^3 y + e^{xy}.$

(b) $\quad z = \log \dfrac{1}{\sqrt{1 + xy}}.$

(c) $\quad z = \log (x^2 y + e^y).$

(d) $\quad z = \dfrac{xy}{x + y}$ for $x \neq -y.$

(e) $\quad z = \dfrac{(1 + x^2)^2}{\sqrt{x^2 + y^2}}, |x| + |y| > 0.$

(f) $\quad z = e^{[1/(x+y)]-(x/y)}$ for $y \neq 0.$

(g) $\quad z = xy^2 + y^2 x.$

(h) $\quad z = [1 - (x^2 + y^2)]^{1/2}, x^2 + y^2 \leq 1.$

(i) $\quad z = e^x + \log y.$

2. Let $z = \sqrt{1 - (x^2 + y^2)}$ when the domain is the disk $x^2 + y^2 \leq 1$. Find the absolute maximum and minimum of the function.

3. Let $z = x(1 - x) + y^2(1 - y)$. Find the absolute maximum if the domain is the square $\{(x, y) \mid 0 \leq |x| \leq 1 \text{ and } 0 \leq |y| \leq 1\}.$

4. Let $z = 1 + x + 2y$. Find the absolute maximum and absolute minimum if the domain is the triangle $\{(x, y) \mid 0 \leq x, 0 \leq y, \text{ and } 2x + y \leq 2\}.$

5. Let $z = x^2 + y^2$, when the domain is the square $\{(x, y) \mid |x| \leq 1 \text{ and } |y| \leq 1\}$. Find the absolute maximum and minimum.

6. Find the distance from the point $(1, 1, 1)$ to the plane $x - 2y + 3z = 10$; that is, minimize the quantity $(1 - x)^2 + (1 - y)^2 + (1 - z)^2$ when $x - 2y + 3z = 10$.

7. The Schlock China Company manufactures two kinds of china, economy and prestige. Let x be the price of the economy line per set and y be the price of the prestige line per set. Then $E(x, y) = 100 - 4x + y$ is the number of economy sets sold as a function of price and $P(x, y) = 80 - y + x$ is the number of prestige sets sold as a function of price. Let $G(x, y)$ be the total gross sales in dollars as a function of the price. Then,

$$G(x, y) = xE(x, y) + yP(x, y).$$

If the production capacity of the company is for all intents unlimited, and if the sole interest of the company is to maximize gross sales, how should they price each line? [*Note:* $G(x, y)$ can be rewritten as $100x + 80y - (y - 2x)^2 - 2xy$ by means of a little algebra. Thus one can conclude that for either x or y (or both) very large, the gross sales will be small, in fact negative. Hence on heuristic grounds one may conclude that an absolute maximum does exist.]

*8. A trough is to be constructed out of wood in the shape indicated in Figure 9.9.1.

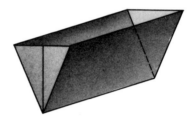

Figure 9.9.1

It is to have ends and a bottom but no top. There are 60 sq ft of lumber available. What is the maximum volume of the trough? [*Hint:* Introduce variables as indicated in Figure 9.9.2. Then the volume $V(x, y, z) = xyz$ and the

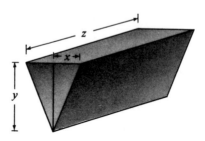

Figure 9.9.2

surface area

$$S(x, y, z) = 2xy + 2\sqrt{x^2 + y^2} \cdot z.$$

Justify these equations. Thus we wish to maximize the quantity $V(x, y, z) = xyz$ subject to the side condition $2xy + 2\sqrt{x^2 + y^2} \cdot z - 60 = 0$. Complete the problem.]

9. The Super Six Motor Company produces three models of their automobile: the Atlas, the Apollo, and the Hercules. Total production is limited in that $2x + y + 5z = 103$ (when x is the number of Atlases per day, y is the number of Apollos per day, and z is the number of Hercules's per day.) The total profit $P(x, y, z)$ is given by the equation

$$P(x, y, z) = x^2 - xy + y^2 + z^2.$$

How many cars per day in each of the model lines should Super Six produce to make their profit largest?

10. Find the least-squares regression line for the data

x	1	3	4	6	9	10
y	-3	2	3	7	12	14

Graph your answer.

surface area

$$S(x, y, z) = 2xy + 2\sqrt{x^2 + y^2} \cdot z.$$

Justify these equations. Thus we wish to maximize the quantity $V(x, y, z) = xyz$ subject to the side condition $2xy + 2\sqrt{x^2 + y^2} \cdot z - 60 = 0$. Complete the problem.]

9. The Super Six Motor Company produces three models of their automobile: the Atlas, the Apollo, and the Hercules. Total production is limited in that $2x + y + 5z = 103$ (when x is the number of Atlases per day, y is the number of Apollos per day, and z is the number of Hercules's per day.) The total profit $P(x, y, z)$ is given by the equation

$$P(x, y, z) = x^2 - xy + y^2 + z^2.$$

How many cars per day in each of the model lines should Super Six produce to make their profit largest?

10. Find the least-squares regression line for the data

x	1	3	4	6	9	10
y	-3	2	3	7	12	14

Graph your answer.

Appendix

TABLE A1. Exponential Functions

x	e^x	e^{-x}	x	e^x	e^{-x}
0.00	1.0000	1.0000	2.3	9.9742	0.1003
0.05	1.0513	0.9512	2.4	11.023	0.0907
0.10	1.1052	0.9048	2.5	12.182	0.0821
0.15	1.1618	0.8607	2.6	13.464	0.0743
0.20	1.2214	0.8187	2.7	14.880	0.0672
0.25	1.2840	0.7788	2.8	16.445	0.0608
0.30	1.3499	0.7408	2.9	18.174	0.0550
0.35	1.4191	0.7047	3.0	20.086	0.0498
0.40	1.4918	0.6703	3.1	22.198	0.0450
0.45	1.5683	0.6376	3.2	24.533	0.0408
0.50	1.6487	0.6065	3.3	27.113	0.0369
0.55	1.7333	0.5769	3.4	29.964	0.0334
0.60	1.8221	0.5488	3.5	33.115	0.0302
0.65	1.9155	0.5220	3.6	36.598	0.0273
0.70	2.0138	0.4966	3.7	40.447	0.0247
0.75	2.1170	0.4724	3.8	44.701	0.0224
0.80	2.2255	0.4493	3.9	49.402	0.0202
0.85	2.3396	0.4274	4.0	54.598	0.0183
0.90	2.4596	0.4066	4.1	60.340	0.0166
0.95	2.5857	0.3867	4.2	66.686	0.0150
1.0	2.7183	0.3679	4.3	73.700	0.0136
1.1	3.0042	0.3329	4.4	81.451	0.0123
1.2	3.3201	0.3012	4.5	90.017	0.0111
1.3	3.6693	0.2725	4.6	99.484	0.0101
1.4	4.0552	0.2466	4.7	109.95	0.0091
1.5	4.4817	0.2231	4.8	121.51	0.0082
1.6	4.9530	0.2019	4.9	134.29	0.0074
1.7	5.4739	0.1827	5	148.41	0.0067
1.8	6.0496	0.1653	6	403.43	0.0025
1.9	6.6859	0.1496	7	1096.6	0.0009
2.0	7.3891	0.1353	8	2981.0	0.0003
2.1	8.1662	0.1225	9	8103.1	0.0001
2.2	9.0250	0.1108	10	22026	0.00005

TABLE A2.   Natural Logarithm Function

(It is possible to find the logarithm of numbers not included in the table. To find the log of 0.04, for example, write

$$\log (0.04) = \log \frac{0.4}{10} = \log 0.4 - \log 10$$

$$= -0.9162 - 2.3026$$

$$= -3.2188.$$

In similar manner,

$$\log 100 = \log 10^2 = 2 \log 10.)$$

x	$\log x$	x	$\log x$	x	$\log x$
0.0	. . . . . .	3.5	1.2528	7.0	1.9459
0.1	-2.303	3.6	1.2809	7.1	1.9601
0.2	-1.609	3.7	1.3083	7.2	1.9741
0.3	-1.204	3.8	1.3350	7.3	1.9879
0.4	-0.916	3.9	1.3610	7.4	2.0015
0.5	-0.693	4.0	1.3863	7.5	2.0149
0.6	-0.511	4.1	1.4110	7.6	2.0281
0.7	-0.357	4.2	1.4351	7.7	2.0412
0.8	-0.223	4.3	1.4586	7.8	2.0541
0.9	-0.105	4.4	1.4816	7.9	2.0669
1.0	0.0000	4.5	1.5041	8.0	2.0794
1.1	0.0953	4.6	1.5261	8.1	2.0919
1.2	0.1823	4.7	1.5476	8.2	2.1041
1.3	0.2624	4.8	1.5686	8.3	2.1163
1.4	0.3365	4.9	1.5892	8.4	2.1282
1.5	0.4055	5.0	1.6094	8.5	2.1401
1.6	0.4700	5.1	1.6292	8.6	2.1518
1.7	0.5306	5.2	1.6487	8.7	2.1633
1.8	0.5878	5.3	1.6677	8.8	2.1748
1.9	0.6419	5.4	1.6864	8.9	2.1861
2.0	0.6913	5.5	1.7047	9.0	2.1972
2.1	0.7419	5.6	1.7228	9.1	2.2083
2.2	0.7885	5.7	1.7405	9.2	2.2192
2.3	0.8329	5.8	1.7579	9.3	2.2300
2.4	0.8755	5.9	1.7750	9.4	2.2407
2.5	0.9163	6.0	1.7918	9.5	2.2513
2.6	0.9555	6.1	1.8083	9.6	2.2618
2.7	0.9933	6.2	1.8245	9.7	2.2721
2.8	1.0296	6.3	1.8406	9.8	2.2824
2.9	1.0647	6.4	1.8563	9.9	2.2925
3.0	1.0986	6.5	1.8718	10.0	2.3026
3.1	1.1314	6.6	1.8871		
3.2	1.1632	6.7	1.9021		
3.3	1.1939	6.8	1.9169		
3.4	1.2238	6.9	1.9315		

TABLE A3.   Integrals (In all entries, C is an arbitrary constant.)

(1) $\displaystyle\int x^r \, dx = \frac{x^{r+1}}{r+1} + C \qquad (r \neq -1)$

(2) $\displaystyle\int \cos x \, dx = \sin x + C$

(3) $\displaystyle\int \sin x \, dx = -\cos x + C$

(4) $\displaystyle\int \sec^2 x \, dx = \tan x + C$

(5) $\displaystyle\int \csc^2 x \, dx = -\cot x + C$

(6) $\displaystyle\int \sec x \tan x \, dx = \sec x + C$

(7) $\displaystyle\int \csc x \cot x \, dx = -\csc x + C$

(8) $\displaystyle\int e^x \, dx = e^x + C$

(9) $\displaystyle\int \frac{dx}{x} = \log |x| + C$

(10) $\displaystyle\int a^x \, dx = \frac{a^x}{\log a} + C$

(11) $\displaystyle\int \log |x| \, dx = x(\log |x| - 1) + C$

(12) $\displaystyle\int \frac{dx}{\sqrt{a^2 - x^2}} = \text{arc sin} \left(\frac{x}{a}\right) + C$

(13) $\displaystyle\int \frac{dx}{a^2 + x^2} = \left(\frac{1}{a}\right) \text{arc tan} \left(\frac{x}{a}\right) + C$

(14) $\displaystyle\int \frac{dx}{\sqrt{x^2 + a^2}} = \log (x + \sqrt{x^2 + a^2}) + C$

(15) $\displaystyle\int \frac{dx}{\sqrt{x^2 - a^2}} = \log (x + \sqrt{x^2 - a^2}) + C$

(16) $\displaystyle\int \frac{dx}{a^2 - x^2} = \frac{1}{2a} \log \left(\frac{a + x}{a - x}\right) + C \qquad (x^2 < a^2)$

(17) $\displaystyle\int \frac{dx}{x^2 - a^2} = \frac{1}{2a} \log \left(\frac{x - a}{x + a}\right) + C \qquad (x^2 > a^2)$

(18) $\displaystyle\int \frac{dx}{x\sqrt{a^2 - x^2}} = -\frac{1}{a} \log \left(\frac{a + \sqrt{a^2 - x^2}}{x}\right) + C \qquad (0 < x < a)$

(19) $\displaystyle\int \frac{dx}{x\sqrt{a^2 + x^2}} = -\frac{1}{a}\log\left(\frac{a + \sqrt{a^2 + x^2}}{|x|}\right) + C$

(20) $\displaystyle\int \frac{x\,dx}{ax + b} = \frac{x}{a} - \frac{b}{a^2}\log|ax + b| + C$

(21) $\displaystyle\int \frac{x\,dx}{(ax + b)^2} = \frac{b}{a^2(ax + b)} + \frac{1}{a^2}\log|ax + b| + C$

(22) $\displaystyle\int \frac{dx}{x(ax + b)} = \frac{1}{b}\log\left|\frac{x}{ax + b}\right| + C$

(23) $\displaystyle\int \frac{dx}{x(ax + b)^2} = \frac{1}{b(ax + b)} + \frac{1}{b^2}\log\left|\frac{x}{ax + b}\right| + C$

(24) $\displaystyle\int \sqrt{a^2 - x^2}\,dx = \frac{x}{2}\sqrt{a^2 - x^2} + \frac{a^2}{2}\arcsin\left(\frac{x}{a}\right) + C$

(25) $\displaystyle\int \sqrt{x^2 + a^2}\,dx = \frac{x}{2}\sqrt{x^2 + a^2} \pm \frac{a^2}{2}\log|x + \sqrt{x^2 + a^2}| + C$

(26) $\displaystyle\int x^2\sqrt{a^2 - x^2}\,dx = -\frac{1}{4}x(a^2 - x^2)^{3/2} + \frac{a^2 x}{8}\sqrt{a^2 - x^2} + \frac{a^4}{8}\arcsin\left(\frac{x}{a}\right) + C$

(27) $\displaystyle\int \sec x\,dx = \log|\sec x + \tan x| + C$

(28) $\displaystyle\int \csc x\,dx = \log|\csc x - \cot x| + C$

(29) $\displaystyle\int e^{ax}\sin bx\,dx = \frac{e^{ax}(a\sin bx - b\cos bx)}{a^2 + b^2} + C$

(30) $\displaystyle\int e^{ax}\cos bx\,dx = \frac{e^{ax}(b\sin bx + a\cos bx)}{a^2 + b^2} + C$

(31) $\displaystyle\int x^n\log x\,dx = x^{n+1}\left[\frac{\log x}{n + 1} - \frac{1}{(n + 1)^2}\right] + C$

(32) $\displaystyle\int \sin^n x\,dx = \frac{-\sin^{n-1} x\cos x}{n} + \frac{n - 1}{n}\int \sin^{n-2} x\,dx + C$

(33) $\displaystyle\int \cos^n x\,dx = \frac{\cos^{n-1} x\sin x}{n} + \frac{n - 1}{n}\int \cos^{n-2} x\,dx + C$

(34) $\displaystyle\int x^n e^{ax}\,dx = \frac{x^n e^{ax}}{a} - \frac{n}{a}\int x^{n-1}e^{ax}\,dx + C$

(35) $\displaystyle\int \frac{dx}{ax^2 + bx + c} = \frac{2}{\sqrt{4ac - b^2}}\arctan\left(\frac{2ax + b}{\sqrt{4ac - b^2}}\right) + C \qquad (b^2 < 4ac)$

(36) $\displaystyle\int \frac{x\,dx}{ax^2 + bx + c} = \frac{1}{2a}\log|ax^2 + bx + c|$

$\displaystyle\qquad\qquad\qquad - \frac{b}{a\sqrt{4ac - b^2}}\arctan\left(\frac{2ax + b}{\sqrt{4ac - b^2}}\right) + C \qquad (b^2 < 4ac)$

Answers to Selected Exercises

CHAPTER 1   **Section 1.1**

1. (a) Rule 5.   (b) Rule 5.   (c) False, $-x < -y$ by Rule 6.   (d) Rule 5.
(e) Rule 7 and then Rule 5.   (f) Rule 7 and then Rule 5.
(g) Rule 6.

3. (a) $[-\frac{1}{2}, \frac{1}{2}]$.   (b) $[-8, 2]$.   (c) $(-\infty, -2) \cup (6, \infty)$.   (d) $(-\infty, 2]$.
(e) $[\frac{3}{2}, \frac{5}{2}]$.   (f) $[-5, 2]$.

4. (a) $(-\infty, -3] \cup [0, \infty)$.   (c) $[-3, 0]$.   (e) $[-\frac{1}{2}, 0]$.   (g) $(0, 1]$.

6. 5 commercials.   **8.** 87.   **11.** (a) Let $a = 1, b = 2, c = -3, d = -2$.

Section 1.2

2. (a) $\sqrt{52}$.   (c) $\sqrt{53}$.   (e) $\sqrt{410}$.   (g) $\sqrt{2}$.

3. (c)

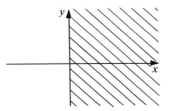

3. (e)

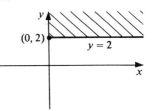

4. (a) $y/x < 0$ when either (1) $y > 0$ and $x < 0$ or (2) $y < 0$ and $x > 0$. Thus, $y/x < 0$ in quadrants II and IV.
(b) $y/x \geq 0$ when either (1) $y \geq 0$ and $x > 0$ or (2) $y \leq 0$ and $x < 0$. Thus, $y/x \geq 0$ in quadrants I and III (including the x axis).

6. By truck, cost is $1,800. By train, cost is $\simeq$$2,000. Cheaper to ship via San Francisco.

7. 54 miles.   **8b.** $x^2 + y^2 = 25$; $(2, 5)$ and $(-1, 4)$ do not lie on circle.

Section 1.3

1. (a) $m = \dfrac{6 - (-1)}{2 - 1} = 7.$ (c) $m = \dfrac{-1 - 0}{\frac{1}{9} - 0} = -9.$ (e) $m = \dfrac{4 + 1}{2 - 3} = -5.$

    (g) $m = \dfrac{6 - 6}{-1 - \frac{7}{3}} = 0.$ (i) $m = \dfrac{0 - 2}{1 - 0} = -2.$

2. In all cases we use the form $y = mx + b$, where m is given in Exercise 1 and b is determined from the fact that the line is to go through the given point.

    (a) $y = 7x + b.$ Since the line goes through $(1, -1)$, $b = -8$ and the desired equation is $y = 7x - 8.$ (c) $y = -9x.$ (e) $y = -5x + 14.$ (g) $y = 6.$
    (i) $y = -2x + 2.$

4. As in Exercise 2, we use the form $y = mx + b.$ (a) $y = 5x + 12.$
    (c) $y = -8x + 26.$ (e) $y = 0.$ (g) $y = \frac{2}{3}x - 5.$ (i) $y = -2x + 17.$

6. \$186,000.

8. (a) $y = 10,000 - 500x.$ (b) $y = \$8,500.$ (c) $y = d - \dfrac{(d - S)}{T} x.$

10. The slope of the new line must be $m = -2.$ Since the line goes through $(7, 11)$, $b = 25$, and the equation is $y = -2x + 25.$

11. (a) $m = -\frac{1}{3}.$ (c) $m = -2.$ (e) $m = +\frac{9}{4}.$

12. (a) $x - y = 1,$ $2x + 4y = 3$ implies $4x - 4y = 4,$ $2x + 4y = 3.$
    Adding, $6x = 7; x = \frac{7}{6}, y = \frac{1}{6}.$
    (c) $9x + 3y = 7,$ $x + 3y = 6$ implies $8x = 1,$ $x = \frac{1}{8},$ $y = \frac{47}{24}.$
    (e) $x - y = 1,$ $\frac{1}{3}x + \frac{2}{3}y = 0$ implies $x = \frac{6}{11},$ $y = -\frac{5}{11}.$

14. $y = -4x - 21.$ 16. $\frac{3}{5}\sqrt{5}.$

Section 1.4

1. (a) The rule is a function. The domain is the set of 300 students and the range a subset of the numbers 0–100.
    (c) Not a function, since one author in A may have written many books in B.
    (e) (i) The rule is a function. Domain is the set of mothers, range the set of eldest children.   (ii) Not a function, as one mother could have many children.
    (g) A function. Domain is the set of people, range is the alphabet.

3. (b) $f(\frac{2}{3}) = \frac{2}{3}; f(2) = 2; f(-\frac{1}{2}) = \frac{1}{2}; f(-\frac{2}{3}) = \frac{2}{3}; f(-\frac{1}{2}) = \frac{1}{2}; f(\frac{1}{3}) = \frac{1}{3}; f(-2) = 1.$

4. (a) The domain of f is the set of real numbers.
    (b) (i) $f(2) = 3.$ (ii) $f(-10) = \frac{9}{11}.$ (iii) $f(a + 2) = (a + 3)/(a + 1),$ $a \neq -1, f(a + 2) = 1$ if $a = -1.$ (iv) $f(a^2) = (a^2 + 1)/(a^2 - 1),$ for $a \neq \pm 1, f(a^2) = 1,$ if $a = \pm 1.$ (v) $f(1/a) = (1 + a)/(1 - a)$ for $a \neq 1,$ $f(1/a) = 1,$ if $a = 1.$

6. $f(x) = x^3.$ $f(\frac{1}{2}x) = \frac{1}{8}x^3; f(x + 1) = (x + 1)^3; f(2x) = 8x^3; f(x + \frac{1}{3}) = (x + \frac{1}{3})^3.$

8. This is not $h(x) = 1$ since there is a point missing.

10. (a)

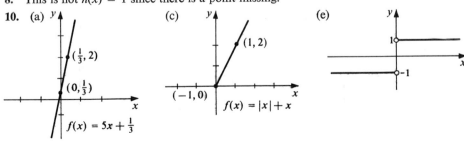

    (a) $(\frac{1}{3}, 2)$, $(0, \frac{1}{3})$, $f(x) = 5x + \frac{1}{3}$
    (c) $(1, 2)$, $(-1, 0)$, $f(x) = |x| + x$
    (e) 1, -1

12. No, in a circle each value of the domain has two corresponding range values. No, a vertical line has infinitely many range values for the one domain value.

14. (a)
$$f(n) = \begin{cases} n & \text{for} & 1 \le n \le 10{,}000 \\ 10{,}000 + 1.5(n - 10{,}000) & \text{for} & 10{,}000 < n \le 20{,}000 \\ 25{,}000 + 2(n - 20{,}000) & \text{for} & 20{,}000 < n. \end{cases}$$

This may be simplified as
$$f(n) = \begin{cases} n & \text{for} & 1 \le n \le 10{,}000 \\ 1.5n - 5{,}000 & \text{for} & 10{,}000 < n \le 20{,}000 \\ 2n - 15{,}000 & \text{for} & 20{,}000 < n. \end{cases}$$

$f(15{,}000) = \$17{,}500; f(20{,}000) = \$25{,}000;$
$f(25{,}000) = \$35{,}000; f(60{,}000) = \$105{,}000.$

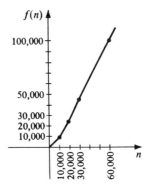

16. (b) For given $s(p) = 30 - 60/(p - 2)$, $s(4) = 0$, $s(6) = 15$, $s(8) = 20$, $s(10) = 22\frac{1}{2}$, $s(17) = 26$. The function gives a good approximation. (c) $s(62) = 29$, $s(122) = 29\frac{1}{2}$. Note that the supply "levels off." As the price increases, the supply increases very little.

Section 1.5

1. (a)(i)   (a)(iv)

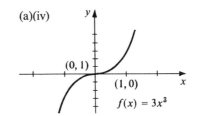

3. The equation $S(t)$ does not describe the given problem. For $t > \frac{5}{2}$, the function is decreasing. It should be increasing. For $t = 10$, given any I, $S(t) = 0$. This is not meaningful.

5. 15 empty seats.   **6.** When there are 40 empty seats, the agency will receive \$3,000.

Section 1.6

1. (a) $f^{-1}(y) = y - 2$.   (c) $f^{-1}(y) = 1/(y - 1)$.   (d) No inverse exists.
(e) $f^{-1}(y) = (\frac{1}{2}y)^{1/2}$.   (f) $f^{-1}(y) = \frac{1}{3}y + \frac{2}{3}$.   (i) $f^{-1}(y) = \frac{1}{5}y - \frac{1}{15}$.

2. (a) Since no two people can have the *same* social security number, the correspondence is one-to-one and f^{-1} exists. f^{-1} is the function that assigns to each *social security number* (considered as a set of a digits) a person.   (b) In this case more than one person in the class may have the same grade. Hence, no inverse function can exist.

Section 1.7

1. (a)

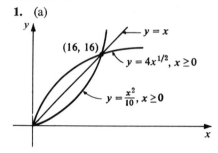

(b)

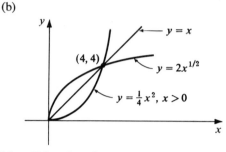

(c) $f(x) = 3x^{1/3}$

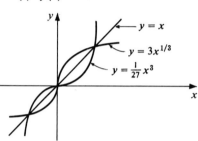

(g) $f(x) = \log_3 2x$

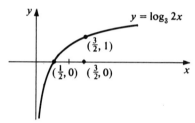

2. (a) $2^2 \cdot 2^3 = 2^5$.   (c) $5^3 \cdot 5^{-4} = 5^{-1}$.   (e) $(2^2)^5 = 2^{10}$.   (f) $6^2 = 2^2 \cdot 3^2$.
    (g) $10^{15} \cdot 10^9 = 10^{24}$.   (i) $5^4/5^3 = 5^1 = 5$.   (j) $10^{-5} \cdot 10^{29} = 10^{24}$.
    (l) $(4^{1/2})^3 = (4^3)^{1/2} = 4^{3/2} = 8$.

3. (a) $x = 2$.   (c) $x = 5$.   (e) $x = 4$.   (g) $x = 2$.   (i) $? = 30$.

4. (a) $x = 343$.   (c) $x = 0.02$.

6. (a) $\log_3 \frac{1}{243} = -5$.   (c) $\log_{1/2} 4 = -2$.   (e) $(0.5)^{-4} = 16$.
    (g) $\log_8 2 = \frac{1}{3}$.

9. (a) $(3^3)^3 = 3^9$ while $3^{(3^3)} = 3^{27}$; hence $3^{3^3} > (3^3)^3$.   (b) $2^{(2^{2^2})}$; *relative sizes:*
$2^{(2^{2^2})} > 2^{222} > 22^{22} > 2^a$ (where $a = [2^{(2^2)}]$) > 2222.

11. $k = 1, b = 7.5, y = 7.5x$.

12. $a \simeq 0.018$ when $P = 800{,}000$, $k \simeq e^{-68}$, $x \simeq 2050$.

13. (i) If $f(x) = a^x$, $f(x + h) - f(x) = a^{x+h} - a^x = a^x(a^h - 1)$.   (ii) If
$f(x) = a^x$, $f(x - 1)f(1) = a^{x-1} \cdot a = a^x = f(x)$.   (iii) If $f(x) = a^x$,
$f(x + h)/f(h) = a^{x+h}/a^h = a^x = f(x)$.

15. (a) If $f(x) = \log_a x$, then $f(x + h) - f(x) = \log_a (x + h) - \log_a x =$
$\log_a [(x + h)/x] = \log_a (1 + h/x)$.   (c) If $f(x) = \log_a x$, then $f(pq) - f(p/q) =$
$\log_a pq - \log_a p/q = \log_a p + \log_a q - \log_a p + \log_a q = 2 \log_a q = 2f(q)$.

Section 1.8

1. (a) $[f + g](x) = 5x + 1$.
      The domain is all real numbers.

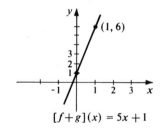

$[f+g](x) = 5x + 1$

(c) $[f \circ g](x) = 4x + 4$. The domain is all real numbers.

3. (a) Yes. $[f + g](x) = f(x) + g(x) = g(x) + f(x) = [g + f](x)$.
   (b) Yes. $[f \cdot g](x) = f(x) \cdot g(x) = g(x) \cdot f(x) = [g \cdot f](x)$.   (c) No (cf. Exercise 2).

4. $[g \circ f](x) = 1 + 3^x$. No. $1 + 3^x \neq 3^{1+x} = 3 \cdot 3^x$.

5. (a) $[g \circ f](x) = e^{\log x} = x$.   (b) $[f \circ g](x) = \log (e^x) = x$.

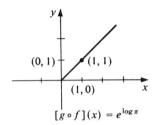

$[g \circ f](x) = e^{\log x}$

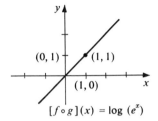

$[f \circ g](x) = \log (e^x)$

6. (a) $h(x) = g[f(x)]$, where $f(x) = x^2 + 7$ and $g(x) = \log x$.
   (c) $h(x) = g[f(x)]$, where $f(x) = x + e^{-x} + 1$ and $g(x) = \log x$.
   (e) $h(x) = g[f(x)]$, where $f(x) = [\log (x^2 + 1)]/(x^4 + 1)$ and $g(x) = x^{1/3}$.
   (g) $h(x) = g[f(x)]$, where $f(x) = 3x$ and $g(x) = 2^x$.
   (i) $h(x) = g[f(x)]$, where $f(x) = \log x$ and $g(x) = \log x$.
   (k) $h(x) = g[f(x)]$, where $f(x) = \sqrt{1 + x}$ and $g(x) = e^x$.
   (m) $h(x) = g[f(x)]$, where $f(x) = x^2 + 2$, and $g(x) = 2^x$.

7. (a)   (b) $f(2) = -2, f(\tfrac{1}{2}) = 2$.

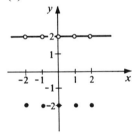

9. (a) By moving the graph of f up 7 units.   (b) Yes. Leave the graph of f fixed
and move the coordinate axis over 3 spaces to the right. $h(x) = [f \circ g]$, where
$g(x) = x - 3$.

12. (a) $g(f(x)) = | |x| - 1|$; this has three "corners." The function
$f(x) = | | |x| - 1| - 1|$ has five "corners."

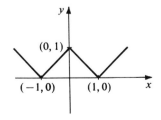

Section 1.9

1. (a) $f(x) = 2 \sin x$.

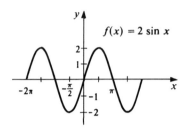

(c) $f(x) = |\sin x|$.

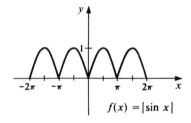

(e) $f(x) = 3 \cos x$.

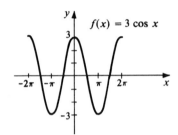

(g) $f(x) = |\cos x|$.

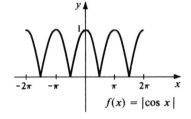

3. (a) $x = \frac{1}{4}[(4n + 1)\pi]$; $n = 0, 1, 2, \ldots, -1, -2, \ldots$.
(b) $x = \frac{1}{2}[(2n + 1)\pi]$; $n = 0, 1, 2, \ldots, -1, -2, \ldots$.
(c) $x = n\pi$; $n = 0, 1, 2, \ldots, -1, -2, \ldots$.

5. $g[f(x)] = 1 + \cos x$, $[f \circ g] = \cos (1 + x)$.

7. $g[f(x)] = \sin x^2$, $[f \circ g](x) = (\sin x)^2$.

Section 1.10

1. (a) (i) $-7 \le x \le 1$; $[-7, 1]$. (iii) $3 \le x \le 7$; $[3, 7]$. (v) $1/|x - 2| > 3$ implies that $|x - 2| < \frac{1}{3}$ and $-\frac{1}{3} < x - 2 < \frac{1}{3}$. Thus, the solution set is $(\frac{5}{3}, \frac{7}{3})$.

3. First write the equation as $y = -\frac{a}{b}x - \frac{c}{b}$, for $b \ne 0$. If $b = 0$ and $a \ne 0$, then the equation of the line is $x = -c/a$. In like manner, if $a = 0$ and $b \ne 0$, the equation is $y = -c/b$. If both $a = 0$ and $b = 0$, then $c = 0$ and, of course, no line exists.
(a) $-a/b = \frac{3}{2}$; c can be anything. (b) $a = \frac{3}{4}b$; $c = -3b$. (c) $c = 0$.
(d) $a + b + c = 0$. (e) $a = 0$. (f) $b = 0$. (g) $-a/b = \frac{2}{3}$, c can be anything. (h) $a/b = \frac{5}{2}$, c can be anything. (i) $b = 1$, $a = -3$, $c = 4$.

5. Area of the triangle is $\frac{1}{2}$ base $\times$ altitude $= 4$.

6. $x = $ cost of A, $600 - x = $ cost of B. $3x + 7(600 - x) \le 3,000$, $x \ge 300$. $300 is the minimum A can cost.

7. The minimum number of miles Truck A must travel is 120.

9. (i) $y = \log (1/x) = -\log x$. (ii) $y = \log |x|$ (since log is defined only for $x > 0$, $\log |x| = \log x$).

10. (b) $e^{\log x} = 100$ implies that $x = 100$. (d) If $\dfrac{e^x + 1}{e^x - 1} = 3$, then

$e^x + 1 = 3e^x - 3$ or $2e^x = 4$, $e^x = 2$ and $x = \log 2$.

11. Domain of f is $(-\infty, \infty)$; domain of g is $(-\infty, 0]$. $(f \circ g)(x) = x$; $[g \circ f](x)$ is *not* defined since $f(x) = x^2$ and $x^2 \notin (-\infty, 0]$. The domain of $(f \circ g)$ is $(-\infty, 0]$ while the range is $[0, \infty)$.

13. (a) $f + g(x) = x^2 + \sqrt{1 - x^2}$, domain is $-1 \le x \le 1$ or $[-1, 1]$.
(c) $f \circ g(x) = (\sqrt{1 - x^2})^2 = (1 - x^2)$; domain is $[-1, 1]$. $g \circ f(x) = \sqrt{1 - x^4}$, domain is $-1 \le x \le 1$.

15. (a) Time from $(0, a)$ to $(x, 0) = (\sqrt{x^2 + a^2})/s$. The time from $(x, 0)$ to $(b, 0) = (b - x)/r$. $T = [(\sqrt{x^2 + a^2})/s] + (b - x)/r$ for $0 \le x \le b$.

(b) $T = \dfrac{\sqrt{x^2 + \frac{9}{16}}}{2} + \dfrac{1 - x}{6} = \dfrac{\sqrt{16x^2 + 9}}{8} + \dfrac{1 - x}{6}$.

x	0	$\frac{1}{4}$	$\frac{1}{2}$	$\frac{3}{4}$	1
T	0.54	0.52	0.54	0.57	0.61

17. The domain is all real $x \ge 0$, and the range is $0 < y \le 1$ for $y = 2^{-x}$. (We cannot consider negative time!)

19. $[f \circ f](x) = \dfrac{[(x + 3)/(2x - 1)] + 3}{2[(x + 3)/(2x - 1)] - 1} = \dfrac{7x}{7} = x = g(x)$.

CHAPTER 2 **Section 2.2**

1.

	$\lim\limits_{x \to a^-} f(x)$	$\lim\limits_{x \to a^+} f(x)$	$\lim\limits_{x \to a} f(x)$
a	yes	yes	no
b	yes	yes	yes
c	yes	yes	yes
d	no	yes	no
e	yes	yes	no
f	yes	yes	yes
g	no	yes	no
h	no	no	no

3. (a) $\lim\limits_{x \to 3^-} f(x) = 3$; $\lim\limits_{x \to 3^+} f(x) = -2$; $\lim\limits_{x \to 3} f(x)$ does not exist.

(c) $\lim\limits_{x \to 3^+} f(x) = 4 = \lim\limits_{x \to 3^-} f(x) = \lim\limits_{x \to 3} f(x)$.

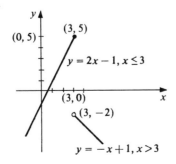

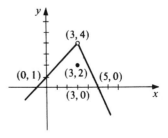

(e) $\lim\limits_{x\to 0^-} f(x) = 0 = \lim\limits_{x\to 0^+} f(x) =$ $\lim\limits_{x\to 0} f(x)$.

(g) $\lim\limits_{x\to -3^-} f(x) = 8$; $\lim\limits_{x\to -3^+} f(x) = -3$; $\lim\limits_{x\to -3} f(x)$ does not exist.

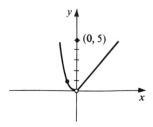

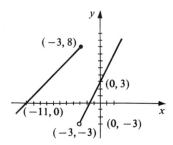

5. (b) $\lim\limits_{x\to 0^+} f(x) = 3$; $\lim\limits_{x\to 0^-} f(x) = 2$. (c) No. $\lim\limits_{x\to 0^+} f(x) \neq \lim\limits_{x\to 0^-} f(x)$.

Section 2.3

1. Yes. 1. 3. 1. 5. 2. 7. $\sqrt{2}$. 9. -2.

11. $\lim\limits_{x\to 1^+} f(x) = \lim\limits_{x\to 1^-} f(x) = 2$, $\lim\limits_{x\to 1} f(x) = 2$. 13. No. 15. No.

17. (a) $\lim\limits_{x\to 0^+} \dfrac{f(x) - f(0)}{x} = 1$, $\lim\limits_{x\to 0^-} \dfrac{f(x) - f(0)}{x} = -1$. (b) No. [See (a).]

Section 2.4

1. (a) 5, P 1, 2, 4, 5. (c) 136, P 1, 3, 4, and 5. (e) 0, P 1, 3, 4, and 5.

3. (a) 4. (b) 5. (c) $\frac{4}{3}$. (d) $\frac{6}{5}$. 5. 1. 7. (a) -1. (b) No.

*9. na^{n-1}. *11. (a) b/d. (b) c/f.

Section 2.5

1. Yes.

3. (b) Continuous at both $x = \frac{1}{2}$ and $x = -\frac{1}{2}$. (c) f is not continuous at 0. Let $f(0) = 0$; then f is continuous there.

5. (a) Yes. $\lim\limits_{x\to 1} f(x) = -2 = f(1)$. (b) No. f is not defined at $x = 1$.

7. (b) Yes. $\lim\limits_{x\to 3} f(x) = 1 = f(3)$. (c) No. We must define $f(1) = 1$ in order for f to be continuous at $x = 1$.

9. No. $\lim\limits_{x\to 0} f(x) = \lim\limits_{x\to 0} 1/x$, which does not exist.

11. $f(x) = \log (x - 1)$ is continuous on $(1, \infty)$. $f(x) = \log (x - b)$ is continuous on (b, ∞).

13. Graph is discontinuous at 1, 2, 3, 4, 5, 6, 7, 8, and 9 ounces.

Section 2.6

1.

	$\lim\limits_{x\to a^-} f(x)$	$\lim\limits_{x\to a^+} f(x)$	$\lim\limits_{x\to a} f(x)$	continuity
2.6.1	yes	yes	no	no
2.6.3	yes	yes	yes	no
2.6.5	yes	no	no	no
2.6.7	yes	no	no	no

2. No, $\lim\limits_{x \to 3} f(x) \neq f(3)$.   **3.** The function is continuous.

5. (b) 0; 1.   (c) Yes, since $f(0) = 0$.   **7.** (a) -3.   (b) 3.

9. (b) f is not continuous at $x = 3$. $\lim\limits_{x \to 3} f(x) = 6 \neq f(3)$. Define $f(3) = 6$.

11. 2.   **12.** $-\frac{1}{9}$.

14. Since f is continuous, it can have no "jumps" in its graph. Thus it must cross the straight line $y = 0$, someplace between $x = 0$ and $x = 1$.

CHAPTER 3   Section 3.1

1. (a) 2, if $x \neq 3$.   (b) Yes. 2.   (c) Yes. $y = 2x + 1$.   **3.** $y = -x + 2$.

5. (a) 0.   (b) 2.   (c) 2.   (d) 1.   (e) 4.   (f) -2.   (g) $\frac{1}{2}$.

7. -2 (or $-2,000$ boxes/cent).   *9. $c = d = 1$.

Section 3.2

1. (b) No. (i) f is *not* continuous at $x = 1$.   (ii) $\lim\limits_{x \to 1} \dfrac{f(x) - f(1)}{x - 1}$ does not exist.

3. (a) (i) -1.   (ii) Does not exist.   (b) Yes. $\lim\limits_{x \to 0} f(x) = 0 = f(0)$.

No. $\lim\limits_{x \to 0^+} \dfrac{f(x) - f(0)}{x}$ does not exist.

5. (b) Yes.   (c) 0; 1. $\lim\limits_{x \to 0^+} \dfrac{f(x) - f(0)}{x} = 1$; $\lim\limits_{x \to 0^-} \dfrac{f(x) - f(0)}{x - 0} = 0$.

$\lim\limits_{x \to 1^+} \dfrac{f(x) - f(1)}{x - 1} = 0$; $\lim\limits_{x \to 1^-} \dfrac{f(x) - f(1)}{x - 1} = 1$.

6. At $x = 1, 2, 3, 4, 5, 6, 7, 8,$ and 9, since the function is not continuous at these points.

8. (b) No. $\lim\limits_{x \to 3} f(x) \neq f(3)$.   (c) No. $\lim\limits_{x \to 3^+} \dfrac{f(x) - f(3)}{x - 3} \neq \lim\limits_{x \to 3^-} \dfrac{f(x) - f(3)}{x - 3}$.

Section 3.3

1. (a) 3; 27.   (b) 0; 1.   (c) $\frac{1}{6}(2^{-5/6})$; $\frac{1}{6}(4)^{5/6}$.   (d) $1/4\sqrt{2}$; $\frac{1}{4}$.
(e) e^2; 1.   (g) 4; 1.   (i) 3; $\frac{9}{2}$.   (k) $-\frac{3}{16}$; $\frac{3}{625}$.

3. $y = \frac{1}{6}x + \frac{3}{2}$.   **4.** $y = (1/e)x$.

6. (a) No.   (b) $f'(1) = 2$; $f'(-1) = 1$; $f'(2) = 4$.

(c) $f'(x) = \begin{cases} 1 & \text{for } x < 0, \\ 2x & \text{for } x > 0. \end{cases}$

7. (a) $21(5)^{20}$.   (c) $e^{-1/2}$.   (e) $\frac{1}{32}$.   **9.** $A = \pi r^2$, $A' = 2\pi r = C$.

11. $V' = 4\pi r^2$; $V'(2) = 16\pi$; $V'(4) = 64\pi$; $V'(6) = 144\pi$; $V'(8) = 256\pi$.

Section 3.4

1. (a) $6x^2 + (1/x)$.   (c) $1 + 2x + 3x^2$.   (e) $-e^x - x^{-2} + (1/x)$.
(g) $x^{-1/2} + 4x^3 + e^x$.   (i) $4x^3 + 6x^2 - 2x$.   (k) $-(1/x^2) + e^x - 20x^9$.
(m) $e^x + 1$.   **3.** $y = 8x - 12$.   **5.** 199,997.   **7.** $20\frac{1}{2}$ million.

9. (a) $R(x) + x(220 - 5x)$; $R'(x) = 220 - 10x$.   (b) $R'(3) = 190$; $R(3) = 615$.

10. (a) $243 + 1/3^3\sqrt{3}$.   (c) $8 + 2e$.   (e) $10 - 5e$.

11. $a_1 + 2a_2x + 3a_3x^2 + \cdots + na_nx^{n-1}$.

Section 3.5

1. (a) $e^x(x^2 + 2x)$ (b) $1/x - x^9 - 10x^9 \log x$. (c) $(3x^2 - x^3)/e^x$.
(d) $(xe^x - 3e^x)/x^4$. (e) $(1 - 5 \log x)/x^6$. (f) $5e^x + 5xe^x \log x + 5e^x \log x$.

(g) $20(\log x)/x$. (h) $2e^{2x}$. (i) $\dfrac{-1}{x^2 \left(1 + \dfrac{1}{x}\right)^2 \left(1 + \dfrac{1}{1 + 1/x}\right)^2} = \dfrac{-1}{(2x + 1)^2}$.

(j) $e^x \left(\dfrac{1}{x} + \log x\right) + \dfrac{25x^2 + 30x + 5}{2\sqrt{x}}$. (k) $\dfrac{e^x}{(1 + e^x)^2}$. (l) $\dfrac{x(2 + x)}{(1 + x)^2}$.

(m) $\dfrac{e^x[(1/x) - \log x] + 1/x}{(1 + e^x)^2}$. (n) $\dfrac{e^x(x - 1)^2}{(1 + x^2)^2}$. (p) $\dfrac{e^x(x + x \log x - 1)}{x(1 + \log x)^2}$.

3. (a) $3e$. (c) 1. (e) $\frac{1}{2}$. 4. 0. $C'(x) = 3x^2 - x + 7$; $C'(10) = 297$.

6. (a) $R(x) = xF(x) = x[(1/x) + \frac{1}{2}x + 0.01x^2] = 1 + \frac{1}{2}x^2 + 0.01x^3$,
$R(10) = 1 + \frac{1}{2}(100) + \frac{1}{100}(1,000) = 61$.
(b) $R'(x) = x + 0.03x^2$, $R'(10) = 10 + 0.03(100) = 13$.
(c) $T(x) = xF(x) - c(x) = 1 + \frac{1}{2}x^2 + 0.01x^3 - 0.30x - 0.001x^2$,
$T(10) = 1 + 50 + 10 - 3 - \frac{1}{10} = 57\frac{9}{10}$,
$T'(x) = x + 0.03x^2 - 0.3 - 0.002x$, $T'(10) = 12\frac{17}{25}$.
(d) Marginal revenue is $x + 0.03x^2$. Marginal cost is $0.30 + 0.002x$. They are equal when
$$x = \dfrac{-499 + \sqrt{258001}}{30} \simeq \dfrac{-499 + 508}{30}; \qquad x \simeq \tfrac{3}{10}.$$

8. (a) $\frac{7}{2}$. (b) -16. (c) $1,055\frac{4}{5}$. (d) $7e + (1/e) + \frac{5}{2}$.
(e) $\dfrac{e^3 + 2e^2 + 2e - 2}{(1 + e)^2}$,

Section 3.6

1. (a) $2(1 + x)$. (c) $x(1 + x^2)^{-1/2}$. (e) $2(1 - 2x)^{-2}$.
(g) $(3x^2 + 2x - 1)e^{(x^3 + x^2 - x)}$. (i) $(2 \log x)/x$.

2. (a) $\dfrac{4x + 1}{2x^2 + x + 1}$. (c) $\dfrac{3x^2 - 4x + 2}{(x + 1)^4}$. (e) $\dfrac{e^x + 1}{e^x + x}$.

(g) $-4(x + \log x)^{-5} \cdot (1 + 1/x)$. (i) $\dfrac{xe^x + e^x}{xe^x} = 1 + \dfrac{1}{x}$.

(k) $\dfrac{2(\log x)}{x} + 3x^2 e^{x^3}$. (m) $\dfrac{4(x + 1)}{(x^2 + 2x + 1)}$.

(o) $e^{[(\log x/x)^{10}]} \cdot 10 \left(\dfrac{\log x}{x}\right)^9 \cdot \dfrac{1 - \log x}{x^2}$. (q) $\dfrac{1}{x \cdot \log x \cdot \log (\log x)}$.

3. $b = 0, \pm 2$. 4. (a) 16. (c) $2e$. (e) $-\frac{64}{1089}$. 5. $x = 0, -1$.

7. $f'(x) = \dfrac{1}{e^x} \cdot e^x = 1$. 9. $y = 4x - 2$. 10. $2x(\sqrt{3x^4 - 1})$.

Section 3.7

1. (a) $6x$. (b) $(x^2 - 1)^{-1/2} - x^2(x^2 - 1)^{-3/2}$. (c) $1/x$.

(d) $e^x \log x + \dfrac{2e^x}{x} - \dfrac{e^x}{x^2}$. (e) $8e^{4x^2}(8x^3 + 8x^2 + 3x + 1)$. (f) $4(x - 1)^{-3}$.

(g) $\dfrac{-(1 + \log x)}{x^2 \log^2 x}$. (h) $4e^{(1-x)/(1+x)}[(1 + x)^{-3} + (1 + x)^{-4}]$.

(i) $-2x(x^2 - 1)^{-4/3} + \frac{16}{9}x^3(x^2 - 1)^{-7/3}$.

(j) $\dfrac{(6x - 3x^4) + e^x[2 + 4x + x^2 - 4x^3 - 2x^4 + x^5 - 2x^2e^x]}{(x^3 + x^2e^x + 1)^2}$.   (k) 2.

(l) $18(x^3 - 9x^2 + 1)[4x^4 - 48x^3 + 135x^2 + x - 3]$.

3. $y' = 2(1 - x)^{-2}$, $y'' = 4(1 - x)^{-3}$, $y''' = 12(1 - x)^{-4}$, $y'''(\frac{1}{2}) = 192$.

5. $\dfrac{-[(f(x))^2f''(x) - 2f(x)(f'(x))^2]}{(f(x))^4} = -\dfrac{ff'' - 2(f')^2}{f^3}$, at those points x, where $f(x) \neq 0$.

Section 3.8

1. (a) $\lim (x^5 - 32)/(x - 2) = \lim (x^4 + 2x^3 + 4x^2 + 8x + 16) = 80$, or note this is
simply the derivative of $f(x) = x^5$ evaluated at $x = 2$. $f' = 5x^4$ and $f'(2) = 5 \cdot 16 = 80$.

(b) $f'(x) = \dfrac{-1}{(x + 2)^2}$, $f'(2) = \dfrac{-1}{4^2} = -\dfrac{1}{16}$.

2. (a) $\dfrac{x^3 - 8x}{(x^2 - 4)^{3/2}}$.   (c) $4x + 3x^2e^x + x^3e^x$.   (e) $\dfrac{5(2x + 2)}{x^2 + 2x + 1}$.

(g) $8/x$.   (i) $2xe^{-x^2} - 2x^3e^{-x^2} - e^{-x} \log x + \dfrac{e^{-x}}{x}$.

3. Slope is $\frac{1}{2}$ and tangent goes through $(1, 0)$. The equation is $2y = x - 1$.

5. $x = 3$.

7. $R(x) = x(5 - \frac{1}{20}x)^2$. Marginal revenue is $R'(x) = (5 - \frac{1}{20}x)^2 - (\frac{1}{10}x)(5 - \frac{1}{20}x)$.
$(5 - \frac{1}{20}x)^2 - (\frac{1}{10}x)(5 - \frac{1}{20}x) = 0$ when $x = 100$. The corresponding fare is 0.

9. (a) 18.   (b) 2.   (c) 2.   (d) $-\frac{5}{4}\sqrt{2}$.

11. $d'(p) = -45 + 2p$ and $d'(25) = 5$.

13. $f(x) = (x - 5)^{-4/5}\left(\dfrac{1}{x} + 2\right)^{1/6}$, $f'(x) = \dfrac{25 - 29x - 48x^2}{30x^2(x - 5)^{9/5}[(1/x) + 2]^{5/6}}$.

15. $y'(x) = \dfrac{2 - 2x^2}{(1 + x^2)^2}$; $y'(x) = 0$ when $x = \pm 1$. $y''(x) = \dfrac{4x^3 - 12x}{(1 + x^2)^3}$; $y''(x) = 0$ when
$x = 0, \pm\sqrt{3}$.

17. $\dfrac{d}{dx}F(ax + b) = F'(ax + b) \cdot a$. Using the chain rule, this equals $aG(ax + b)$ since
$F'(x) = G(x)$.

CHAPTER 4   Section 4.1

1. (a) Local and abs max at $x = 0$. No min.   (c) Local and abs max at $x = 0$.
No min.   (e) Local max at $x = -1, 1$, abs max at $x = -1, 1$. Local min at
$x = 0$, abs min at $x = 4$.   (g) Local max at $x = -1, 1, 2$. Local min at $x = 0$,
abs min at $x = 0$.

2. (a) Max at $x = 10$, min at $x = 0$.   (c) Max at 2, min at -1.   (e) Max at
-3, min at 5.   (g) Max at 4, min at 1.   (i) Max at $-1 + \sqrt{2}$, min at $-1 - \sqrt{2}$.
(k) Max at $+ 1$, min at -1.   (n) Max at 1, min at 4.   (p) Max at 1, min at 0.

4. $x = 10$, $y = 10$.

Section 4.2

1. 64.   3. Income is \$9,000 and fee is \$3.00 per member.

5. 4 in. pieces to make a square of sides 4 in.   7. $8\sqrt{2}$.   9. \$8.   11. $3'' \times 6''$.

13. $\frac{8''}{3} \times \frac{32''}{3} \times \frac{32''}{3}$.   15. 3,456 cu in.   17. $500\pi\sqrt{3}/9$.   19. Area is 1.

Section 4.3

1. (a) $f' > 0$, $-2 < x < 1$; $f' < 0$, $1 < x < \frac{7}{2}$; $f'' > 0$, $x \in (-1, 0) \cup (2, \frac{7}{2})$; $f'' < 0$, $x \in (-2, -1) \cup (0, 2)$.   (c) $f' > 0$, $x \in (-1, \frac{1}{2}) \cup (1, 2)$; $f' < 0$, $x \in (\frac{1}{2}, 1) \cup (2, 4)$; $f'' < 0$, $1 < x < 3$.

2. (a) Inc, $x > \frac{5}{2}$; dec, $x < \frac{5}{2}$; concave up for all x.   (c) Inc, $x \in (-\infty, -2) \cup (\frac{4}{3}, \infty)$; dec, $-2 < x < \frac{4}{3}$; concave up, $x > -\frac{1}{3}$; concave down, $x < -\frac{1}{3}$.   (e) Inc, $x < \frac{1}{2}$; dec, $x > \frac{1}{2}$; concave down for all x.   (g) Inc, $|x| > 1$; dec, $-1 < x < 1$; concave up, $0 < x$; concave down, $x < 0$.   (i) Inc, $x > -2$; dec, $x < -2$; concave up for all x.   (k) Inc, $x \in (-\infty, -\frac{1}{3}) \cup (3, \infty)$; dec, $x \in (-\frac{1}{3}, 3)$; concave up, $x > \frac{2}{3}$; concave down for $x < \frac{2}{3}$.   (m) Inc, $-\dfrac{1}{\sqrt{2}} < x < \dfrac{1}{\sqrt{2}}$; dec, $x \in \left(-\infty, -\dfrac{1}{\sqrt{2}}\right) \cup \left(\dfrac{1}{\sqrt{2}}, \infty\right)$; concave up, $x \in (-\sqrt{\frac{3}{2}}, 0) \cup (\sqrt{\frac{3}{2}}, \infty)$; concave down, $x \in (-\infty, -\sqrt{\frac{3}{2}}) \cup (0, \sqrt{\frac{3}{2}})$.   (v) Inc, $x > 1$; dec, $x < 1$; concave down on domain.

4. $b \geq \frac{3}{2}$.   **5.** (a) $-\frac{1}{2}$.   (c) $\frac{8}{27}$.   **7.** $E(p) = \alpha$.

Section 4.4

1. (a) Inc, $x > \frac{5}{2}$; dec, $x < \frac{5}{2}$; concave up for all x; min $(\frac{5}{2}, -\frac{1}{6})$.   (c) Inc, $x \in (-\infty, -2) \cup (\frac{4}{3}, \infty)$; dec, $-2 < x < \frac{4}{3}$; concave up, $x > -\frac{1}{3}$; concave down, $x < -\frac{1}{3}$; max, $(-2, 13)$; min, $(\frac{4}{3}, -5\frac{14}{27})$.   (e) Inc, $x \in (-\infty, -1) \cup (6, \infty)$; dec, $-1 < x < 6$; concave up, $x > \frac{5}{2}$; concave down, $x < \frac{5}{2}$; max, $(-1, 18)$; min, $(6, -325)$.   (g) Inc, $|x| < 1$; dec, $|x| > 1$; concave up, $x < 0$; concave down, $x > 0$; max, $(1, 6)$; min, $(-1, 2)$.   (i) Inc, $x \in (-\infty, -1) \cup (3, \infty)$; dec, $-1 < x < 3$; concave up, $x > 1$; concave down, $x < 1$; max, $(-1, 6)$; min, $(3, -26)$.   (k) Inc, $0 < x < \frac{2}{3}$; dec, $x \in (-\infty, 0) \cup (\frac{2}{3}, \infty)$; concave up, $x < \frac{1}{3}$; concave down, $x > \frac{1}{3}$; max, $(\frac{2}{3}, \frac{4}{27})$; min, $(0, 0)$.   (m) Inc, $x \in (-\infty, 0) \cup (2, \infty)$; dec, $x \in (0, 2)$; concave up, $x > 1$; concave down, $x < 1$; min, $(2, 0)$; max, $(0, 4)$.   (v) Inc, $x > \frac{1}{2}\sqrt[3]{4}$; dec, $x < \frac{1}{2}\sqrt[3]{4}$; concave up, $x \in (-\infty, -1) \cup (0, \infty)$; concave down, $-1 < x < 0$; max, none; min, $(\frac{1}{2}\sqrt[3]{4}, \frac{3}{2}\sqrt[3]{2})$.

4. At $x = 1/e$.

Section 4.5

1. (a) Local min at $x = 3$.   (b) Local max at $x = -4$, local min at $x = 3$.   (c) No max or min.   (e) Local max at $x = 0$.   (g) Local min at $x = \sqrt[3]{2}$.   (i) Local max at $x = -\sqrt{\frac{1}{3}}$, local min at $x = +\sqrt{\frac{1}{3}}$.   (k) No max or min.

2. (a) Min, $(1, 0)$; max, $(-1, 4)$.   **3.** $\sqrt{50}$.   **5.** $\frac{1}{8}\sqrt{15}$.

7. $a = 4$, $b = 6$, $d = -5$, $c = 0$.

10. Inc for $0 < x < e$; dec for $x > e$; concave up for $x > e^{3/2}$; concave down for $x < e^{3/2}$. As $x \to \infty$, $f(x) \to 0$.

11. (a) Max at $x = 3$, min at $x = 0$.   (c) No max, no min.   (e) Min at $x = 0$, max at $x = 3$.   (g) Min at $x = \frac{1}{2}$, no max.

Section 4.6

1. 12.   **3.** $x = \frac{1}{2}$.   **5.** $x = y = 6$.   **7.** 0.2 sec at age 20.

9. (a) 10.   (b) $x = 10$ yields min cost.   **11.** $(2/\sqrt{2})e^{-1/2}$.

13. $(4\sqrt{3} + \frac{5}{2})(\frac{10}{3}\sqrt{3} + 3)$.   **15.** $x = \sqrt{BD/C}$.   **17.** $50' \times 100'$.

19. $x = 30$.   **21.** $\frac{4}{9}\sqrt{3}$.

Section 4.7

1. (a) Inc for $x < -5 \cup x > 1$; dec for $-5 < x < 1$; concave up for $x > -2$; concave down for $x < -2$; max at $x = -5$; min at $x = 1$. (c) Inc for $x < 0$; dec for $x > 0$; concave up for all x; max at $x = 0$; min at $x = \pm 1$. (e) Inc for $x < 0$; dec for $x > 0$; concave up for $x > 3/\sqrt{2} \cup x < -3/\sqrt{2}$; concave down for $-3/\sqrt{2} < x < 3/\sqrt{2}$; max at $x = 0$.

2. Max occurs at $x = \sqrt{2}$; max total revenue is $4\sqrt{2}$.

3. $T(x) = \dfrac{8x}{4 + x^2}$, max $T(x) = 2$. **5.** $22,500.

6. The inflection point is at $x = 11.67$. This is the point of diminishing returns.

7. (a) $y' = 0$ and $y'' = 2$. By the second derivative test (b), we know y has a local min at x_0. (b) $y' = 0$ and $y'' = 0$. By the second derivative test (c), we have no information. (c) $y' = 0$ and $y'' = -2$. By the second derivative test (a), x_0 is a local max.

9. Length = width = height = $\sqrt[3]{12}$. **10.** $A = 1,250$. **11.** $x = 2$.

12. (a) $x = 6$. (b) $9 per item. (c) $34 a week.

CHAPTER 5 **Section 5.1**

1. (a) $\frac{1}{4}x^4 + C$. (b) $-(1/x) + C$. (c) $2x^{1/2} + C$. (d) $e^x + C$.

(e) $\frac{5}{2}x^2 + C$. (f) $\begin{cases} -\dfrac{1}{3x^3} + C_1 & \text{for } x > 0, \\[2mm] -\dfrac{1}{3x^3} + C_2 & \text{for } x < 0. \end{cases}$

2. (a) $\frac{2}{5}x^{5/2} + C$. (c) $\frac{1}{3}t^3 + \frac{1}{2}t^2 + C$. (e) $\frac{2}{3}x^{3/2} - \frac{2}{5}x^{5/2} + C$.
(f) $\frac{1}{2}x^2 + C$. (h) $\frac{1}{2}x^2 + C$. (j) $\frac{3}{2}x^2 - 5e^x + C$. (l) $\frac{1}{6}x^6 - \frac{2}{7}x^{7/2} + C$.

3. (a) $h' = -x(1 - x^2)^{-1/2}$. (b) $\sqrt{1 - x^2} + C$.

5. (a) $\frac{1}{2}x^2 - (1/x) + C$. (c) $\frac{1}{3}x^3 - \frac{2}{3}x^{3/2} + C$. (d) $\frac{2}{3}t^{3/2} + 2t^{1/2} + C$.
(f) $(at^{9001}/9001) + C$. **6.** (a) $g'(x) = (2x + 1)^{1/2}$. (b) $\frac{1}{3}(2x + 1)^{3/2} + C$.

Section 5.2

1. (a) $\frac{1}{3}x^3 - \frac{3}{2}x^2 + 9x + C$. (c) $5x - 2x^2 + C$.

2. $2x^{3/2} + \frac{10}{7}x^{7/2} + \frac{19}{20}x^{20/19} + C$. **3.** $-x^{-1} - \frac{3}{2}x^{-2} + C$.

5. $-\frac{3}{2}x^{2/3} + \frac{6}{5}x^{5/6} - \frac{9}{8}x^{8/9} + C$. **7.** $-2e^{-x} + e^x + C$.

9. $\log|x| - (3/x) + \frac{3}{4}x^{4/3} - 3x^{1/3} + C$. **11.** $\frac{1}{5}x^5 - \frac{4}{3}x^3 + C$.

13. (a) $f(x) = x^2 - \frac{3}{2}x^4$. (b) $x^3 - \frac{7}{2}x^2 + 2x + 3$.
(c) $f(x) = 2x^4 - x^2 - 20$. **15.** $f'(x) = \frac{1}{2}x^2 + x + \frac{1}{2}, f(x) = \frac{1}{2}(\frac{1}{3}x^3 + x^2 + x)$.

Section 5.3

1. (a) $\int 2x(x^2 + 1)\, dx = \int f'(x)[f(x)]\, dx = \frac{1}{2}[f(x)]^2 + C = \frac{1}{2}(x^2 + 1)^2 + C$.
(b) $f(x) = x^4 + 3x^3 + 1; \frac{1}{2}(x^4 + 3x^3 + 1)^2 + C$. (d) $f(x) = e^{x^2+1}; e^{x^2+1} + C$.
(f) $f(x) = x + e^x; \log|x + e^x| + C$. (h) $f(x) = x^2 + 2x + 1$;
$\frac{1}{2}\log|x^2 + 2x + 1| + C$.

2. (a) $dy = (3x^2 + 2e^{2x})\, dx$. (c) $dy = 2xe^{(x^2+1)}\, dx$. (e) $du = \dfrac{x\, dx}{\sqrt{x^2 + 1}}$,

(g) $dy = 4x^3\, dx$. (i) $dy = dx/x$. **3.** $\frac{3}{8}x^{8/3} + \frac{6}{7}x^{7/6} - \dfrac{1}{x} + C$.

5. $\frac{3}{4}(x^2 + 2x + 3)^{2/3} + C$. **7.** $\frac{2}{5}u^{5/2} - \frac{2}{3}u^{3/2} + C, u = (x + 1)$.

9. $\frac{2}{3}u^{3/2} + 18u^{1/2} + C, u = (x - 9)$. **11.** $\frac{1}{3}(\log |x|)^3 + C$.

13. $\log (\log |x|) + C$. **15.** $(\log |x|)^2 + C$. **17.** $2e^{\sqrt{x}} + C$.

19. $-\frac{1}{8}[\log (1 - x^2)]^2 + C$. **21.** $\frac{1}{3}e^{x^3} + (x^2 + 1)^{3/2} + C$.

24. $f(x) = \frac{1}{3}e^{x^3+3x} + \frac{2}{3}$. **26.** $f(x) = 3x - 5$. **28.** $f(x) = -\frac{2}{63}(x^2 - 1)^2 + \frac{128}{63}$.

Section 5.4

1. (o) $(\frac{1}{4} - 0) + (\frac{1}{4})^2(\frac{1}{2} - \frac{1}{4}) + (\frac{1}{2})^2(\frac{3}{4} - \frac{1}{2}) + (\frac{3}{4})^2(1 - \frac{3}{4})$.

3. (a) $\frac{1}{6}[2^3 + (\frac{13}{6})^3 + (\frac{14}{6})^3 + (\frac{15}{6})^3 + (\frac{16}{6})^3 + (\frac{17}{6})^3]$.

  (c) $\frac{1}{6}[(\frac{25}{12})^3 + (\frac{27}{12})^3 + (\frac{29}{12})^3 + (\frac{31}{12})^3 + (\frac{33}{12})^3 + (\frac{35}{12})^3]$.

Section 5.5

1. (a) $\frac{3}{2}$. (c) $\log 4$. (e) $\frac{28}{3}$. (g) 1. (i) $2(e - e^{-3/2})$.

  (k) $\log 2 + \frac{1}{2}$. (m) $\log 3$. (o) $-\frac{1}{2}e^{-4} + \frac{1}{2}$.

3. (a) $\frac{19}{9}$. (c) $\frac{4}{9}(\sqrt[4]{27} - \sqrt[4]{8})$. (e) $\frac{1}{6}$. (f) $4 - \frac{1}{4}[(1/e) + 1]^4$.

  (h) $\frac{1}{2}e(e^3 - 1)$. (j) $\frac{1}{4}[(\log 5)^2 - (\log 2)^2]$.

5. Yes, since $F(x) = -\frac{1}{2}x^2$ on $[-1, 0)$; $F(x) = \frac{1}{2}x^2$ on $(0, 1]$; and when $x = 0$, $F(x) = 0$. $\int_{-1}^{1} f(x)\,dx = 1$. **7.** $F'(t) = f(t) = 2t$.

Section 5.6

1. (a) $\frac{4}{3}$. (c) 8. (e) e^{-1}. **3.** $\log 3$. **5.** $\frac{64}{3}$. **7.** $1 + e$.

9. $\frac{4}{3}$. **11.** $b = -2$.

13. (a) $\int_{-a}^{a} \sqrt{a^2 - x^2}\,dx$ gives us the area of the semicircle. Thus, area and the integral equal $\frac{1}{2}\pi a^2$. (b) $\frac{1}{4}\pi a^2$.

15. (a) $F'(x) = \sqrt{1 - x^2}$. (b) Write $u(x) = x^2$. If $G(u) = \int_0^u \sqrt{(1 - t^2)}\,dt$,

then $\dfrac{dF}{dx} = \dfrac{dG}{du}\dfrac{du}{dx} = 2x\sqrt{1 - u^2} = 2x\sqrt{1 - x^4}$.

Section 5.7

1. Yes, about 4 years. **3.** (a) 16. (b) $69\sqrt{3}$. **5.** $P_s = 12$.

7. Average value $= \dfrac{1}{t_2 - t_1} \displaystyle\int_{t_1}^{t_2} f'(t)\,dt = \dfrac{f(t_2) - f(t_1)}{t_2 - t_1}$. **9.** $a\sqrt{3} - \frac{1}{3}a$.

11. Approximately 6 years. **13.** Approximately $1,213. **15.** After 4 years.

Section 5.8

1. (a) $2e^x(x - 1) + C$. (c) $x^2(\frac{1}{2} \log |x| - \frac{1}{4}) + C$.

  (e) $e^x(x^3 - 3x^2 + 6x - 6) + C$. (g) $\frac{2}{3}x^{3/2} (\log |x| - \frac{2}{3}) + C$.

2. (a) $2[(e - 2)/e]$. (c) $\frac{2048}{105}$.

5. Let $u = x, dv = g'(x)\,dx; du = dx, v = g(x)$. **7.** $\log b/a$.

Section 5.9

1. (a) $\frac{5}{3}x^3 + x^{-9} - 15x^{-2/3} + C$. (c) $(1/10e^9)(e^{20} - 1)$.

  (e) $-(1 - x^2)^{1/2} + C$. (g) $\frac{1}{48}(4x^2 - 9)^6 + C$.

  (i) $2(x - 3)^{3/2} + \frac{2}{5}(x - 3)^{5/2}$. (k) $2 \log 2 - \frac{3}{4}$.

  (m) $-(3 \log |x| + 1)/9x^3 + C$. (o) $2e^4(2e^2 - 1)$. (q) $\frac{1}{3}(\sqrt{11} - \sqrt{2})$.

3. $\frac{4}{15}$. **5.** $\frac{26}{3}$.

7. $s = 5x + 10 \log x + \frac{1}{2}x^2 \log x - \frac{1}{4}x^2$. $s(1) = 4\frac{3}{4}, s(5) = \frac{75}{4} + \frac{45}{2} \log 5$. The change in sales is $s(5) - s(1) = 14 + \frac{45}{2} \log 5$.

9. $\int_{-1}^{1} (x^3 - x)\,dx = 0$. **11.** $330. **13.** 6.9%.

Section 5.11

1. $\frac{16}{3}a^3$.   3. $\frac{1}{2}\sqrt{3}$.

5. (a) $\frac{1}{2}\pi$.   (b) $\frac{111}{4}\pi$.   (c) $\frac{1}{3}\pi(2\sqrt{2} - 1)$.   (d) $\pi(\frac{43}{2}\sqrt{2} - \frac{115}{4})e^{2\sqrt{2}} + \frac{1}{4}\pi e^2$.
   (e) $\pi(2\log^2 2 - 4\log 2 + 2)$.   (f) $\frac{1}{123}\pi(6^{41} - 5^{41})$.

CHAPTER 6   Section 6.1

1. (a) $\dfrac{-y}{x}$.   (b) $\dfrac{-6xy - 4y^2}{3x^2 + 8xy}$.   (c) $-ye^x$.   (d) $-\dfrac{1}{x\log x}$.

   (e) $\dfrac{y(-y^3 - 5x^5 \log 2y)}{x(x^5 + 3y^3 \log x)}$.   (f) $\dfrac{-y(y + 3x)}{x(x + 3y)}$, $x \neq 0$, $x + 3y \neq 0$.

   (g) $-2xy$.   (h) $-\dfrac{9x}{4y}$, $y \neq 0$.   (i) $-\dfrac{b^2}{a^2}\dfrac{x}{y}$.

   (j) $\dfrac{y}{y - x}$.   (k) $-\dfrac{x^2 - ay}{y^2 - ax}$.

3. $A'(2AB + C^2) + B'(2BC + A^2) + C'(2CA + B^2) = 0$.

5. $u'(x)e^{u(x)} + [v'(x)]/[v(x)]$.

7. $u'[2v^2 + 1/(u - 2w) + e^{w^2 + u}] + v'[4uv] + w'[2/(2w - u) + 2we^{w^2 + u}]$.

9. $(y + a) = -\frac{1}{4}(x - 2a)$.   11. $x = y = 1/\sqrt{2}$.

Section 6.2

1. (b) 0; no, $y'' = 2$.

2. (a) $\dfrac{1}{6x}$.   (c) $\dfrac{1}{2(x + e^x)}$.   3. (a) $2(x + 1)^{1/2}$.   (b) $\frac{2}{9}$.

5. f^{-1} exists and $(f^{-1})'(x) = -\sqrt{1 - x^2}/x$.   7. $\frac{1}{12}$.

9. $F'(x) = 1/(1 + x^2)$, $(F^{-1})^1(y) = 1 + y^2$.

11. $y' = \dfrac{y - y\log(xy) + e^y[\log(xy)]^2 - ye^x[\log(xy)]^2}{e^x[\log(xy)]^2 - xe^y[\log(xy)]^2 - x + x\log(xy)}$.

Section 6.3

1. (a) $12 - 2e^4$.   (b) When $0 = 3t^2 - 2e^{2t}$.

3. (a) 104.   (b) $24e^{36} - 1$.   (c) -8.   (d) $\frac{22}{9}$.   (e) $\frac{3}{2}\sqrt{2}$.   (f) $-\frac{7}{2}$.

5. (a) $4{,}800$.   (c) $360{,}000$.   (e) $-4{,}800$, speed $4{,}800$.

7. At $h = 60$, $dh/dt = 3/25\pi$. At $h = 30$, $dh/dt = 12/25\pi$.

9. $dA/dt = \frac{4}{5}(225\pi)^{1/3}$ sq ft/sec.   11. $s(t) = -3t^2 + 25t$, $s(\frac{25}{6}) = \frac{625}{12}$.

13. $a(t) = 12t^2$, $s(t) = t^4 + 5t + 2$.   15. $(10/\pi)^{1/2}$ yd/min; $(5/6\pi)^{1/2}$ yd/min.

Section 6.4

1. (a) 1.   (b) 1.   (c) 0.   (d) 0.   (e) Undefined.   3. 0.   5. 1.

Section 6.5

1. (a) $-5\sin 5x$.   (b) $-\dfrac{1}{2\sqrt{1 - x}} \cdot \cos\sqrt{1 - x}$.   (c) $-\dfrac{\cos(1 - x)}{2\sqrt{\sin(1 - x)}}$.

   (d) $3(\cos 3x)e^{\sin 3x}$.   (e) $-2\tan 2x$.   (f) $\dfrac{3(x + 5)^2}{\cos^2(x + 5)^3}$.   (g) $\dfrac{e^x}{\cos^2 e^x}$.

(h) $\dfrac{2}{(1 - x)^2 \sin [(1 + x)/(1 - x)] \cos [(1 + x)/(1 - x)]}$.

(i) $-\dfrac{-(\sin x)[\cos (\cos x)]}{\sin (\cos x)}$.　(j) $\dfrac{e^x(\sin x - 2 \cos x)}{(\sin x)^3}$.　(k) $\dfrac{1}{x \cos^2 (\log x)}$.

(l) $(\cos [\sin (\sin x)])(\cos (\sin x))(\cos x)$.　(m) $\dfrac{e^{\tan x}[2(1 + \sin x) - \cos^3 x]}{2(\cos^2 x)(1 + \sin x)^{3/2}}$.

3. $\dfrac{d}{dx} (\cos x) = \dfrac{d}{dx} \sin (x + \tfrac{1}{2}\pi) = \cos (x + \tfrac{1}{2}\pi) = -\sin x.$

7. Let $y = \arccos x$; then $\cos y = x.$　$-\sin y \dfrac{dy}{dx} = 1$　or　$\dfrac{dy}{dx} = -\dfrac{1}{\sqrt{1 - x^2}}$.

Since $\sin^2 y + \cos^2 y = 1$, $\sin y = \sqrt{1 - x^2}$.

8. (a) $\dfrac{2}{\sqrt{2 - x^2}}$.　(c) $\dfrac{1}{2\sqrt{x}(1 + x)}$.　(f) $\dfrac{2x \cos x^2}{1 + (\sin x^2)^2}$.

10. $d\theta/dt = \tfrac{5}{148}.$　**11.** $dx/dt = 800\pi$ ft/min.

Section 6.6

1. (a) $\dfrac{dy}{dx} = \dfrac{2\sqrt{y}(1 - 2y)}{4x\sqrt{y} - 2\sqrt{y} + 1}$.　(b) $\dfrac{dy}{dx} = \dfrac{x - xy^2}{x^2y - y}$.

(c) $\dfrac{dy}{dx} = \dfrac{5x^4 - 4(x + y)^3 - 4(x - y)^3}{4(x + y)^3 - 4(x - y)^3 - 5y^4}$.

(d) $\dfrac{dy}{dx} = \dfrac{y(3x^2 - 1 - \log xy - 2xye^{x^2 y})}{x(xye^{x^2 y} + 1)}$.　(e) $\dfrac{dy}{dx} = \dfrac{y - 3x^2}{3y^2 - x}$.

(f) $\dfrac{dy}{dx} = -\dfrac{y^2}{x^2}$.　(g) $\dfrac{dy}{dx} = -\dfrac{ye^y}{x[ye^y \log xy + e^y - 2y^2]}$.　**3.** $r' = \tfrac{3}{16}.$

5. $-92\tfrac{4}{13}$ mi/hr.　**7.** $s(t) = 12t - \tfrac{1}{2} \cdot 9t^2, 0 = 12t - 9t, s(\tfrac{4}{3}) = 12 \cdot \tfrac{4}{3} - \tfrac{9}{2}(\tfrac{4}{3})^2 = 8.$

8. (a) $\tfrac{1}{4}.$　(c) $\infty.$　(f) $1/e.$

9. (a) Yes; $-x^2.$　(b) Yes; $\tfrac{1}{3}.$　(c) No.　(d) Yes; $1/4x^3.$　(e) No.

11. (a) $v(t) = -pt + 88, s(t) = -\tfrac{1}{2}pt^2 + 88t.$ $p \simeq 38.72 \simeq 39.$　(b) $s(t) = 25$ ft.

CHAPTER 7　Section 7.1

1. (a) $\tfrac{1}{5} \sin 5x + C.$　(b) $-\tfrac{1}{2} \cos x^2 + C.$　(c) $-\tfrac{1}{2} \log \cos 2x + C.$
(d) $\tfrac{1}{2} \sin^2 x + C$ or $-\tfrac{1}{2} \cos^2 x + C.$　(e) $\tfrac{1}{3} \sin^3 x + C.$　(f) $e^{\tan x} + C.$
(g) $-\cos x \log |\cos x| + \cos x + C.$　(h) $-\tfrac{1}{2}x \cos 2x + \tfrac{1}{4} \sin 2x + C.$

3. 1.　**4.** $\tfrac{1}{2}\pi - 1.$　**6.** $\tfrac{1}{4}\pi^2.$　**8.** $\tfrac{3}{2}.$

Section 7.2

1. (a) $x\sqrt{x^2 + \tfrac{9}{4}} + \tfrac{9}{4} \log |2x + 2\sqrt{x^2 + \tfrac{9}{4}}| + C.$　(b) $-\tfrac{1}{3} \log \dfrac{3 + \sqrt{9 - 2x^2}}{x\sqrt{2}} + C.$

(c) $\left(\text{Let } u^2 = \dfrac{x - 1}{x + 2}\right), \sqrt{(x - 1)(x + 2)} - \tfrac{3}{2} \log \dfrac{\sqrt{x - 1} + \sqrt{x + 2}}{\sqrt{x - 1} - \sqrt{x + 2}} + C.$

(d) $e^{8x}(\tfrac{1}{8}x^3 - \tfrac{3}{64}x^2 + \tfrac{3}{256}x - \tfrac{3}{2048}) + C.$　(e) $u = e^x; \dfrac{1}{\sqrt{15}} \arctan (\sqrt{\tfrac{3}{5}} e^x).$

(f) $\tfrac{1}{2}x^4 e^{2x} - x^3 e^{2x} + \tfrac{3}{2}x^2 e^{2x} - \tfrac{3}{2}xe^{2x} + \tfrac{3}{4}e^{2x} + C.$

(g) $\displaystyle\int \dfrac{x \, dx}{\sqrt{x^2 - 9}} - \int \dfrac{3 \, dx}{\sqrt{x^2 - 9}} = (x^2 - 9)^{1/2} - 3 \log \left(\dfrac{x + \sqrt{x^2 - 9}}{3}\right) + C.$

(h) $-\frac{1}{3}\log\left(\dfrac{3 + \sqrt{9 - x^2}}{x}\right) + C.$ (i) $\frac{1}{5}x - \frac{4}{25}\log|5x + 4| + C.$

(j) $-\frac{1}{4}x(25 - x^2)^{3/2} + \frac{25}{8}x\sqrt{25 - x^2} + \frac{625}{8}\arcsin\frac{1}{5}x + C.$

3. (a) $[1/(2\sqrt{2})][x\sqrt{x^2 + \frac{3}{2}} - \frac{3}{2}\log(x + \sqrt{x^2 + \frac{3}{2}})] + C.$ (b) $\frac{1}{2}\log\frac{5}{3}.$
   (c) $\frac{3}{16}\arcsin x^2 - \frac{3}{16}x^2\sqrt{1 - x^4} - \frac{1}{8}x^6\sqrt{1 - x^4} + C.$ 5. $\pi(\frac{1}{4}e^2 - \frac{3}{4}).$

Section 7.3

1. (a)

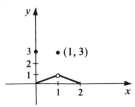

(b) 1, $f(x)$ is bounded and there is only one point of discontinuity.

3. (a)

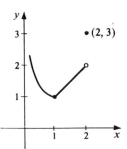

(b) $3\frac{1}{2}.$

4. (a) $\frac{6}{5}.$ (c) $-1.$ (e) Diverges.
5. (a) 1.   (b) 4.   (c) $-\frac{1}{4}.$ (d) Divergent.
6. (a) Diverges.   (c) $-1.$ (e) Diverges.
7. (a) For $n < 1$, $1/(1 - n).$ (b) For $n > 1$, $1/(n - 1).$
9. $A = \lim\limits_{N \to \infty} \int_1^N (1/x)\, dx = \lim\limits_{N \to \infty} (\log N - \log 1) = \infty,$ diverges.
   $V = \lim\limits_{N \to \infty} \int_1^N \pi(1/x^2)\, dx = \pi.$

11. $k = \frac{3}{16},\ \mu = \frac{3}{4},\ \sigma^2 = \int_0^2 (x - \frac{3}{4})^2(4 - x^2)\, dx.$ 13. $\dfrac{1}{3^{p-1}} - \dfrac{1}{4^{p-1}}.$

Section 7.4

1. Concave up, area greater; down, less.   2. 1.1949.   4. $\frac{1}{4}\pi.$
5. $A(x) = 22.33$ sq ft, $V = 117{,}902.4$ cu ft.

Section 7.5

1. (a) $\frac{1}{6}\log\left(\dfrac{3 + x}{3 - x}\right).$ (b) $\frac{1}{2}\log\left(\dfrac{x + \sqrt{x^2 + \frac{9}{4}}}{2}\right) + C.$

   (c) $\sqrt{x^2 + 2x - 3} + \log\left(\dfrac{x + \sqrt{x^2 + 2x - 3}}{2}\right) + C.$

   (d) $\sqrt{x^2 + 9} + 2\log\left(x + \sqrt{x^2 + 9}\right) + C.$

(e) $\frac{1}{2}\sqrt{3}\ x\sqrt{x^2 + \frac{5}{3}} + \frac{5}{6}\sqrt{3}\ \log|\sqrt{3}\ x + \sqrt{3x^2 + 5}|.$

(f) Let $u = e^x + 1$, $-\log\dfrac{e^x + 1}{e^x}$.

(g) Let $u = 2x^2 + 4x + 1$; $\frac{1}{4}(2x^2 + 4x + 1)[\log(2x^2 + 4x + 1) - 1] + C.$

(h) $\dfrac{1}{6(3x + 6)} + \dfrac{1}{36}\log\left|\dfrac{x}{3x + 6}\right| + C.$

(i) $-\frac{1}{4}x(25 - x^2)^{3/2} + \frac{25}{8}x\sqrt{25 - x^2} + \frac{625}{8}$ arc sin $\frac{1}{5}x + C.$

(j) $\log\left(e^x + \sqrt{e^{2x} + 9}\right) + C.$

2. (a) No.　　(c) $\frac{1}{3}$.　　(e) 6.　　3. $A = \int_0^\infty xe^{-x^2}\ dx = \frac{1}{2}.$

5. This is $F'(2)$, where $F'(x) = e^{-x^2}$. $\therefore F'(2) = e^{-4}.$

7. (a) 1.114 approximately.　　(b) 1.108 approximately.

CHAPTER 8　　Section 8.1

1. $y(x) = \frac{1}{4}x^4 + C_1x + C_2$, then $y'(x) = x^3 + C_1$, $y''(x) = 3x^2.$

3. $dy/dx = C_1e^x + C_2(e^x + xe^x)$, and $d^2y/dx^2 = C_1e^x + C_2e^x + C_2(e^x + xe^x).$
$(d^2y/dx^2) - [2(dy/dx)] + y = e^x(C_1 + 2C_2 + C_2x) - 2e^x(C_1 + C_2 + C_2x) + e^x(C_1 + C_2) = e^x(C_1 + 2C_2 + C_2x - 2C_1 - 2C_2 - 2C_2x + C_1 + C_2x) = e^x(0) = 0.$

5. (a) $y = \frac{1}{5}x^5 - x^3 + x + 1.$　　(b) $y = \frac{1}{6}x^6 + e^x + 2.$　　(c) $y = -\frac{1}{2}e^{-x^2} + 1.$
　　(d) $y = x^2[(\log x) - \frac{1}{4}] - \frac{1}{2}x^2 + x - \frac{1}{4} = \frac{1}{2}x^2(\log x - \frac{3}{2}) + x - \frac{1}{4}.$

7. (a) $y' = \frac{3}{2}x^2 + x + C_1$ and $y = \frac{1}{2}x^3 + \frac{1}{2}x^2 + C_1x + C_2.$
　　(b) $f(x) = \frac{1}{2}x^3 + \frac{1}{2}x^2 + 2x + 1.$　　(c) $f(x) = \frac{1}{2}x^3 + \frac{1}{2}x^2 + \frac{1}{2}x.$

Section 8.2

1. (a) $y = Ce^{x^3/3}.$　　(b) $y\ dy = x\ dx$, then $\frac{1}{2}y^2 = \frac{1}{2}x^2 + C_1$ or $y^2 = x^2 + C$,
$y = \sqrt{x^2 + C}$ is a set of solutions. $y = -\sqrt{x^2 + C}$ is also a set of solutions.
(c) $(y - y^2)\ dy = (x + x^2)\ dx$, then $\frac{1}{2}y^2 - \frac{1}{3}y^3 = \frac{1}{2}x^2 + \frac{1}{3}x^3 + C_1$, $3y^2 - 2y^3 = 3x^2 + 2x^3 + C.$　　(d) $e^y\ dy = e^x/(1 + e^x)\ dx$, $e^y = \log(1 + e^x) + C$,
$y = \log[\log(1 + e^x) + C].$　　(e) $y = 4 + Ce^{x^4/4}.$　　(f) $y = \pm\sqrt{\log x^2 + C_1}$
for $\log x^2 + C_1 > 0.$

3. $s = C/(A + P)^B.$　　5. $p(s) = 10(1 - e^{-s}).$

Section 8.3

1. (a) $y = e^{-x} + Ce^x.$　　(c) $y = \frac{1}{2}x^3 + Cx.$　　(e) $y = xe^{1/x} + Ce^{1/x}.$
　　(f) $y = \frac{1}{2}\log x + C/\log x.$　　(g) $y = \frac{1}{3} + Ce^{-x^3}.$

2. (a) $y = 1 + e^{-x^2}.$　　(c) $y = \frac{1}{3}x^2 + \frac{2}{3}x.$　　(e) $y = \frac{1}{2}x^2e^{-x^2} + e^{-x^2}.$

3. $K = [(b - c) + cx] - (b - c)e^{-x}.$

5. (a) $Y = y^2$; then $dY/dx = 2y\ dy/dx.$ Then $2xy\ dy/dx + y^2 = x$ becomes
　　　$x\ dY/dx + Y = x.$　　(b) The general solution is $y = \pm\sqrt{\frac{1}{2}x + C/x}.$

Section 8.4

1. $N(t) = 2{,}000e^{kt}$; $k = \log\frac{3}{2}$, $N(4) = 10{,}125.$

3. $N(t) = 300e^{0.06t}.$ Thus $N(10) = 300e^{0.6} = 300(1.8221) = \$546.63.$

5. $dy/dx = -a$, $y(x) = -ax + C$. $y(12) = 82$, then $82 = -12a + C$. $y(0) = 100$, then $100 = C$ and $a = \frac{3}{2}$. Then $y(x) = -\frac{3}{2}x + 100$. At 2 years, $x = 24$. $y(24) = -\frac{3}{2}\cdot 24 + 100 = 64$. After x months, $y(x) = -\frac{3}{2}x + 100$. $y = 0$ when $x = 66\frac{2}{3}$, then $66\frac{2}{3}$ months is the maximum length of service of the waiters.

7. $N(t) = N_0 e^{Kt}$, where $N(t)$ is the amount in t years. $N_0 = \$2,000$ and $K = 0.07$. Then $N(t) = 2,000e^{0.07t}$. In 25 years, $N(25) = 2,000e^{0.07 \cdot 25} = (2,000)(5.7618) = \$11,523.60$. Find t when $N(t) = \$20,000$. $20,000 = 2,000e^{0.07t}$, $10 = e^{0.07t}$, then $\log 10 = 0.07t$, and $2.3026 = 0.17t$, $t = 32.89$. In 33 years the amount will be $20,000.

9. $dN/dt = -KN$, $N = N_0 e^{-Kt}$, $\frac{1}{2}N_0 = N_0 e^{-Kt}$, $\frac{1}{2} = e^{-K \cdot 5600}$. $-\log 2 = -K \cdot 5,600$, $K = \log 2/5,600$, $N = N_0 e^{-t(\log 2/5600)} = N_0 e^{-t(0.000124)}$. $N(500) = N_0 e^{-500 \cdot 0.000124} = N_0 e^{-0.062} = N_0 \cdot 0.93$. 93% remains and 7% disintegrates. $N(50,000) = N_0 e^{-5000 \cdot 0.000124} = N_0 e^{-0.062} = N_0(0.5381)$. 54% remains.

11. $N = N_0 e^{kt}$; $\frac{1}{4}$.

14. Eighty-nine machines added at the end of the year.

Section 8.5

1. (a) $y = \frac{1}{12}x^4 + \frac{1}{20}x^5 + 3e^x + C_1 x + C_2$.   (b) $y = e^{-x^2/2}\int e^{x^2/2}\, dx + Ce^{-x^2/2}$.

   (c) $y = x \log x + 2x^{3/2} + xC$.   (d) $y = \frac{3}{2}x^4 + x^2 C$.   (e) $y = e^{\sqrt{2\log xC}}$.

   (f) $y = \dfrac{1}{1-x}[\frac{1}{2}x^2 + x + C]$.   (g) $\frac{1}{3}y^3 - \frac{1}{2}y^2 - 2y = \frac{1}{3}x^3 + \frac{1}{2}x^2 + Cx$.

   (h) $y = \frac{2}{3}ax - 1/x + C/\sqrt{x}$.

3. $y = -e^{-x^2}/2x + C/x$; as $x \to 0$, $y \to 0$.

5. $dN/dt = -KN^2(t)$, $dN/[N^2(t)] = -K\,dt$ or $[N^{-1}(t)]/-1 = -Kt - C$, $-1/[N(t)] = -Kt - C$ or $N(t) = 1/(Kt + C)$. When $t = 0$, $N_0 = 1$. $1 = 1/C$, $C = 1$. When $t = 1$, $N(t) = \frac{1}{2}$, $\frac{1}{2} = 1/(K + C)$, $K = 1$. The solution is $N(t) = 1/(1 + t)$.

7. (a) Present value is $606.50.   (b) It takes 13.9 years to double its value.

9. $N = 1/(1 + Ke^{-at})$.

CHAPTER 9   Section 9.1

1. (a)

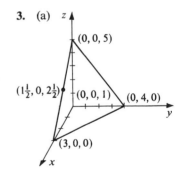

   (b) No.   (c) Yes.

Section 9.2

1. No.

2. The equation of a torus is $z^2 + (\sqrt{x^2 + y^2} - c)^2 = a^2$; z is not a function.

5. (a) 3.　　(b) 4.　　(c) 27.　　(d) 4.　　(e) 27.　　(f) 6.

6. (a)

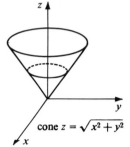

cone $z = \sqrt{x^2 + y^2}$

(c)

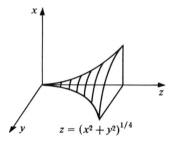

$z = (x^2 + y^2)^{1/4}$

(d)

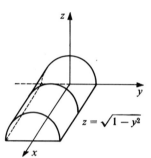

$z = \sqrt{1 - y^2}$

(f)

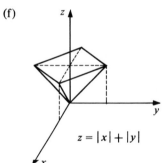

$z = |x| + |y|$

(j)

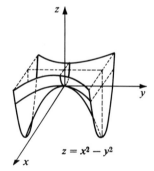

$z = x^2 - y^2$

Section 9.3

2. No.　　**3.** $f(0, 0) = 0$.　　**5.** No.　　$\lim_{(x,y)\to(0,0)} f(x, y)$ does not exist.

Section 9.4

1. $\partial z/\partial x = 1 + 2xy + 9x^2y^2$. $\partial z/\partial y = x^2 + 6x^3y$, at $(8, -1)$ $\partial z/\partial y = -3{,}008$.

2. (a) $\partial z/\partial x = 2ye^{2xy}$; $\partial z/\partial y = 2xe^{2xy}$.　　(b) $2x$; $3y^2$.　　(c) y^2; $2xy + 1$.
(d) $1/y$; $-x/y^2$.　　(e) $6x^5y^5$; $5y^4x^6$.　　(f) $\partial z/\partial x = \log y$; $\partial z/\partial y = x/y$.
(g) $\partial z/\partial x = 4/y$; $\partial z/\partial y = -(4x/y^2) + 2y$.
(i) $\partial z/\partial x = 2yx/(x^2 + y)$; $\partial z/\partial y = \log (x^2 + y) + y/(x^2 + y)$.

3. $\partial f/\partial x = 1/\sqrt{y^2 + 1}$. At $(1, 2)$, $\partial f/\partial x = 1/\sqrt{5}$.

4. $\partial z/\partial x|_{(x_0, y_0)} = A$, $\partial z/\partial x|_{(x_0, y_0)} = B$.

5. $\partial^2 z/\partial x^2 = 2y^5 + e^{x+y}$; $\partial^2 z/\partial x^2|_{(1,0)} = e \cdot \partial^2 z/\partial y^2 = 20x^2y^3 + e^{x+y}$;
$\partial^2 z/\partial y^2|_{(-1, 1)} = 21$; $z_{xy}|_{(2,3)} = 1{,}620 + e^5$.

7. $\partial^2 f/\partial x^2 + \partial^2 f/\partial y^2 = e^x \cos y - e^x \cos y = 0.$

9. $\partial^2 f/\partial x\, \partial y = 2x, \partial^2 f/\partial x^2 = 2y. \therefore x(\partial^2 f/\partial x^2) - y(\partial^2 f/dx\, dy) = 2xy - 2xy = 0.$

10. $\partial z/\partial x = 2, \partial z/\partial y = 0.04y. \partial z/\partial x = 2$ means that for every dollar the customer is willing to put down on the house the banker will loan \$2. $\partial z/\partial y = 0.04y$ means that for every dollar in the customer's salary the banker will loan an additional 4¢.

12. (a) $\partial f/\partial x = -2; \partial f/\partial y = 1; \partial g/\partial x = 1; \partial g/\partial y = -1.$ f and g are competitive.
   (b) $\partial f/\partial x = -2; \partial f/\partial y = 1; \partial g/\partial x = -2; \partial g/\partial y = -3.$ f and g are neither competitive nor complementary.

13. (a) $\partial z/\partial u = 1, \partial z/\partial v = 1$ at $u = 1, v = 1.$
   (b) $\partial z/\partial u = 0, \partial z/\partial v = 13$ at $u = 1$ and $v = \frac{1}{2}.$
   (c) $\partial z/\partial u = \partial z/\partial v = 1$ at $u = 1 = v.$

Section 9.5

1. (a) $\partial f/\partial x = 3x^2 + 1.$ $3x^2 + 1$ not 0. Hence $\partial f/\partial x \neq 0.$
   (c) $\partial f/\partial x = 2x + 4y + 1.$ $\partial f/\partial y = 2y + 4x.$ Solution $(\frac{1}{6}, -\frac{1}{3}).$
   (e) $\partial f/\partial x = 2x + 24y.$ $\partial f/\partial y = 16y + 24x.$ Solution $(0, 0).$
   (g) $\partial f/\partial x = 3x^2 - 3y^2.$ $\partial f/\partial y = -6xy + 3y^2.$ Solution $(0, 0).$

2. $\partial f/\partial x = \partial f/\partial y = 0$ at the origin; but f does not have a relative maximum or minimum there.

3. (a) $\partial f/\partial x = -2xe^{-(x^2+y^2)}.$ $\partial f/\partial y = -2ye^{-(x^2+y^2)}.$ Solution $(0, 0).$ This point is an absolute maximum.
   (c) $\partial f/\partial x = e^{-xy}(1 - xy - y^2).$ $\partial f/\partial y = e^{-xy}(1 - x^2 - xy).$ Solution $(\frac{1}{2}\sqrt{2}, \frac{1}{2}\sqrt{2})(-\frac{1}{2}\sqrt{2}, -\frac{1}{2}\sqrt{2}).$ These are both saddle points.   6. Saddle point.

Section 9.6

1. The maximum occurs at $(\frac{1}{2}, \frac{1}{2}).$ The minimum occurs at $(0, 0), (1, 0), (0, 1), (1, 1).$
   (b) $\partial f/\partial x = \partial f/\partial y = 0$ at $(3, -2).$ $f(3, -2) = -22$ is minimum value.   (c) $\partial f/\partial x = \partial f/\partial y = 0$ at $(\frac{23}{17}, -\frac{13}{17}).$ This is a saddle point.   (d) $f(0, 0)$ is a maximum on the interval. The minimum then occurs on the boundary.   (e) Maximum at $(3, 2),$ minimum at $(0, 5).$ 2. f has a maximum at $f(\frac{1}{2}\sqrt{2}, 0).$

3. There is no maximum, there is no minimum.

4. Minimum cost for $x = 2\sqrt[3]{10}, y = 3\sqrt[3]{10}, z = \sqrt[3]{10}.$

5. $x = 2\sqrt[3]{150}, y = 2\sqrt[3]{150}, z = \sqrt[3]{150}.$ 6. $y = 2$ and $x = \log 4{,}000 - 1 \approx 7.3.$

7. $x = 23, y = 41, u = 36$ gives minimum fuel costs.

Section 9.7

1. Maximum occurs at $x = \frac{1}{3}\sqrt{3}, y = \frac{1}{3}\sqrt{3}, z = \frac{1}{3}\sqrt{3}.$

2. Maximum volume when $x = \frac{4}{3}\sqrt{15}, y = \frac{2}{3}\sqrt{15}, z = \frac{10}{3}\sqrt{15}.$

3. Minimum occurs at $x = 2\sqrt[3]{150} = y, z = \sqrt[3]{150}.$

5. Maximum profit when $x = 14, y = 4,$ and $z = 5.$ 7. $50^0.$

Section 9.8

1. The regression line has the equation $y = x.$

2. Regression line has the equation $15y + 21x + 8 = 0.$

3. Regression line has the equation $47y = -57x + 102.$

4. Regression line has the equation $10y = 23x - 3.$ If advertising expenditures are 6, then sales are approximated by 13.5 units per month. If advertising expenditures are set at 10 units per month, then sales will be about 22.7 units per month.

Section 9.9

1. (a) $\partial z/\partial x = 6x^2y + ye^{xy}$; $\partial z/\partial y = 2x^3 + xe^{xy}$.
   (d) $\partial z/\partial x = y^2/(x + y)^2$; $\partial z/\partial x = x^2/(x + y)^2$.
   (g) $\partial z/\partial x = 2y^2$; $\partial z/\partial y = 4xy$.   (i) $\partial z/\partial x = e^x$; $\partial z/\partial y = 1/y$.

2. The absolute maximum is 1 at $(0, 0)$ and the absolute minimum occurs at points on the boundary $x^2 + y^2 = 1$. It is 0.

3. $z(\frac{1}{2}, \frac{2}{3}) = \frac{43}{108}$ is absolute maximum. The absolute minimum is at $(0, 0)$, $(0, 1)$, $(1, 0)$, $(1, 1)$.

4. Absolute minimum at $(0, 0)$ is 1. Absolute maximum at $(0, 2)$ is 4.

5. Absolute minimum occurs at $(0, 0)$ and is 0. The absolute maximum is 2 and occurs at $(1, 1)$, $(-1, -1)$, $(1, -1)$, and $(-1, 1)$.

6. Minimum occurs at $(\frac{11}{7}, -\frac{1}{7}, \frac{10}{7})$, and is $\frac{4}{7}\sqrt{14}$.

7. An absolute maximum does occur when $x = 30$, $y = 70$.

8. The maximum volume is $20\sqrt{5}$.

9. Maximum occurs when $x = 10$, $y = 8$, and $z = 15$.

10. The equation of the regression line is $123y = 325x - 1,070$.

Index

Note: t after entry indicates table.